Innovative Millet Processing

Samuel Jaddu • Rama Chandra Pradhan
Editors

Innovative Millet Processing

Harnessing Novel Technologies for Nutritional Excellence

Springer

Editors
Samuel Jaddu
Department of Food Process Engineering
National Institute of Technology
Rourkela, Odisha, India

Rama Chandra Pradhan
Professor & Associate Dean (Research)
Department of Food Process Engineering
National Institute of Technology
Rourkela, Odisha, India

ISBN 978-3-031-99863-8 ISBN 978-3-031-99864-5 (eBook)
https://doi.org/10.1007/978-3-031-99864-5

This Springer imprint is published by the registered company Springer Nature Switzerland AG
The registered company address is: Gewerbestrasse 11, 6330 Cham, Switzerland

Preface

Millets, once considered ancient grains, are now regaining global recognition for their exceptional nutritional value, resilience to climate change, and potential in ensuring food and nutritional security. As consumer preferences shift toward healthier and more sustainable food systems, the role of millets in modern diets has grown significantly. However, to unlock their full potential, innovative and scientifically sound processing technologies are essential.

This book, *Emerging Technologies in Millet Processing: Bridging Tradition and Innovation*, offers a comprehensive exploration of both conventional and cutting-edge methods used in the processing of millets. From age-old traditional practices to state-of-the-art techniques such as cold plasma, pulsed electric field, ultrasound, and radiofrequency heating, the chapters delve into the science, engineering, and applications behind each method.

The compilation begins by laying the foundation with an overview of current millet processing technologies and then progressively covers advanced approaches like high-pressure processing, ohmic heating, and extrusion. Unique emphasis is placed on the comparative analysis of traditional versus modern methods, ensuring that readers appreciate both the cultural heritage and technological evolution of millet processing. In addition, the book examines how these methods influence nutritional quality and functional properties, offering evidence-based insights valuable for food scientists, technologists, agronomists, and policymakers.

This book is the result of collaborative efforts from experts in food engineering, cereal chemistry, and emerging food technologies, each contributing their knowledge to address the multifaceted challenges and opportunities in millet utilization. It is intended for researchers, academicians, industry professionals, and students who seek to understand and implement sustainable, efficient, and nutritionally beneficial processing strategies for millets.

As the International Year of Millets (2023) has sparked renewed interest in these humble grains, we hope this work will serve as a timely and valuable resource, supporting continued innovation in this field and helping elevate millets to their rightful place in global food systems.

We gratefully acknowledge all contributors, reviewers, and institutions that made this publication possible.

Rourkela, Odisha, India

Samuel Jaddu
Rama Chandra Pradhan

Contents

1 Introduction to Millet Processing Technologies: Current Scenario, Scope, and Prospects

Bindusri Naraharasetti, Snehasis Chakraborty, Kaliramesh Siliveru, and P. V. Varaprasad

1.1 Introduction

Millets are the grain from the Poaceae family and are considered as the oldest crops. These are grown in the semiarid tropical regions of Africa and Asia and are a major source of food for people in the regions. Millets come in a variety of small-seeded varieties, including barnyard millet (*Echinochloa utilis*), foxtail millet (*Setaria italica*), little millet (*Panicum sumatrense*), kodo millet (*Paspalum setaceum*), proso millet (*Penicum miliaceum*), and pearl millet (*Pennisetum glaucum*). In addition to barley (*Hordeum vulgare*), oats (*Avena sativa*), and maize (*Zea mays*), they are referred to as coarse cereals (Saleh et al. 2013). Pearl millet (*Pennisetum glaucum*) and finger millet (*Eleusine coracana*) are the two millets most frequently used for food and feed. Climate change, water scarcity, population growth, rising food prices, and other socioeconomic repercussions are predicted to significantly threaten agriculture and food security in the twenty-first century, especially for the poorest people in desert and subarid countries (Hassan et al. 2021). These consequences make it difficult for scientists to investigate the viability of creating, processing, and using substitute, sustainable food sources to eliminate penury and malnutrition. The chapter aims to clarify the economic and nutritional benefits of millet, how millet will be the solution to the food security issues and discusses summaries of both conventional and modern millet processing techniques, encompassing both traditional methods. It looks at how various processing techniques affect millet's nutritional value and components, including carbohydrates, protein, ash, and mineral content.

B. Naraharasetti · S. Chakraborty · K. Siliveru (✉)
Department of Grain Science and Industry, Kansas State University, Manhattan, KS, USA
e-mail: kaliramesh@ksu.edu

P. V. Varaprasad
Feed the Future Innovation Lab for Collaborative Research on Sustainable Intensification, Kansas State University, Manhattan, KS, USA

Department of Agronomy, Kansas State University, Manhattan, KS, USA

S. Jaddu, R. Chandra Pradhan (eds.), *Innovative Millet Processing*,
https://doi.org/10.1007/978-3-031-99864-5_1

It addresses the main problems facing the millet processing sector as well as possible areas of future growth.

1.2 Global Millet Production

FAO estimates that in 2020, millet was produced on 74.00 million hectares, producing 89.17 million metric tons of grain worldwide. With finger millet, foxtail millet, proso millet, barnyard millet, little millet, and kodo millet, sorghum and pearl millet account for approximately 90% of the world's millet. More than 93 countries worldwide grow millet, including pearl millet and minor millet (Singh et al. 2023). China, Nigeria, and India have been the top producers of millet. Between 2016 and 2020, millet production grew from 10.28 million tons to 12.49 million tons. India's share of global millets output increased from 37.34% in 2016 to over 41% in 2020. Asia has made the largest contribution to the millet production pool, accounting for 80.46% in 2016 and 79.36% in 2020 (Harish et al. 2024).

Sorghum, pearl millet, finger millet, foxtail millet, kodo millet, proso millet, barnyard millet, and little millet are the eight-millet species that are typically grown in India using rainwater. Additionally, at least four to five types of millet are grown in each of the millet-growing regions, either as a major crop or as an allied crop in conjunction with oilseeds, pulses, spices, and condiments. For example, sorghum and pearl millet are allied and primary crops, respectively, in the desert regions of Rajasthan. Similarly, sorghum is grown as a major crop in Telangana, Andhra Pradesh, Maharashtra, and portions of Central India, while in other Southern regions, it is regarded as a fodder crop. Similarly, finger millet is a minor crop in Telangana but a major crop in Gujarat and Tamil Nadu (Dayakar Rao et al. 2017).

The global consumption of millet has decreased by over 1%, and it is anticipated to increase between 2019 and 2024. Due to supply and demand issues like urbanization, government policies, and rising earnings, millet's significance as a staple grain has decreased over the past 20 years. These days, high-yielding cultivars and technological advancements increase the output of millets. Over the past 60 years, the area planted to small millets has shrunk from 8 million hectares in 1949–1950 to 1.8 million hectares in 2017–2018. Similar trends were seen in small millet production, which decreased from 4 million tons to 2.44 million tons in the corresponding year. The majority of the losses were in all other small millet varieties as opposed to finger millet. The area under millets in India decreased from 51.44 lakh hectares (1950–1955) to 6.23 lakh hectares (2015–2019), with a 16.21% annual loss in millet area; similarly, millet production decreased from 21.13 lakh tons to 4.01 thousand tons, with a 13.58% annual loss (Gowri et al. 2020). The millet production data are shown in Figs. 1.1 and 1.2.

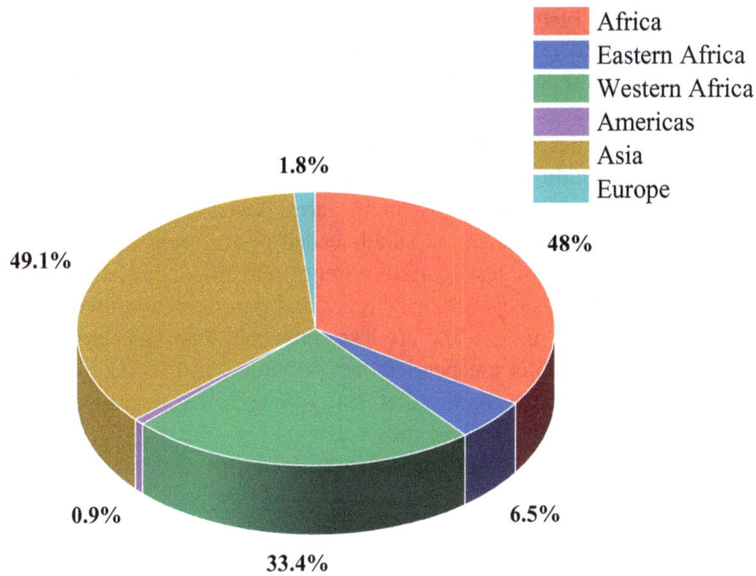

Fig. 1.1 Global production of millets (2018–2020). (Source: Deevi et al. 2024)

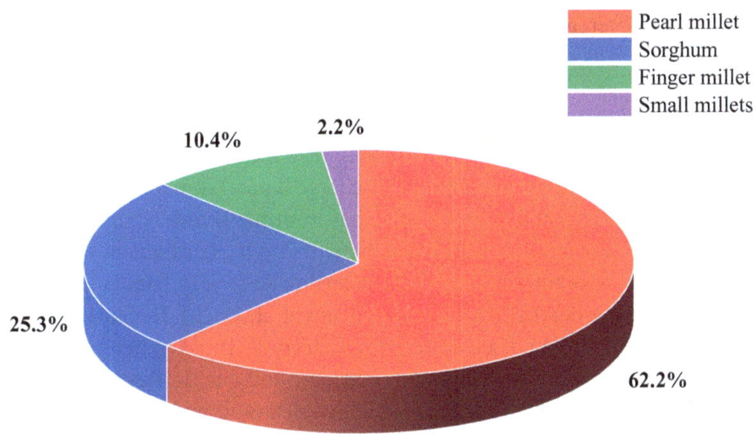

Fig. 1.2 Production of millets in India (2022–2023). (Source: Sendhil et al. 2023)

1.3 Types of Millets and Their Nutritional Attributes

A millet is an ovoid or globose grain of the caryopsis type, typically gray, white, yellow, brown, or purple (Fig. 1.1). The endosperm, germ, and pericarp make up the grain. The pericarp is made up of three layers: the epicarp (outer), mesocarp (middle), and endocarp (inner), with the mesocarp lacking starch granules. During milling, the pericarp breaks at the cross and tube cell layers. Beneath the endocarp lies the seed coat, and the main portion of the kernel, the endosperm, is made up of an aleurone layer (rich in minerals, B vitamins, and oil) and two zones: corneous (dense with starch granules) and floury (contains more starch and protein bodies, and is more amenable to enzyme digestion). The floury endosperm has bigger, spherical starch granules than the corneous zone (Serna-Saldivar and Espinosa-Ramírez 2019). The embryonic axis and scutellum, which are abundant in lipids, proteins, enzymes, and minerals, make up the germ. Based on grain size, nutritional value, and environmental adaptation, small-seeded cereal crops called millets are separated into primary varieties, including major and minor millets. Because of their economic value, major millets are bigger and dominate worldwide farming; minor millets are smaller in size and are less consumed than their larger counterparts. Under the large millet are pearl millet, sorghum, and finger millet. Minor millets include foxtail millet, proso millet, barnyard millet, kodo millet, small millet, brown top millet, and buckwheat millet (Bhatt et al. 2022). These millets provide a varied nutritional profile and flexibility to fit different agroclimatic environments.

1.3.1 Pearl Millet

Pearl millet (*Pennisetum glaucum*), also referred to as bulrush millet, is a small-grain C4 tropical cereal grass cultivated widely in arid and semiarid regions. In Nigeria, it is known as gero, in India as bajra, in Niger as hegni, in Mali as sanyo, in Sudan as dukhon, and in Namibia as mahangu (Taylor 2016). High temperatures, high salt content, low pH, and dry conditions are all favorable for pearl millet. Due to its ability to tolerate adverse growth circumstances, it may survive in regions where other crops cannot survive. India is the largest pearl millet producer (Dayakar Rao et al. 2017). Nutritionally, pearl millet has multiple health advantages, making it an excellent supplement to a variety of diets. Its high fiber content (11.4%), iron content (0.8 g/kg), and zinc content (0.31 g/kg) make it very beneficial in treating anemia by increasing hemoglobin levels (Kulthe et al. 2016). Because it is gluten-free, those who have celiac disease can benefit from it. It has anticancer properties that may help inhibit tumor development. Omega-3 fatty acids, flavonoids, and phenolics all work to prevent DNA scission, LDL cholesterol oxidation, and the growth of HT-29 adenocarcinoma cells (Krishnan and Meera 2018). Additionally, it is also rich in phosphorus, which is necessary for bone growth, development, and repair, and crucial for the formation of ATP, the energy currency of the body, contributing to overall bone health. Additionally, pearl millet's ability to turn the stomach alkaline helps prevent and reduce the effects of stomach ulcers (Cavazos et al. 2013). The grain is favorable for heart health due to its lignin and phytonutrients, which serve as potent antioxidants. Pearl millet's hypoallergenic properties make it a safe and digestible option for infants, lactating mothers, the elderly, and those recovering from illness (Rai et al. 2008). Pearl millet can be cooked similarly to rice or ground into flour for making light porridge or gruel, often consumed with sugar and milk. Alternatively, it can be prepared as a thick porridge and served with soup or stew. In

Ghana, it is used for malting, with the malt then utilized to produce a local beer known as pito (Baryeh et al. 2002).

1.3.2 Finger Millet

Finger millet (*Eleusine coracana)* is commonly referred to as ragi and mandua in India; kaddo in Nepal; fingerhirse in Germany; koracan in France; bulo in Uganda; kambale, lupoko, mawele, amale, and bule in Zambia; poho, rapoko, zviyo, njera, and mazhovole in Zimbabwe; finger millet, African millet, and koracan in England; dagussa, tokuso, and barankiya in Ethiopia; and wimbi and mugimbi in Kenya. It serves as a staple food in various regions of eastern and central Africa, as well as in India (Singh et al. 2012). It grows better in regions with higher rainfall (600–1200 mm), especially acid soils, and reaches maturity in 100–130 days. Being able to adapt to various agroclimatic conditions is one of this millet's key characteristics, and makes it is the most productive of all the millets (Gupta et al. 2017). The chemical composition of finger millet reveals that its total carbohydrate content is 66.8%, while its protein content exhibits significant variation, from 7.3%. Finger millet contains approximately 3.64 g/kg of calcium and 0.0542 g/kg of iron. It helps to reduce calcium and iron deficiency, which will lead to anemia (Singh et al. 2012). Finger millet is abundant in bioactive compounds, especially polyphenols, which offer antioxidant, anticancer, antidiabetic, and anti-inflammatory benefits. Dietary fiber supports healthy digestion, helps manage weight, and lowers cholesterol levels. It helps to reduce oxidative stress (Devi et al. 2014). It is used in both natural and malted forms to prepare a variety of meals. Porridge, puddings, pancakes, biscuits, roti, bread, noodles, and other snacks are made from millet's grains, which are then turned into flours. In addition, when malted, it is a nutritious food for babies (Gull et al. 2014) (Table 1.1).

1.3.3 Proso Millet

Proso millet (*Panicum miliaceum)* is also known as gijang in Korea, kibi and mijo in Spain, panic millet in France, broomcorn millet in China, common millet in the USA, barri in India, broomtail millet, kashfi millet, red millet, white millet, brown millet, and Chinese millet (Das et al. 2019). Grown in a range of soil types and hostile growth conditions, proso millet is salt, alkali, cold, and drought-resistant, growing widely in India, China, Nepal, Africa, Russia, Ukraine, Belarus, the Middle East, Turkey, and Romania (Zhang et al. 2014). A nutrient-dense grain, proso millet is high in vitamins, minerals, dietary fiber, polyphenols, and proteins. It has 11% protein, 3.5% fat, 9.0% crude fiber, 3.6% ash, 70.1% carbohydrate, 9.0% total dietary fiber, and 13.3% total phenolic components. The high lecithin found in millet helps the neurological health system (Kalinova et al. 2007). Rich in minerals (P, Ca, Zn, Fe), vitamins (niacin, B-complex vitamins, folic acid), and essential amino acids (methionine and cysteine), it contains iron, zinc, calcium, and phosphorus in

Table 1.1 Nutritional composition of different millet

Millet	Protein (%)	Carbohydrate (%)	Fat (%)	Crude fiber (%)	Calcium (g/kg)	Phosphorus (g/kg)	Energy(kcal)	Reference
Pearl millet	10.4	61.7	5	11.4	0.02	2.96	363	Kulthe et al. (2016)
Finger millet	7.3	66.8	1.3	3.6	3.64	2.83	320	Singh & Raghuvanshi (2012)
Proso millet	11.0	70.0	3.5	9.0	0.14	2.06	341	Nithyashree & Vijayalaxmi (2022) Kalinova et al. (2007)
Kodo millet	8.3	66.6	1.4	9.0	0.15	1.88	353	Bunkar et al. (2021)
Foxtail millet	11.7	75.3	3.5	2.2	0.47	2.90	349	Yang et al. (2022)
Barnyard millet	10.5	68.8	3.6	13.6	0.22	2.81	398	Kim et al. (2011)
Little millet	8.4	74.7	2.1	3.2	0.15	1.88	329	Srilekha et al. (2019)
Sorghum	10.4	70.7	3.1	2.0	0.27	2.22	329	Abah et al. (2020)
Teff	13.3	73.1	2.4	3.0	1.80	4.29	367	Rishitha et al. (2023)
Brown top millet	11.6	71.3	1.9	8.0	0.28	2.76	368	Bhavani et al. (2024)

weight ranges of 0.0512 g/kg, 0.0440 g/kg, 0.14 g/kg, and 2.02 g/kg, respectively (Nithyashree & Vijayalaxmi 2022). Consuming more proso millet and its derivatives is associated with a decreased chance of developing chronic illnesses such as elevated blood cholesterol, heart disease, type II diabetes, and liver damage. Flavonoids and polyphenols offer a healthy diet for those with celiac disease, direct scavenging of reactive oxygen species (ROS), and anti-apoptotic action that reduces the incidence of eye problems (Agarwal et al. 2019).

Proso millet is eaten as a sweet dish by adding apples during cooking and honey during cooling in Germany. It is eaten as a traditional food, as porridge in many areas, with milk and sugar in Russia. In China, porridge is eaten with a combination of beans and sweet potato or squash (Santra et al. 2013).

1.3.4 Kodo Millet

Kodo millet (*Paspalum scrobiculatum*), which is also referred to as Indian crown grass, cow grass, or rice grass, is cultivated in West Africa, India, Pakistan, the Philippines, Indonesia, Vietnam, and Thailand. Different regional names for it include kodo in Bengali, kodra in Gujarati and Marathi, kodon in Hindi, varagu in Tamil, and arikelu in Telugu (Deshpande et al. 2015). Kodo millet is a nutrient-dense grain, with 353 kcal per 100 grams. Its total is 66.6% carbs, 9% fiber, 2.4% minerals, 1.4% fat, 8.3% protein (mostly glutelin), and 2% ash. Its iron content is around 0.039 g/kg (Bunkar et al. 2021). This millet is sometimes called nutria cereal because of its profusion of vitamins, minerals, and nutrients, including sulfur. Though it is somewhat low in tryptophan, it also boasts important amino acids such as valine, threonine, and lysine. Its great lecithin content and simple digestion help to boost nervous system wellness. Particularly beneficial for postmenopausal women managing high blood pressure or cholesterol. Kodo millet is gluten-free and appropriate for those with gluten sensitivity (Dayakar Rao et al. 2017). Kodo millet has beneficial plant compounds, including hydroxybenzoic and hydroxycinnamic acids and flavonoids (such as quercetin, luteolin, and catechin), which promote immunity and provide antioxidant advantages (Dey et al. 2022). Flexible and used in both traditional and modern cuisine, kodo millet flour can be found in laddoos, idli, dosa, chapati, biscuits, bread, and cookies. Gluten-free bread is also produced by using hydrocolloids to replicate its texture. When combined with other cereals, its nutritive value rises even more to produce products like biscuits. Often ground into flour in India, it is becoming more and more common in multigrain, gluten-free recipes (Deshpande et al. 2015).

1.3.5 Foxtail Millet

Foxtail millet (*Setaria italica*) is recognized as a major millet in terms of world output and ranks as the sixth highest-yielding grain. It is one of the easiest cereal grains to raise and among the oldest crops worldwide (Sharma et al. 2018). It cannot resist waterlogging yet is rather tolerant of a drought. Its early development lets it survive some drenches. Its fast growth makes it a good temporary capture crop. It may flourish at several heights, soil types, and temperatures. Humans eat it as well as caged birds and poultry that feed on its grain (Liu et al. 2019). Foxtail millet flour provides 2.21% of crude fiber, 75.33% of carbohydrates, 3.48% of fat, and 11.7% of protein. It had 0.47 g/kg of calcium, 0.0459 g/kg of iron, and 0.023 g/kg of zinc (Yang et al. 2022). Like other millets, it has a laxative action that helps maintain a healthy digestive system. It also supports bowel movements and facilitates digestion. It is also a good source of crude fiber. Furthermore, shown to have various health benefits is foxtail millet, which prevent cancer and hypoglycemia and have hypolipidemic effects (Abedin et al. 2022). Comprising a grain that is both gluten-free and quite nutritious, foxtail millet is being employed in a variety of dietary products, including composite flours, unfortified weaning mixes, and composite breads. Other flours used to enhance the nutritional profile and create items with acceptable rheological and sensory properties are roasted soybean flour, malted barnyard millet flour, and skim milk powder. Excellent taste acceptability, improved shelf life, and a richness of minerals make foxtail millet a vital component in fermented foods such as soybean paste and probiotic dahi. Foxtail millet bran oil has a great oxidative stability and is a good source of tocopherols (Arora et al. 2023).

1.3.6 Barnyard Millet

An ancient crop, barnyard millet (*Echinochloa esculenta*/*Echinochloa frumentacea*) is cultivated in warm regions, especially in Asia, including countries like China, Japan, Korea, and India. It's the fourth most often grown minor millet and provides nourishment for many low-income communities. India leads the global production in barnyard millet output with 0.146 million hectares under cultivation and 0.147 million tons produced (Renganathan et al. 2020). The grains weigh 0.186 g per kilogram, with 10.5% protein, 68.8% carbohydrates, and 13.6% fiber. Reducing phytic acid levels (3.30–3.70 mg per 100 g) improves mineral absorption and qualifies as a good option for those with anemia (Kim et al. 2011). Naturally, barnyard millet is a great source of easily digested protein and provides both soluble and insoluble fibers. Its great fiber count helps to release sugar gradually into the bloodstream, therefore preserving steady blood sugar levels. The main fatty acids in barnyard millet are linoleic acid, palmitic, and oleic acids. It also generates resistant starch, which reduces blood sugar and fat levels and might help control heart disease and diabetes (Dayakar Rao et al. 2017). The grains are dehulled, boiled, and consumed like rice, making barnyard millet a popular alternative to rice in the Indian Himalayan region. Madira ki kheer, a sweet porridge from Uttarakhand, is a well-known dish created with it. It is used in classic southern Indian meals, including chakli, dosa, and idli. Products including cookies, candies, noodles, rusk, ready-mixes, and popping snacks also contain barnyard millet. However, greater industry participation is required for large-scale production and commercialization to satisfy worldwide demand (Sood et al. 2015).

1.3.7 Little Millet

Indian millet (*Panicum sumatrense*) is another name for little millet, which is indigenous to India. China, East Asia, India, Malaysia, and the Caucasus are the primary growing regions. Little millet may grow in both tropical and temperate regions. Little millet, often referred to as Tribal millet and King of Cereals, is more of a nonperishable grain that can be kept for a long time and encourages consumers to eat healthily, which is utilized mostly by tribals. It had 8.42% protein, 2.10% fat, 1.75% ash, 74.75% carbohydrate, 3.20% crude fiber, and 351.65 kcal (Srilekha et al. 2019). Nutraceuticals such as phenolics, lignans, resistant starch, sterols, phytates, and gamma-aminobutyric acid (GABA) are also abundant in little millet. These bioactive nutraceuticals' additive and synergistic actions in millet-based diets may have many positive health consequences (Guha et al. 2015). It contains fiber, which helps to reduce body fat, and it is a rich source of phosphorus. Its low-calorie content and antioxidants support a balanced diet and weight, which can aid in weight loss. Healthy polyunsaturated fatty acids make up a large portion of the fats found in little millets. Little millets include flavonoids that are crucial for the immune system and self-defense (Indirani et al. 2021).

1.3.8 Sorghum

Sorghum (*Sorghum bicolor*) is cultivated in warmer climates worldwide. Following wheat, maize, rice, and barley, it is quantitatively the fifth-largest and most significant cereal grain in the world. Mostly planted in semiarid regions of the globe, sorghum is a tropical grass. Sorghum is grown in Uganda, Kenya, Tanzania, Burundi, and Zambia. It mostly spreads throughout Africa from West Africa south of the Sahara through Sudan, Ethiopia, and Somalia, West Africa, and Upper Egypt. This crop is rather important in the USA, France, Italy, Venezuela, central and northern China, India, Pakistan, Thailand, Australia, and the drier parts of Argentina and Brazil. While pearl millet is better suited to dry, hot temperatures, the crop has spread to drier parts of the world and thrives in dry, cool weather (Leder et al. 2004). Sorghum has 10.4% protein, 3.1% fat, 1.6%ash, 2.0% crude fiber, 70.7% carbohydrate, 329 kcal, 0.27 g/kg calcium, 0.054 g/kg iron, 0.0038 g/kg thiamin, 0.0015 g/kg riboflavin, and 0.043 g/kg niacin (Abah et al. 2020). Sorghum grain offers a range of potential health benefits, primarily due to its rich content of phenolic compounds, including condensed tannins and 3-deoxy anthocyanidins. These have significant antioxidant properties, which aid in the reduction of free radicals and oxidative stress, important anti-inflammatory effects by stopping the synthesis of pro-inflammatory molecules, which is essential in avoiding inflammation-related disorders. Particularly in colon, breast, and leukemia tumors, sorghum phenolics cause death and stop the spread of cancer cells. By lowering the absorption and production of cholesterol, sorghum helps to improve heart health by promoting cardiovascular health (Xiong et al. 2019). A variety of value-added products with sorghum are kisra, khamir, ogi-baba, pito, bouza, kunu-zaki, mahewu, etc. Beer and injera are two fermented food and drink items of sorghum. Ethiopians eat injera, a big, round, fermented bread that resembles pancakes. Morvite is a soft porridge precooked by adding vitamins, citric acid, and sugar, which is called a product of King Food in South Africa (part of the Tiger Foods group) (Rashwan et al. 2021).

1.3.9 Teff

Teff (*Eragrostis tef*) is known as a drought-resistant, resilient growing warmer seasons and a staple food to Ethiopia and Eritrea. Teff can flourish in a variety of environments, including waterlogging and drought. It is grown in the UK, Australia, China, India, North America, and other African nations (Barretto et al. 2021). Teff has 8.98% protein, 73.1% carbohydrate, 2.38% fat, 3.0% crude fiber, 1.80 g/kg, and 4.29 g/kg. It has 367 kcal in the grain (Rishitha et al. 2023). 8% of the teff grain is composed of complex carbohydrates. Teff is classified as a starchy cereal due to its roughly 73% starch content. The percentage of amylose varied between 20% and 26%. Teff typically contains between 8% and 11% crude protein. According to the fractional protein composition of teff, prolamins make up a minor component (12%), while glutelins (45%) and albumins (37%) are the main protein storage (Zhu et al. 2018). Teff shows anti-oxidative properties, can raise hemoglobin levels in the body, and can help prevent malaria, anemia, and diabetes. Having a high calcium content prevents obesity and osteoporosis. Teff has an iron content that makes it an effective method to combat iron deficiency without any side effects of supplements (Gebru et al. 2020). Teff is used to make a lot of traditional products such as injera, kitta (unleavened bread), atmit or muk (gruel), arake (distilled beverage), and shamit (local beer) (Gebremariam et al. 2014). Teff's amino acid content is well-balanced. A key limiting amino acid in cereals, lysine, is present in rather high levels in teff. More isoleucine, leucine, valine, tyrosine, threonine, methionine, phenylalanine, arginine, alanine, and histidine teff also possesses than other cereals. Another important feature is teff's gluten-free nature. Unsaturated fatty acids, primarily oleic acid (32.4%) and linoleic acid (23.8%), are abundant in teff grains. Increased teff consumption is anticipated to result in a higher dietary fiber intake and related health advantages (Baye et al. 2014).

1.3.10 Brown Top Millet

Brown top millet (*Brachiaria ramose*), also called korale or karlakki in Karnataka, and andakorra in Andhra Pradesh, is an underused millet. It is an annual grass with tiny seeds, farmed in China, Australia, Arabia, Western Asia, and Africa as a grain crop with its origins in Southeast Asia. Brown top millet grows in Karnataka, Tamil Nadu, Andhra Pradesh, and north-central India (Nagaraju et al. 2020a, b). Energy levels were 368.62 kcal. The amounts of fat, crude fiber, and carbohydrates are 71.32%, 8.06%, and 1.89%, respectively. Protein is 11.64%. It includes vital macro- and micronutrients such as phosphorus (2.76 g/kg), potassium (0.60 g/kg), magnesium (0.945 g/kg), calcium (0.28 g/kg), iron (0.0772 g/kg), zinc (0.025 g/kg), manganese (0.0199 g/kg), and sodium (0.076 g/kg) (Bhavani et al. 2024). It includes quinones, flavonoids, tannins, and resins, all of which have potential health advantages. As a result, it is an excellent alternative for treating lifestyle-related disorders. Brown top millet provides a greater concentration of phosphorus than other nutrients, which is necessary for healthy bones and teeth. Its low glycemic index (40–50 GI) results in a gradual increase in blood glucose levels after meals. This is particularly helpful for diabetics and those with impaired glucose tolerance. For blood pressure regulation, glucose management, and insulin efficiency, brown top millet's high magnesium content is essential (Singh et al. 2022a, b). Brown top millet grains from non-shattering kinds are used to make porridge, unleavened bread, or cooked whole grains (similar to rice). Brown top millet is often mixed into flour and used to produce bread (roti, dosa), or polished and cooked to make gruel (rice, anna, and kheer) (Maitra et al. 2020).

1.4 Economic Importance of Millets

Millets offer a promising solution, as they exhibit high photosynthetic efficiency, a short growing period, robust dry matter production, and exceptional resistance to heat and drought. With their great photosynthetic efficiency, short growing period, strong dry matter generation, and extraordinary resilience to heat and drought, millets present a promising option not just to improve food security but also to make a valued and affordable commodity. They can also adjust quickly to acidic, saline, or tainted with aluminum soils. These characteristics make millet a major crop for implementing climate-resilient agricultural methods and mitigating climate change (Wilson et al. 2022). While maintaining appropriate calorie output and land area, expanding the farming of coarse grains, including millets and sorghum, can help boost nutritional security, improve climate resilience, and reduce energy, water, and greenhouse gas emissions. Realizing these advantages depends on being able to move from rice to coarse grains. Ensuring food security for the increasing population and solving world hunger are among the most important issues facing us today. Projections indicate that between 2 and 3 billion people will suffer from hunger and nutritional insecurity if present food production patterns continue, and it is anticipated that by 2025, there will be 9 billion people on the planet (Singh et al. 2022a, b). With micronutrient shortages, sometimes known as hidden hunger, the world confronts both agricultural and nutritional issues that pose a major threat. Given the increased use of irrigated land, this problem demands more focus on drylands to increase grain output. But low productivity of drylands makes it difficult to generate premium crops. Food security is under increasing risk as the variety of crop species which are essential to world food supplies declines. Although global evaluations of this trend are few, historical data shows changes in the variety, abundance, and composition of crops in national food systems. The proportional contributions of several food commodities have become more equal over time, therefore lessening the dominance of several main staples. With an increase in the importance of internationally significant cereal and oil crops and a decline in other cereals, oils, and starchy roots, national food supplies have thus become increasingly comparable. This emphasizes how urgently programs with a nutrition focus are needed to increase food security (Khoury et al. 2014). Rich in secondary metabolites with several applications in biology, millet grains phenols and flavonoids, two bioactive elements found in these grains, have antifungal and antibacterial action (Nithiyanantham et al. 2019). Some high-income nations are realizing millet grains have great potential for bioethanol generation and biofilm formation. Processing millet straw into biomethane allows one to use it as an absorbent for contaminated wastewater. This emphasizes the possibility for effective and ecologically friendly millet use with little formation of waste (Zhu et al. 2014).

1.5 Current Scenario of Millet Processing

The Food and Agriculture Organization (FAO) has highlighted that conventional food processing methods such as decortication, roasting, malting, germination, and milling are commonly used to enhance the edible, nutritional, and sensory qualities of millets. However, industrial processing methods for millet are less developed compared to other cereals, leading to certain inevitable negative effects on their properties. Despite these challenges, millets hold significant potential as a source of nutritious, high-value food items, catering to both traditional consumers and emerging markets with diverse varieties (Sarita et al. 2016).

1.5.1 Physical Processing

Physical processing of millets, such as dehulling and germination, is required to enhance the digestibility and usability in food processing applications. It has a significant impact on the millet flour's ultimate quality.

1.5.1.1 Dehulling

Dehulling treatments play a crucial role in loosening the bond between the seed hull and cotyledons, facilitating easier hull removal through mechanical action and minimizing breakage to improve product quality. In millet processing, dehulling is a crucial step involving removing the bran and hard outer coating of cellulosic material adhered to the grain's surface. The decortication process greatly impacts the nutritive and biochemical characteristics of millet grains because of the irregular distribution of nutrients in the grain. Decortication, aimed at removing antinutrients from the cell wall, significantly influences grain nutrient properties (Joshi and Rao 2024).

1.5.1.2 Milling

Grain is ground during milling to separate the germ, bran, and endosperm, reducing particle size to produce refined flour. In millet processing, milling, whether manual or mechanical, improves the digestibility of carbohydrates and protein by up to 80%, primarily by removing antinutritional factors such as phytic acid and tannins during bran separation (Yousaf et al. 2021). Figure 1.2 explains the process of manufacturing millet flour from the grains. The millet flour-making process begins with cleaning the grains through sieving. Next, the grains are conditioned with water in a 3:10 ratio to increase their moisture content, preparing them for milling. In some cases, the bran is separated using mechanical abrasive decortication, but this step is not always needed, as the whole millet flour is preferred in some situations. Afterward, the grains pass through the first break roller milling. The flour fraction obtained through the first break roller milling is passed through the sifting and purification process. Flour with a particle size smaller than 212 μm is classified as the first break flour. The remaining flour continues through the reduction roller milling process, where it is further milled and then again sifted and purified. The flour from the first and second fractions is then combined. Traditional methods like pestle and mortar, rubbing with gunny bags, hand and leg pounding, and stone grinding were often used for dehulling and pearling but resulted in significant milling waste. Early mechanized approaches, which employed batch-type abrasive polishers designed for cereals and pulses, had low dehulling efficiency and often damaged millet grains, proving ineffective for millet processing (Mittal et al. 2024). With the expansion of millet cultivation, specialized machinery has emerged to improve processing efficiency. For instance, the Vivek thresher-cum-pearler, developed by PHET, VPKAS in Uttarakhand, can process 60 kg per hour, while the Central Institute of Agricultural Engineering (CIAE) in Bhopal introduced a dehusking machine capable of handling 100 kg per hour for grains with 10–12% moisture content (AK Gupta et al. 2018). Modified rice milling machines and centrifugal shellers are also used to process small millets, although the husk fraction in processed millet products can range from 1.5% to 29.3%. Challenges in millet processing include the grains fibrous seed coat, pigmentation, astringent flavor, and the limited shelf life of processed products. Techniques such as pearling, debranning, and chemical treatments can enhance nutritional value and consumer acceptability (Jaybhaye et al. 2014). Still, limited access to sophisticated milling processes keeps millet's application for flour-based goods limited. Further developments and broad industry acceptance of contemporary milling processes are required to satisfy growing demand and release millet's full nutritious value (Fig. 1.3).

1.5.2 Biological Processing

Fermentation and malting have been traditional methods used to create indigenous fermented and malted products, which have improved nutritional, health, and sensory qualities. Fermented grains, enriched with bioactive components like γ-aminobutyric acid and natural phenolics, exhibit diverse antioxidants and anticancer effects. Furthermore, fermentation has been shown to increase antioxidant effects, including free radical scavenging, reducing, and metal-chelating abilities, by boosting antioxidant phenolics and peptides (Gan et al. 2017). The observed

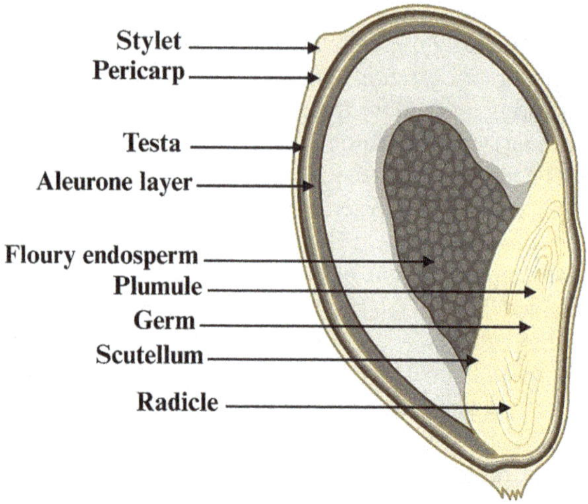

Fig. 1.3 Cross-sectional view showing the general appearance of millets

reduction in the antinutrient content could be attributed to the metabolic activities of microbes and enzymes during the fermentation and malting processes (Akinola et al. 2017).

1.5.2.1 Germination

Germination is a traditional processing technique that enhances the nutritional and functional qualities of grains, therefore producing lighter, more delicate flours fit for use in baby and health food recipes. Furthermore, these surgeries improve the bioavailability of nutrients by reducing antinutrient levels in grains. Increased enzyme activity and the use of fats as fuel across the process cause pearl millet's protein content to climb by as much as 50% during germination to drop (Yousaf et al. 2021). In pearl millet (*Pennisetum typhoides*), germination lowers tannin levels from 1.6% to 0.83%. It also helps break down proteins (14%–26%) and starch (from 86% to 112%). This process reduces phytic acid, tannins, and polyphenols, among other nutrients; these compounds can create complexes that impede protein absorption. Moreover, by reducing the antinutrient value, germination raises the in vitro extractability and bioavailability of important minerals, including zinc, iron, and calcium. The fact that both pearl and finger millets have exhibited similar benefits emphasizes the need for germination to increase the nutritional possibilities of various grains (Sarita et al. 2016). Pre-treatment research is still lacking, nevertheless, since millet-specific milling methods have only recently been developed.

1.5.3 Thermal Processing

Thermal treatments, including heating, blanching, and roasting, have been extensively researched to determine their impact on the nutrients and antinutrients present in PMF. These methods apply heat to the grain, substantially altering the chemical structure of nutrients and antinutrients.

1.5.3.1 Extrusion

Extrusion cooking is a quick, high-temperature processing technique; moist, soft food products are fed into an extruder to reach specified temperature, pressure, and residence time parameters (Sawant et al. 2013). This method uses mechanical and heat energies to gelatinize and rapidly break down complex starch and protein structures though variations in screw speed, die, feed composition, moisture content, and cooking temperature along the extruder (Nagaraju et al. 2020a, b) allow for a range of bespoke products. In one experiment extrusion conditions were optimized using a die diameter of 3 mm, a screw speed of 350 rpm, and dual heating zones set at 110 °C and 90 °C. The amount of tiny millet in the mix raised the extrudates' moisture content (4.45–5.7%), ash content (0.75–1.45%), protein content (3.25–4.9%, bulk density, and hardness (0.11–0.24 N) of the extrudates increased when the percentage of tiny millet in the mix was raised from 0% to 20%. Higher millet proportion lowered the carbohydrate content (89.65–84.42%), the expansion ratio (4.01–2.63), and the calorie value (378.98–365.73 kcal). Extrudates with favorable sensory qualities were created by the study using a composite flour composed of trace amounts of millet, rice, and maize in a 10:45:45 ratio (Saini et al. 2018). Good inactivating enzymes and reducing antinutritional components, #extrusion cooking also reduces high temperatures and feed moisture levels of extrusion aid to promote starch gelatinization, hence boosting water absorption and product expansion. By itself, improved texture, higher nutritional quality, and appealing sensory characteristics created by careful control of extrusion conditions and ingredient ratios promote consumer acceptance. Extruded millet-based creative, value-added goods (Yadav et al. 2022) are produced extensively in the food industry using extrusion technologies.

1.5.3.2 Infrared Technology

Infrared technology is the application of infrared radiation, a kind of electromagnetic radiation, in the food industry. The spectrum has a visible light-to-microwave spectrum, which can be divided into three wavelength categories: near-IR (0.78–1.4 µm), mid-IR (1.4–3.0 µm), and far-IR (3.0–1000 µm) (Vantansever et al. 2012). IR radiation creates heat by vibrating water molecules when it passes through food. Among the various benefits this approach provides are shorter heating periods, equal heat distribution, lower energy use, fast heat transmission, and better product quality. By constricting cytoplasmic membranes, breaking down mesosomes, and causing intracellular contents to spill, the thermal energy from IR can destroy microbial cells. Furthermore, by inhibiting lipid hydrolysis, IR therapy can inactivate lipase, thereby lowering the generation of free fatty acids (Jahan et al. 2024). Additionally, quite successful in preventing germination and extending millet shelf life is infrared therapy. Usually, grains are handled under controlled dosages in either packaged or bulk systems using IR. For millet grains that are whole or dehulled, IR is a safe postharvest technique (Dekka et al. 2023) since gamma irradiation at dosages of 0.75 kGy or above greatly suppresses fungal development.

1.5.3.3 Microwave Technology

Microwave energy ranging between 300 MHZ and 300 GHZ is extensively applied to lower the moisture content and antinutritional components and improve the antioxidant qualities of the grains. Reducing antinutrients during microwave processing results in peptide bond hydrolysis, covalent bond disruption, and disulfide bond modification. At the same time, the production of free radicals leads to the reduction of inositol and its phosphate, and heat-sensitive proteins lead to a lowering of the amount of phytic acid and trypsin inhibitors (Jahan et al. 2024). For instance, microwave treatment of pearl millet at 900 W for 40–100 s significantly reduced tannins and inactivated lipase and free fatty acids.# At 18% grain moisture and 100-second exposure, leads to 92.9% reduction in lipase activity, extending the shelf life of microwave-treated pearl millet flour to 30 days with minimal free fatty acid increase, compared to a 10-day shelf life for untreated pearl millet flour (Yadav et al. 2012). Microwave heating also complements soaking as a method to leach and hydrolyze antinutrients. In sorghum, phenolic content and antioxidant activity increased from 47.1 to 50.8 mg GAE/g and 40.9% to 59.1%, respectively, while fungal growth was decreased by 33.4% and 26.2%, respectively, by microwave treatment at 500 and 350 W (Dekka et al. 2023). Overall, microwave processing is a fast, reliable, efficient, and eco-friendly method of food processing, offering improved nutritional quality and extended shelf life for grains.

1.5.3.4 Ohmic Heating

Ohmic heating utilizes the alternative electrical current to produce the internal heat according to Joule's law. It enables uniform and rapid heating throughout the process. The material gives advantages to the food processing (Knirsch et al. 2010). Ohmic heating overcomes the challenge associated with kodo millet having a dense husk and small grain. Ohmic heating improves the milling properties of kodo millet by increasing grain hardness and reducing processing losses. Kodo millet, after being soaked (30% moisture), was parboiled for 20 min at 14 V/cm. This treatment improved head grain yield by 28.903% (Kabui et al. 2024). Ohmic heating did not soften grain texture as the conventional parboiling methods, and it resulted in harder grains with increased lightness, which is beneficial for the small millets. In the future, it increases the shelf life of millet by 40 days without compromising the sensory qualities. Ohmic heating preserves the protein content compared to conventional paraboiling (Jahan et al. 2024). Ohmic heating is an eco-friendly and energy-efficient technology that aligns with sustainable processing goals. This technology has significant potential in optimizing millet grain quality.

1.5.4 Nonthermal Processing

Millet-based goods are being revolutionized by modern processing technologies like ultrasound processing, ohmic heating, high-pressure processing (HPP), and microwave heating. By inactivating bacteria without heat, these technologies increase shelf life and guarantee quick cooking and drying with little loss of nutrients. It enhances nutrient bioavailability and grinding efficiency. These creative techniques enable the global rebirth of millets as a sustainable and nutrient-dense dietary staple while also improving product quality and satisfying the growing demand for nutritious, minimally processed goods.

1.5.4.1 High-Pressure Processing

High-pressure processing imparting 100–800 MPa pressure, temperature ranges freeze the food item to above 100 °C temperature. It has advantages including structural damage prevention and nutritional preservation, therefore preserving food quality by reducing nutrient loss (Farkas et al. 2000). For instance, foxtail millet subjected to 60 minutes at 60 °C and 400 MPa revealed notable changes in protein concentration, total phenolic content, and antioxidant activity. HPP also lowers antinutritional elements such as phytic acid and tannins, so improving the digestibility and nutritional worth of the grain (Sharma et al. 2018). HPP can also improve millet starch's functional qualities. Treatments at 600 MPa for 30 min boosted granule porosity, therefore enhancing water-holding capacity by 0.66–2.19% and so improving hydration. In millet starch, HPP also lowered syneresis and retrogradation, so improving textural stability a benefit particularly relevant for goods needing a long shelf life. Excellent for usage in bread products, sauces, and other processed meals, HPP-treated starch has enhanced flexibility and hydration qualities (Mirzababaee et al. 2022). Being a nonthermal technique, HPP not only preserves vital nutrients but also reduces antinutritional compounds, therefore increasing the usage of millet in many different kinds of food compositions. Essential in modern food processing processes, this sustainable processing method maximizes millet-based goods by raising processing efficiency, nutritional quality, and functional performance.

1.5.4.2 Cold Plasma Technology

Cold plasma is a nonthermal processing technology that employs reactive species—including free radicals, ions, atoms, and electrons in ionized gas—to influence the physical, chemical, and biological properties of food (Doddabematti Prakash et al. 2023). A plasma state, which is defined by electrical conductivity and fluidlike behavior, is created when gases are heated or subjected to electromagnetic fields. It is especially helpful for grain-based goods because the reactive species produced create an oxidative environment that permits surface changes and microbial and enzyme inactivation (Chakraborty et al. 2024). For instance, Charu et al. (2024) explored the effect of cold plasma treatment on microbial reduction of the Dhanashakti pearl millet variety. They achieved a log reduction from 7.61 to 1.78 log CFU/g (5.82 log reduction) and from 6.43 to 1.95 log CFU/g (4.47 log reduction) in the aerobic mesophilic count and total yeast count, respectively, after a cold plasma treatment at 2 kV/5 min. Sarkar et al. (2023) exposed PMF to cold plasma (20–30 kV for 10–20 min) and analyzed its antinutrients. The sample exposed at 30 kV for 20 min showed the highest reduction in antinutrient properties. The reduction of tannin and phytate content was 17% and 27%, respectively.

Another research on little millet flour treated with multipin electrical discharge atmospheric cold plasma (24 W, 30 min) demonstrated improved functional properties, such as increased water and oil absorption, swelling capacity, and solubility index. Meanwhile, viscosity, dispersibility, and color showed slight reductions, suggesting the flour's suitability for various applications (Jaddu et al. 2022). Another investigation on plasma-treated kodo millet starch (10–30 kV, 10–30 minutes) found enhanced water and oil absorption capabilities (1.43–1.55 g/g and 1.01–1.34 g/g, respectively), but decreased amylose content (31.06%–25.94%) and pH (7.5–6.9). These improvements make the starch suited for thickening or oil-absorbing uses in food products, as well as increased hydration and structural properties (Sonkar et al. 2023). Cold plasma treatment enhances the nutritional and functional properties and reduces the antinutrient properties. Cold plasma-treated flour has the potential to make the new formulation in bakery items and porridges.

1.5.4.3 Ultrasound Technology

Ultrasound technology is a nonthermal technology that utilizes sound waves that produce mechanical energy, resulting in compression and expansion cycles. It operates beyond the human hearing above 20KHz. High-energy ultrasounds (20–100 kHz) induce the mechanical, physical, and chemical transformation in food through cavitation and generate 1000 atm pressure and 5000 k temperature. Low-intensity ultrasound (5–10 MHz) is a nondestructive processing though it preserves product quality and safety (Nazari et al. 2018). For instance, applying high-intensity ultrasound (80% amplitude for 12.5 min) to proso millet significantly enhanced its nutritional value. Antioxidant activity increased by 15%, phenolic content by 16%, and dietary fiber content improved, with water-soluble and ethanol-insoluble fibers rising by 38% (Mustac et al. 2019). In brown top millet, ultrasound treatment at 20 kHz and 200 W (applied in intervals over 20 min) brought major improvements in cooking quality. It reduced crystallinity from 59.78% to 22.84%, improved water absorption from 164% to 179.2%, and cut cooking time dramatically from 18.8 min to just 7.9 min. The treatment also lowered gelatinization temperature and enhanced viscosity, resulting in a softer texture. However, it did come with some trade-offs, as total phenolic content and antioxidant activity decreased TPC dropped from 1129.2 to 648 mg GAE/g, and antioxidant activity fell from 62.3% to 44.69% (Mounika et al. 2024). Finger millet also benefited from ultrasound-assisted hydration (66% amplitude for 26 minutes at a 1,3 grain-to-water ratio). This method not only sped up processing but also significantly reduced antinutritional factors; phytate levels dropped by 66.98% and tannins by 62.83%. The treatment improved starch properties, increasing water-binding capacity and solubility while enhancing the grain's structure and functionality compared to traditional soaking methods (Yadav et al. 2021). Ultrasound technology improves texture, shortens processing times, and enhances functional properties. It is particularly useful in bakery applications (Fig. 1.4).

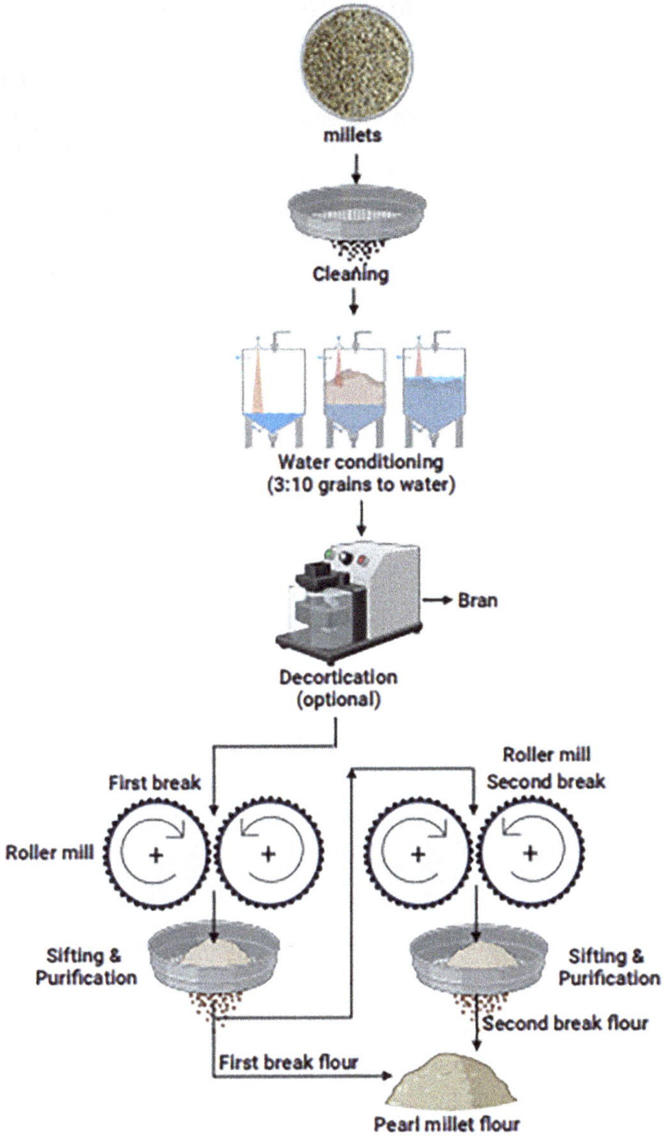

Fig. 1.4 Flow chart showing the procedure to obtain flour from pearl millet grains

1.6 Challenges in the Millet Processing Industry

Numerous difficulties impede the millet processing sector. A high-fiber content, a wide range in size and bonding, and a lack of specific millet processing technology all have an impact on the millet's effectiveness and output. Millet has a shorter shelf life than cereals and other grains such as maize and rice. Aside from that, antinutrient agents reduce the nutrient characteristics of millet. Other major economic and logistical issues include inconsistency in supply and demand, competition from other cereal grains, and a lack of government backing. Another source of concern is the lack of processing and preservation processes. In terms of the consumer, lack of awareness, bitter aftertaste, nonpalatability, and appearance are the concerns that suppress the growth of millet processing technology. However, postharvest losses also occur due to inadequate storage facilities. By addressing those issues, millet processing sectors will gain its full attention.

1.7 Scope, Prospects, and Research Directions

Nowadays, people are switching to healthier grains; millets are expected to be in high demand globally by 2025 due to their low glycemic index and nutraceutical properties. The nature of millets to grow in harsh environments, low soil, and drought conditions also make them a sustainable source for the upcoming generations. India is one of the major producers of millet globally.

There is a lot of scope in the millet processing industry. The sprouting and fermentation of millet improve the nutrient content, such as vitamins, minerals, and antioxidants, and reduce the compounds of antinutrient factors. These processes will help to enhance the bioactive peptides and probiotics. The fortification of millet-based goods with vitamins and minerals such as iron, zinc, and vitamin B complex helps to address the micronutrient deficits. Incorporating the new millet processing technology to overcome the drawbacks of the millets, like shelf life and antinutrient properties, and the lack of proper milling methods in the food industry, will enhance the demand for the Indian market and its products.

There are several opportunities for developing and standardizing value-added products through ready-to-eat protein bars and protein mixes by combining millets with other protein-rich foods like legumes and nuts, cookies, deserts and ready-to-cook foods such as idli, poha, doa mix, etc. by the addition of a minute fraction of other cereal grains. At the same time, new product developments in the bakery field enhance the value of millets globally and strengthen the socioeconomic conditions of the Indian market. Apart from that, future research on specific processing technologies to inactivate enzyme activity and antinutrient factors is the major drawback to the millet processing sections.

1.8 Conclusion

The growing interest in nutritious and sustainable food sources has created the demand for millet-based foods and related processing methods. Traditional methods such as roasting and hand pounding lead to color change, heat-sensitive nutrient loss, shorter shelf life, and inconsistent texture. Modern processing methods like cold plasma technology and ultrasound technology will inactivate the enzyme, which leads to the spoilage of the millets, improve the functionality, and preserve the heat-sensitive nutrients. Millets have significant potential to address future food security needs, particularly in areas prone to climate unpredictability. Continuous study and development of millet processing technology could help contribute to a healthy diet to the future world while maintaining sustainable agriculture practices.

References

Abah CR, Ishiwu CN, Obiegbuna JE, Oladejo AA (2020) Sorghum grains: nutritional composition, functional properties and its food applications. Eu J Nutr Food Saf 12(5):101–111

Abedin MJ, Abdullah ATM, Satter MA, Farzana T (2022) Physical, functional, nutritional and antioxidant properties of foxtail millet in Bangladesh. Heliyon 8(10):e11186

Agarwal S, Chauhan ES (2019) The nutritional composition, various processing and health benefits of Proso millet: a review. Res J Pharm Technol 12(8):4013–4017

Akinola SA, Badejo AA, Osundahunsi OF, Edema MO (2017) Effect of preprocessing techniques on pearl millet flour and changes in technological properties. Int J Food Sci Technol 52(4):992–999. https://doi.org/10.1111/ijfs.13363

Arora L, Aggarwal R, Dhaliwal I, Gupta OP, Kaushik P (2023) Assessment of sensory and nutritional attributes of foxtail millet-based food products. Front Nutr 10:1146545

Barretto R, Buenavista RM, Rivera JL, Wang S, Prasad PV, Siliveru K (2021) Teff (Eragrostis tef) processing, utilization and future opportunities: a review. Int J Food Sci Technol 56(7):3125–3137

Baryeh EA (2002) Physical properties of millet. J Food Eng 51(1):39–46. https://doi.org/10.1016/S0260-8774(01)00035-8

Baye K (2014) Teff: nutrient composition and health benefits, vol 67. Intl Food Policy Res Inst

Bhatt D, Fairos M, Mazumdar A (2022) Millets: nutritional composition, production and significance: a review. J Pharm Innov 11:1577–1582

Bhavani P, Nandini C, Maharajan T, Ningaraju TM, Nandini B, Parveen SG et al (2024) Browntop millet: an overview of breeding, genetic, and genomic resources development for crop improvement. Planta 260(1):10

Bunkar DS, Goyal SK, Meena KK, Kamalvanshi V (2021) Nutritional, functional role of kodo millet and its processing: a review. Int J Curr Microbiol App Sci 10(01):1972–1985

Cavazos A, Gonzalez de Mejia E (2013) Identification of bioactive peptides from cereal storage proteins and their potential role in prevention of chronic diseases. Compr Rev Food Sci Food Saf 12(4):364–380. https://doi.org/10.1111/1541-4337.12017

Chakraborty S, Pulivarthi MK, Raj AS, Prakash SD, Bommina H, Siliveru K (2024) Inactivation of lipase and lipoxygenase in whole wheat flour using atmospheric cold plasma and steam treatments: kinetics, mechanism, and impact on its compositional properties. J Cereal Sci 117:103889

Charu C, Vignesh S, Chidanand DV, Mahendran R, Baskaran N (2024) Impact of cold plasma on pearl and barnyard millets' microbial quality, antioxidant status, and nutritional composition. Food Humanity 2:100238

Das S, Khound R, Santra M, Santra DK (2019) Beyond bird feed: Proso millet for human health and environment. Agriculture 9(3):64

Dayakar Rao B, Bhaskarachary K, Arlene Christina GD, Sudha Devi G, Vilas AT, Tonapi A (2017) Nutritional and health benefits of millets. ICAR_Indian Institute of Millets Research (IIMR) Rajendranagar, Hyderabad, 2

Deevi KC, Swamikannu N, Padmanabhan J (2024) Global millet trends, outlook, challenges, and opportunities. The Pearl Millet Genome:1–14

Dekka S, Paul A, Vidyalakshmi R, Mahendran R (2023) Potential processing technologies for utilization of millets: an updated comprehensive review. J Food Process Eng 46(10):e14279

Deshpande SS, Mohapatra D, Tripathi MK, Sadvatha RH (2015) Kodo millet-nutritional value and utilization in Indian foods. J Grain Proc Storage 2(2):16–23

Devi PB, Vijayabharathi R, Sathyabama S, Malleshi NG, Priyadarisini VB (2014) Health benefits of finger millet (Eleusine coracana L.) polyphenols and dietary fiber: a review. J Food Sci Technol 51:1021–1040

Dey S, Saxena A, Kumar Y, Maity T, Tarafdar A (2022) Understanding the antinutritional factors and bioactive compounds of kodo millet (Paspalum scrobiculatum) and little millet (Panicum sumatrense). J Food Qual 2022(1):1578448. https://doi.org/10.1155/2022/1578448

Doddabematti Prakash S, Siliveru K, Zheng Y (2023) Emerging applications of cold plasma technology in cereal grains and products. In: Trends in food science and technology, vol 141. Elsevier Ltd. https://doi.org/10.1016/j.tifs.2023.10417

Farkas DF, Hoover DG (2000) High pressure processing. J Food Sci 65:47–64

Gan RY, Li HB, Gunaratne A, Sui ZQ, Corke H (2017) Effects of fermented edible seeds and their products on human health: bioactive components and bioactivities. Compr Rev Food Sci Food Saf 16(3):489–531. https://doi.org/10.1111/1541-4337.12257

Gebremariam MM, Zarnkow M, Becker T (2014) Teff (Eragrostis tef) as a raw material for malting, brewing and manufacturing of gluten-free foods and beverages: a review. J Food Sci Technol 51:2881–2895

Gebru YA, Sbhatu DB, Kim KP (2020) Nutritional composition and health benefits of teff (Eragrostis tef (Zucc.) Trotter). J Food Qual 2020(1):9595086

Gowri MU, Shivakumar KM (2020) Millet scenario in India. Econ Aff 65(3):363–370

Guha M, Sreerama YN, Malleshi NG (2015) Influence of processing on nutraceuticals of little millet (Panicum sumatrense). In: Processing and impact on active components in food. Academic Press, pp 353–360

Gull A, Jan R, Nayik GA, Prasad K, Kumar P (2014) Significance of finger millet in nutrition, health and value added products: a review. Magnesium (mg) 130(32):120

Gupta SM, Arora S, Mirza N, Pande A, Lata C, Puranik S et al (2017) Finger millet: a "certain" crop for an "uncertain" future and a solution to food insecurity and hidden hunger under stressful environments. Front Plant Sci 8:643

Gupta AK, Nayak P, Patil M (2018) Processing of minor millets: a review. Int J Farm Sci 8(1):50–58

Harish MS, Bhuker A, Chauhan BS (2024) Millet production, challenges, and opportunities in the Asia-pacific region: a comprehensive review. Front Sustain Food Syst 8:1386469

Hassan ZM, Sebola NA, Mabelebele M (2021) The nutritional use of millet grain for food and feed: a review. Agric Food Secur 10:1–14. https://doi.org/10.1186/s40066-020-00282-6

Indirani K (2021) Review on nutritional profiles and health benefits of little millets–India. Int J Res Eng Sci 9(11):07–11

Jaddu S, Pradhan RC, Dwivedi M (2022) Effect of multipin atmospheric cold plasma discharge on functional properties of little millet (Panicum miliare) flour. Innovative Food Sci Emerg Technol 77:102957

Jahan A, Gnananethri CJ (2024) Novel thermal technologies on the storage of pearl millet grain and flour

Jaybhaye RV, Pardeshi IL, Vengaiah PC, Srivastav PP (2014) Processing and technology for millet based food products: a review. J Ready To Eat Food 1(2):32–48

Joshi J, Rao PS (2024) Characterization and multivariate analysis of decortication-induced changes in pearl millet. J Food Compos Anal 125:105788. https://doi.org/10.1016/j.jfca.2023.105788

Kabui KK, Athmaselvi KA (2024) Ohmic heating parboiling as a novel pretreatment to enhance the milling properties of Kodo millet. J Cereal Sci 10:104049

Kalinová J (2007) Nutritionally important components of proso millet (Panicum miliaceum L.). FoodReview 1(1):91–100

Khoury CK, Bjorkman AD, Dempewolf H, Ramirez-Villegas J, Guarino L, Jarvis A et al (2014) Increasing homogeneity in global food supplies and the implications for food security. Proc Natl Acad Sci 111(11):4001–4006

Kim JY, Jang KC, Park BR, Han SI, Choi KJ, Kim SY et al (2011) Physicochemical and antioxidative properties of selected barnyard millet (Echinochloa utilis) species in Korea. Food Sci Biotechnol 20:461–469

Knirsch MC, Dos Santos CA, de Oliveira Soares AAM, Penna TCV (2010) Ohmic heating–a review. Trends Food Sci Technol 21(9):436–441

Kulthe AA, Thorat SS, Lande SB (2016) Characterization of pearl millet cultivars for proximate composition, minerals and anti-nutritional contents. Adv Life Sci 5(11):4672–4675

Léder I (2004) Sorghum and millets. Cultivated plants, primarily as food sources 1:66–84

Liu TY, Ye N, Song T, Cao Y, Gao B, Zhang D et al (2019) Rhizosheath formation and involvement in foxtail millet (Setaria italica) root growth under drought stress. J Integr Plant Biol 61(4):449–462

Maitra S (2020) Potential horizon of brown-top millet cultivation in drylands: a review. Crop Res 55(1and2):57–63

Mirzababaee SM, Ozmen D, Hesarinejad MA, Toker OS, Yeganehzad S (2022) A study on the structural, physicochemical, rheological and thermal properties of high hydrostatic pressurized pearl millet starch. Int J Biol Macromol 223:511–523

Mittal V, Kumar S, Gupta R, Panghal A, Kumar N (2024) Review on millets dehulling and pearling technologies and machinery. J Food Process Eng 47(5):e14611

Mounika A, Shanmugam A, Rao TJ, Sinija VR, Rawson A, Baskaran N (2024) Evaluation of quality characteristics of ultrasound–treated browntop millet grains. Chem Eng Process-Proc Intens 203:109919

Mustač NČ, Voučko B, Novotni D, Drakula S, Gudelj A, Dujmić F, Ćurić D (2019) Optimization of high intensity ultrasound treatment of Proso millet bran to improve physical and nutritional quality. Food Technol Biotechnol 57(2):183

Nagaraju M, Ramachandra M, Nagarathna SB, Kalpana B, Palanimuthu V, Darshan MB (2020a) Physical properties of an underutilized crop: Browntop millet (Urochloa ramose). Int J Chem Stud 8(6):192–197

Nagaraju M, Tiwari VK, Sharma A (2020b) Effect of extrusion on physical and functional properties of millet based extrudates: a review. J Pharm Phytochem 9(6):1850–1854

Nazari B, Mohammadifar MA, Shojaee-Aliabadi S, Feizollahi E, Mirmoghtadaie L (2018) Effect of ultrasound treatments on functional properties and structure of millet protein concentrate. Ultrason Sonochem 41:382–388

Nithiyanantham S, Kalaiselvi P, Mahomoodally MF, Zengin G, Abirami A, Srinivasan G (2019) Nutritional and functional roles of millets—a review. J Food Biochem 43(7):e12859

Nithyashree K, Vijayalaxmi KG (2022) Effect of popping on nutrients and anti-nutrients in Proso millet. Mysore J Agric Sci 56(4)

Rai KN, Gowda CLL, Reddy BVS, Sehgal S (2008) Adaptation and potential uses of sorghum and pearl millet in alternative and health foods. Compr Rev Food Sci Food Saf 7(4):340–352. https://doi.org/10.1111/j.1541-4337.2008.00049.x

Rashwan AK, Yones HA, Karim N, Taha EM, Chen W (2021) Potential processing technologies for developing sorghum-based food products: an update and comprehensive review. Trends Food Sci Technol 110:168–182

Rateesh Krishnan RK, Meera MS (2018) Pearl millet minerals: effect of processing on bioaccessibility 55:3362. https://doi.org/10.1007/s13197-018-3305-9

Renganathan VG, Vanniarajan C, Karthikeyan A, Ramalingam J (2020) Barnyard millet for food and nutritional security: current status and future research direction. Front Genet 11:500

Rishitha P, Vani SN (2023) Nutritional profile and health benefits of Teff: a review. Pharma Innov J 12:945–949
Saini R, Yadav KC (2018) Development and quality evaluation of little millet (Panicum sumatrense) based extruded product. JPharm Phytochem 7(3):3457–3463
Saleh AS, Zhang Q, Chen J, Shen Q (2013) Millet grains: nutritional quality, processing, and potential health benefits. Compr Rev Food Sci Food Saf 12(3):281–295. https://doi.org/10.1111/1541-4337.12012
Santra DK, Rose DJ (2013) Alternative uses of proso millet. University of Nebraska–Lincoln Extension. Retrieved from https://extensionpublications.unl.edu/assets/pdf/g2217.pdf
Sarita ES, Singh E (2016) Potential of millets: nutrients composition and health benefits. J Sci Innov Res 5(2):46–50
Sarkar A, Niranjan T, Patel G, Kheto A, Tiwari BK, Dwivedi M (2023) Impact of cold plasma treatment on nutritional, antinutritional, functional, thermal, rheological, and structural properties of pearl millet flour. J Food Process Eng 46(5):e14317
Sawant AA, Thakor NJ, Swami SB, Divate AD, Vidyapeet BS (2013) Physical and sensory characteristics of ready-to-eat food prepared from finger millet based composite mixer by extrusion. Agric Eng Int CIGR J 15(1):100–105
Sendhil R, Joseph J, Akhilraj M, Sivasakthi Devi T, Swaminathan N, Pouchepparadjou A (2023) Status of millets in India: trends and prospects. Sensitizing the Millet Farming, Consumption and Nutritional Security-Challenges and Opportunities. Eds, 5–15.
Serna-Saldivar SO, Espinosa-Ramírez J (2019) Grain structure and grain chemical composition. In: Sorghum and millets. AACC Int Press, pp 85–129
Sharma N, Goyal SK, Alam T, Fatma S, Chaoruangrit A, Niranjan K (2018) Effect of high pressure soaking on swater absorption, gelatinization, and biochemical properties of germinated and non-germinated foxtail millet grains. J Cereal Sci 83:162–170
Singh P, Raghuvanshi RS (2012) Finger millet for food and nutritional security. Afr J Food Sci 6(4):77–84
Singh RP, Qidwai S, Singh O, Reddy BR, Saharan S, Kataria SK et al (2022a) Millets for food and nutritional security in the context of climate resilient agriculture: a review. Int J Plant & Soil Science:939–953
Singh S, Suri S, Singh R (2022b) Potential and unrealized future possibilities of browntop millet in the food sector. Front Sustain Food Syst 6:974126. https://doi.org/10.3389/fsufs.2022.974126
Singh S, Yadav RN, Tripathi AK, Kumar M, Kumar M, Yadav S et al (2023) Current status and promotional strategies of millets: a review. Int J Environ Climate Change 13(9):3088–3095
Sonkar S, Jaddu S, Pradhan RC, Dwivedi M, Seth D, Goksen G et al (2023) Effect of atmospheric cold plasma (pin type) on hydration and structure properties of kodo-millet starch. LWT 182:114889
Sood S, Khulbe RK, Gupta AK, Agrawal PK, Upadhyaya HD, Bhatt JC (2015) Barnyard millet—a potential food and feed crop of future. Plant Beeding 134(2):135–147
Srilekha K, Kamalaja T, Uma Maheswari K, Neela Rani R (2019) Nutritional composition of little millet flour. Int Res J Pure Appl Chem 20(4):1–4
Taylor JRN (2016) Millet pearl: overview. In: Encyclopedia of food grains, vol 1. Elsevier, pp 190–198. https://doi.org/10.1016/B978-0-12-394437-5.00011-5
Vatansever F, Hamblin MR (2012) Far infrared radiation (FIR): its biological effects and medical applications: Ferne Infrarotstrahlung: Biologische Effekte und medizinische Anwendungen. Photonics & Laser Med 1(4):255–266
Wilson ML, Van Buren R (2022) Leveraging millets for developing climate resilient agriculture. Curr Opin Biotechnol 75:102683
Xiong Y, Zhang P, Warner RD, Fang Z (2019) Sorghum grain: from genotype, nutrition, and phenolic profile to its health benefits and food applications. Compr Rev Food Sci Food Saf 18(6):2025–2046
Yadav DN, Anand T, Kaur J, Singh AK (2012) Improved storage stability of pearl millet flour through microwave treatment. Agric Res 1:399–404

Yadav S, Mishra S, Pradhan RC (2021) Ultrasound-assisted hydration of finger millet (*Eleusine Coracana*) and its effects on starch isolates and antinutrients. Ultrason Sonochem 73:105542

Yadav GP, Dalbhagat CG, Mishra HN (2022) Effects of extrusion process parameters on cooking characteristics and physicochemical, textural, thermal, pasting, microstructure, and nutritional properties of millet-based extruded products: a review. J Food Process Eng 45(9):e14106

Yang T, Ma S, Liu J, Sun B, Wang X (2022) Influences of four processing methods on main nutritional components of foxtail millet: a review. Grain Oil Sci Technol 5(3):156–165

Yousaf L, Hou D, Liaqat H, Shen Q (2021) Millet: a review of its nutritional and functional changes during processing. Food Res Int 142:110197. https://doi.org/10.1016/j.foodres.2021.110197

Zhang L, Liu R, Niu W (2014) Phytochemical and antiproliferative activity of proso millet. PLoS One 9(8):e104058

Zhu F (2014) Structure, physicochemical properties, and uses of millet starch. Food Res Int 64:200–211

Zhu F (2018) Chemical composition and food uses of teff (*Eragrostis tef*). Food Chem 239:402–415

Ultrasound in Millet Processing: Applications, Advantages, and Future Implications

2

Tanvir Ahmed, Md Suzauddula, Khadiza Akter, and Monir Hossen

2.1 Introduction

Millets are among the traditional cereal grains known to humanity, with a historical and nutritional footprint, particularly in arid and semiarid regions of Africa and Asia, including China and India. These small-seeded, round-shaped cereals belong to the *Poaceae* family and are valued for their nutritional content and agro-industrial relevance. The diverse types of millets, varying in cultivation areas, size, shape, and color, are uniquely suited to thrive in harsh climatic conditions, making them a vital food source in regions where other crops may fail. Globally, millets rank as the sixth-highest yielding grain, with a total production of approximately 33 million tons in 2018. India stands out as the largest producer, followed by Sudan and Niger (Yousaf et al. 2021). Millets are nutrient-dense, containing 65–75% carbohydrates, 7–12% proteins, 2–5% fats, and 8–15% fiber. They are rich in minerals like calcium (10–348 mg/100 g), iron (2.2–17.7 mg/100 g), zinc (0.4–2.8 mg/100 g), and phosphorus (189–293 mg/100 g), as well as vitamins such as thiamine (0.15–0.60 mg/100 g), niacin (0.89–4.6 mg/100 g), and riboflavin (0.09–0.28 mg/100 g), making them a vital component for global nutrition and health (Meherunnahar et al. 2024; Jacob et al. 2024). Millets are consumed in various forms worldwide, from congee and porridge in China, Poland, and Russia to thin pancakes in Sudan and chapatis in India and Pakistan. They are also used in

T. Ahmed (✉)
Department of Food Engineering & Tea Technology, Shahjalal University of Science & Technology, Sylhet, Bangladesh

M. Suzauddula
College of Health and Human Sciences, Kansas State University, Kansas, USA

K. Akter · M. Hossen
Department of Microbiology, Noakhali Science and Technology University, Noakhali, Bangladesh

S. Jaddu, R. Chandra Pradhan (eds.), *Innovative Millet Processing*,
https://doi.org/10.1007/978-3-031-99864-5_2

commercial food products like millet noodles, millet juice, and puffed millet. The rich nutritional profile of different millet types, such as foxtail millet, proso millet, and pearl millet, underscores their potential in addressing nutritional deficiencies (Akanbi et al. 2019). The presence of bioactive compounds in millets contributes to their health benefits, including antioxidant and antimicrobial activities. Recent research highlights the potential of millets in mitigating hidden hunger and enhancing food security.

One of the primary challenges in traditional millet processing is the labor-intensive nature of the methods used. Techniques such as hand pounding and manual dehulling are not only time-consuming but also physically demanding. This labor-intensive process can limit the quantity of millet that can be processed, making it difficult for small-scale farmers to scale up production and meet market demands. Additionally, the reliance on manual labor often leads to inconsistent processing outcomes, with variations in the degree of dehulling and milling affecting the quality and uniformity of the final product (Sabuz et al. 2023). Another significant challenge is the high postharvest losses associated with traditional processing methods. Millets, being small grains, are susceptible to losses during the various stages of processing, including threshing, winnowing, and milling. The lack of efficient storage and handling practices exacerbates these losses, leading to a reduction in the available quantity of millet for consumption and sale. These losses not only impact the food security of farming communities but also affect their economic stability by reducing potential income from millet sales (Sivarajasekar 2023). Nutritional degradation is also a notable issue with traditional millet processing techniques. The processes of hand pounding and manual dehulling often result in the removal of the bran layer, which contains a significant portion of the grain's fiber, vitamins, and minerals. This loss diminishes the nutritional value of the millet, undermining its health benefits (Singh et al. 2024). Furthermore, the absence of advanced techniques to minimize nutrient loss during processing means that traditionally processed millet often falls short of delivering its full nutritional potential.

Nonthermal technologies, particularly ultrasound, offer potential solutions to these challenges by leveraging mechanical energy rather than heat, thereby preserving the grain's integrity and nutritional value. The effect of specific ultrasound frequencies on cell wall disruption is a multifaceted topic, with different frequencies and intensities leading to varied outcomes in cell disruption processes. For *A. niger*, cell disruption occurred more rapidly at 2 MHz compared to 10 MHz, while the opposite was observed for *P. chrysogenum* (Srinivasa Reddy and Linga Reddy 2015). Low-frequency ultrasound, such as 35 kHz, can lead to cell inactivation and disruption in *Microcystis aeruginosa*. Initial exposure results in slight decreases in chlorophyll content and physiological activity, but prolonged exposure leads to significant cell rupture and nutrient release (Tan et al. 2018). Ultrasound technology employs high-frequency sound waves to create cavitation bubbles within the processing medium. When these bubbles collapse, they generate intense localized energy that can effectively disrupt cell structures (Ahmed et al. 2024a). Such a mechanism leads to the formation of free radicals such as hydrogen atoms and hydroxyl radicals. These radicals can initiate various chemical reactions, including the oxidation of lipids, which is evidenced by changes in physicochemical parameters like peroxide and acid values of subjected goods (Pingret et al. 2012). In the context of millet processing, this nonthermal disruption aids in the dehulling and milling processes. By breaking down the outer husk and reducing particle size, ultrasound makes it easier to separate the edible parts of millet without applying excessive heat, which can degrade sensitive nutrients such as vitamins and amino acids (Ahmed et al. 2024b). This preservation of nutrients is crucial for maintaining the health benefits of millet. Moreover, ultrasound technology can significantly reduce the levels of anti-nutritional factors present in millet. These factors, such as phytic acid and tannins, can impair the absorption of essential minerals and proteins. The mechanical action of ultrasound helps in breaking down these compounds or facilitating their removal during processing. As a result, the bioavailability of nutrients in millet is enhanced, making the grain more beneficial for human consumption (Ahmed et al. 2022). This aspect is essential for populations relying on millet as a staple food, ensuring they receive maximum nutritional benefit. Additionally, ultrasound processing can improve the digestibility of millet. The physical effects of ultrasound, such as increased hydration and swelling of starch granules, promote better enzymatic action during digestion. This results in more efficient breakdown and absorption of millet's carbohydrates and proteins, improving its overall nutritional profile. The enhanced digestibility also makes millet a more potential ingredient for a variety of food products, including gluten-free options and health foods aimed at individuals with specific dietary needs (Yadav et al. 2021). This chapter aims to explore the potential of ultrasound technology in millet processing, addressing the traditional challenges faced in this domain. This chapter will also highlight how ultrasound processing can significantly improve efficiency and yield. Detailed discussions will focus on the specific applications and advantages of ultrasound technology, providing insights into its impact on both millet grains and millet-based products. Furthermore, the chapter will delve into future implications, offering a comprehensive understanding of how ultrasound can improve millet processing, ultimately enhancing its utilization and nutritional benefits.

2.2 Current Challenges in Millet Processing

Despite numerous benefits, millet processing faces challenges that hinder its widespread adoption and commercialization. These challenges span technological, economic, sociocultural, and environmental domains, affecting millet-based products' efficiency, cost, and marketability. The urgency and significance of understanding and addressing these issues is crucial to harnessing the full potential of millet as a sustainable food source.

2.2.1 Technological Challenges

2.2.1.1 Inconsistent Grain Quality

Millets are cultivated in diverse regions such as India, Mali, Nigeria, and Niger, each characterized by unique climatic conditions. This geographical diversity results in significant variability in millet grain characteristics, including size, shape, and moisture content. For instance, a study revealed that foxtail millet grains exhibited variations in size when transported between regions. These morphological changes may be influenced by factors such as environmental conditions, field practices, seedling competition, sowing depth, the presence of immature grains, and altitude (Sun et al. 2024). Moreover, another study has shown that the nutritional quality of foxtail millet varies significantly across different production areas, influenced by environmental factors such as geography, climate, and soil (Ma et al. 2024). Processing millets poses several challenges due to the differences in size among various millet types and the low shelf life of processed millets. This variability presents a significant challenge for processing, as equipment must be adjusted to handle grain size, hardness, and moisture content differences. For example, pearl millet (*Pennisetum glaucum*) has a tightly attached pericarp and a larger germ-to-endosperm ratio, making it more complicated and challenging to dehull. In contrast, finger millet (*Eleusine coracana*) has a loosely connected, sack-like pericarp that easily detaches, resulting in a softer texture and easier dehulling (Hassan et al. 2021). Additionally, millets are susceptible to insect infestations, moisture damage, and fungal contamination, further affecting grain quality and processed products. These factors necessitate advanced storage and processing techniques to maintain the quality and extend the shelf life of millet-based products. The inconsistency in grain quality due to these technological challenges remains a significant barrier to the efficient processing and commercialization of millets.

2.2.1.2 Lack of Advanced Processing Equipment

A significant obstacle in millet processing is insufficient suitable infrastructure and equipment. Conventional processing techniques, such as stone grinders, wooden staffs for pounding, and traditional stone mills, require substantial manual labor, are time-consuming, and have lower levels of efficiency. While still in use in many regions, these conventional methods are inadequate for meeting modern food production's demands. The diverse nature of millet grains—varying in size, shape, and husk content—means that no single dehuller can handle all types effectively. The inefficiency of current machinery is notable, with only 70–80% of the grain successfully collected; the remainder is either damaged or still has its outer shell intact (Gudadhe et al. 2023). This inefficiency not only reduces yield but also affects the quality of the final product. Furthermore, despite advancements in dehulling and milling technologies, the high cost of modern equipment remains a substantial barrier, particularly for small-scale farmers and processors. The initial investment required for advanced processing equipment is often prohibitive, limiting access to the latest technologies that could improve efficiency and product quality. Moreover, the technology used is often outdated or not well-suited to industry needs, making it difficult to produce high-quality products in large quantities. The lack of skilled workers to operate the necessary equipment further hampers efficiency. Many processing units are operated by workers without formal training in modern machinery, leading to suboptimal operation and equipment maintenance. This skill gap exacerbates the challenges associated with adopting advanced processing technologies.

2.2.1.3 High Postharvest Losses

Postharvest losses of millets are significant, particularly in sub-Saharan Africa, where agricultural productivity is already low. These losses occur at various stages of the postharvest process, including harvesting, drying, threshing, winnowing, processing, bagging, storage, and transportation. Each stage presents unique challenges that contribute to the overall loss of millet grains. During harvesting, improper timing and techniques can lead to substantial losses. In some African countries, for instance, losses for sorghum and millet during harvesting, drying, and threshing stages range from 11.3% to 12.2% (Abass et al. 2014). These losses are often due to a lack of access to modern harvesting equipment, leading farmers to rely on manual methods that are less efficient and more prone to causing grain damage. Drying is another critical stage where significant losses can occur. Inadequate drying methods, such as sun drying on bare ground, expose grains to contamination and damage. Moisture retention in improperly dried grains can lead to mold growth and spoilage, reducing the quality and market value of the millet (Rasika et al. 2024). The threshing process, which involves separating the grain from the chaff, is often performed using rudimentary tools and techniques, resulting in incomplete threshing and grain losses. Winnowing, the process of cleaning the grain by removing the chaff, is similarly affected by traditional methods that are less effective than modern mechanical winnowers. Inefficient winnowing leads to the retention of impurities in the grain, impacting the quality and safety of the final product. Processing and bagging also contribute to postharvest losses. Inadequate milling techniques can lead to incomplete dehulling and grain breakage, while poor-quality bags and improper bagging techniques can cause spillage and contamination (Chege et al. 2024). Storage poses a significant challenge due to the susceptibility of millets to pests and diseases. Pests such as shoot flies, stem borers, and grain weevils and diseases like downy mildew and grain mold exacerbate the losses. These pests and diseases can infest stored grains, leading to considerable damage and spoilage (Joshi et al. n.d.). Transportation is the final stage where losses can occur. Poor infrastructure, rough handling, and inadequate transportation facilities can physically damage the grains during transit. This damage reduces the quantity of marketable grain and affects its quality, leading to further economic losses for farmers.

2.2.2 Economic Challenges

2.2.2.1 High Processing Costs

Millet processing involves several essential steps to prepare the grains for consumption, including cleaning, dehulling, polishing, and milling. These processes aim to remove nonedible parts such as husks and seed coats, making the grains suitable for secondary processing into products like breakfast cereals and malted drinks (Gowda et al. 2022). However, the processing of millets is significantly cost-intensive due to several factors. Firstly, reliance on labor-intensive methods due to the lack of mechanized processing facilities increases labor expenses, especially in regions where traditional techniques like manual dehulling and milling are predominant. Secondly, the substantial investment in specialized millet processing equipment adds to operational costs. Thirdly, inadequate postharvest infrastructure, including drying and storage facilities, increases losses and additional expenses.

2.2.2.2 Market Access and Demand

Millet farmers face significant financial challenges due to limited market access, inadequate transportation, and insufficient storage facilities. Often perceived as traditional crops with low market appeal, millets typically fetch lower prices, discouraging investment in their cultivation. Farmers, frequently outside cooperatives or associations must sell to intermediaries at reduced rates, further eroding profitability. Limited access to markets and processing facilities lowers selling prices, relegating millets to home consumption or animal feed. The absence of established standards, certifications, branding, packaging, efficient distribution networks, promotional efforts, and comprehensive market information further exacerbates these challenges, hindering farmers' ability to capitalize on value-addition opportunities.

2.2.3 Sociocultural Challenges

2.2.3.1 Consumer Preferences

Millet consumption is declining globally, particularly in India, due to changing food habits and urbanization, where preferences increasingly favor refined and processed foods. Millets face reduced consumer acceptance due to their relatively bitter taste components, extended cooking times, and coarse textures. In India, per capita millet consumption plummeted from 30.94 kg per annum in 1960 to just 3.87 kg per annum in 2022, driven by these shifts and population growth outpacing production (Singh and Singh 2024). The popularity of polished grains such as rice and wheat has overshadowed millets, often perceived as coarse or less desirable. Western dietary influences have also contributed to the perception that refined grains are more prestigious and accessible to prepare. Studies emphasize that factors like health benefits, affordability, palatability, effective marketing strategies, and accessibility through diverse distribution channels influence consumer preferences for millet-based products. Increasing awareness and promoting millets as nutritious, healthy alternatives are critical to improving food security and combating malnutrition.

2.2.3.2 Perception and Awareness

There is often a lack of awareness about millet's nutritional and environmental benefits among consumers and policymakers, impacting funding and support for millet processing initiatives. A study in Coimbatore city revealed growing awareness about millets' health benefits, but barriers such as limited availability, lack of convenience, and established dietary habits persist (Jeyalakshmi 2024). Similarly, Kane-Potaka et al. (2021) found health concerns drive early millet adoption, yet flavor preferences and home consumption habits hinder broader acceptance. Additionally, another study has shown that only 41% of caregivers for children with type 1 diabetes mellitus were aware of millets, with barriers including non-palatability and high costs (Verma et al. 2024). Furthermore, the decline in millet cultivation is primarily due to farmers' limited awareness of the benefits and importance of growing millet.

2.2.4 Environmental and Policy Challenges

2.2.4.1 Climate Change

Climate change poses significant challenges to millet production, particularly in regions such as the Sahel, India, and sub-Saharan Africa. Increased temperatures, droughts, desertification, dry spells, and rainfall variability adversely affect millet yields and growth. In the Central Sahel, millet production is projected to decline by 21%, 20%, and 18% in northern, central, and western Mali, respectively, by 2030. Additionally, due to climate change, pearl millet production in sub-Saharan Africa is expected to decrease by 17% by 2050 (Abubakar et al. 2023). Another study reported that a 1% increase in average yearly temperature in India can lead to a 3.06% decrease in millet yield, while a 1% increase in CO_2 emissions per capita results in a 0.42% gain in yield (Jadaun and Khan 2023). Precipitation and rural population positively impact millet output in the long run, but temperature has a negative effect.

2.2.4.2 Policy Support

Inadequate policy support and funding for millet cultivation and processing have hindered the growth of the millet industry. Research efforts to improve millet varieties, productivity, and processing technologies are underfunded compared to refined cereals like rice and wheat. Additionally, refined cereals are more accessible at subsidized prices through various public feeding and nutrition programs, while millets are not. Consumer preferences have shifted toward more convenient and tastier foods, often influenced by Western culture or the misconception that millet is "poor man's food" (Singh and Vemireddy 2003). The lack of diversified millet-specific processing technologies and equipment further complicates the situation. The absence of a Minimum Support Price for millets has slowed their area expansion, and the lower profitability of millets, due to lower yields and environmental susceptibility, has impacted their commercialization.

2.2.4.3 Sustainability Concerns

Sustainable processing practices are essential for ensuring millet cultivation's environmental and economic viability. Challenges such as the lack of improved millet seeds, water stress during crop establishment, and high input costs for seeds, fertilizers, and pesticides pose significant obstacles to achieving sustainability in the millet industry. Emphasizing sustainable cultivation practices, including organic farming, agroforestry, and the promotion of seed diversity, plays a crucial role in enhancing resilience to pests, diseases, and climate variability. These practices minimize environmental impact and contribute to long-term agricultural productivity and the preservation of natural resources, thereby supporting the sustainable growth of the millet sector.

2.3 Improvement of Millet Processing Using Ultrasound Processing Technology

Ultrasound technology has emerged as a promising tool for enhancing various stages of millet processing, offering benefits such as improved efficiency, reduced processing time, and enhanced product quality. The application of ultrasound in food processing involves using high-frequency sound waves to induce physical and chemical changes in the food matrix. This section explores how ultrasound technology can be applied to improve different aspects of millet processing.

2.3.1 Dehulling

Ultrasound-assisted dehulling has emerged as a promising technique for enhancing millet processing efficiency and quality compared to traditional methods. Conventional dehulling often involves mechanical abrasion or impact, resulting in significant grain damage, nutrient loss, and uneven dehulling. In contrast, ultrasound treatment exploits the cavitation phenomenon, generating microbubbles imploding and creating localized high-pressure and temperature zones within the grain. This disruption weakens the bond between the husk and the kernel, facilitating easier removal (Fig. 2.1). A study reported that hydrothermal treatment combined with ultrasound improved the dehulling efficiency of pearl millet by 20% in terms of seed coat removal and reduced crude fiber content by 13.7% (Lazaro and Favier 2005). The benefits of ultrasound-assisted dehulling are manifold. Firstly, it achieves higher dehulling efficiency with minimal grain breakage, preserving the integrity and nutritional value of the millet. Secondly, the process is relatively gentle, reducing energy consumption and processing time compared to abrasive methods. Moreover, ultrasound treatment can improve the functional properties of millet flour by modifying starch granule size and distribution, enhancing its water-holding capacity and pasting properties. Additionally, ultrasound-assisted dehulling can

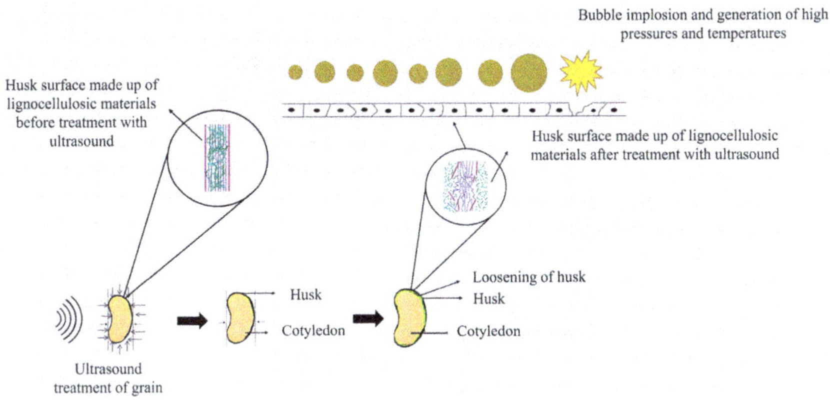

Fig. 2.1 A schematic diagram of the ultrasound-assisted dehulling process of food sample. (Adapted from Chang et al. (2024))

contribute to a cleaner production environment by reducing dust and chaff generation.

2.3.2 Milling

One of the primary advantages of ultrasound-assisted milling is its ability to enhance the efficiency of the milling process. Conventional methods often struggle with the hardness and small size of millet grains, leading to lower throughput and higher energy consumption. In contrast, ultrasonic waves create cavitation bubbles in the milling medium, which collapse and generate intense local pressure and temperature. This action effectively breaks down the millet grains more quickly and uniformly, reducing processing time and energy requirements (Fig. 2.2) (Joshi et al. 2023). In addition, traditional milling can cause significant nutrient loss due to the high mechanical forces and heat generated during the process. A study reported that water-holding capacity of ultrasonicated *kodo* millet flour (259.01 ± 6.0 g/100 g DW) and little millet flour (238.13 ± 3.0 g/100 g DW) significantly surpassed that of chemically treated samples, followed by germination of *kodo* millet flour (212.53 ± 6.8 g/100 g DW) and little millet flour (200.76 ± 2.9 g/100 g DW). Similarly, the oil holding capacity of ultrasonicated *kodo* millet flour (227.23 ± 9.8 mL/100 g DW) and little millet flour (207.5 ± 8.51 mL/100 g DW) exhibited a significantly higher value compared to chemically treated samples, followed by germination of *kodo* millet flour and little millet flour. Ultrasonication potentially disrupts starch granules, influencing the reconstitution characteristics of millet flour. Furthermore, ultrasonicated *kodo* millet flour (801.01 ± 18.1%) demonstrated superior dispersibility compared to chemically treated samples, followed by germination of *kodo* millet flour (749.95 ± 12.7%) (Dey et al. 2024; Sharma et al.

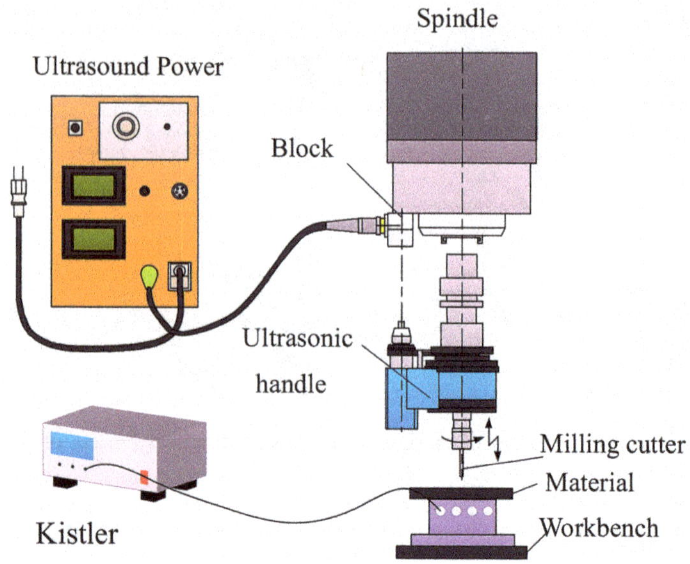

Fig. 2.2 A schematic diagram of ultrasonic vibration-assisted milling. (Adapted from Yang et al. (2021))

2021). Ultrasound-assisted milling operates at lower temperatures and uses less mechanical force, which helps retain essential nutrients such as proteins, vitamins, and minerals (Singh et al. 2024). A study compared the effect of the wet milling method and ultrasound-associated isolation on isolated starch chemical composition. They found that the total starch of wet milling was 87.82%, whereas ultrasound-associated isolation found 92.89% (Bouhallel et al. 2021). This preservation of nutritional content is crucial to millet, valued for its high nutritional profile and health benefits. Furthermore, ultrasonic waves' precise and gentle milling action results in finer and more uniform particle sizes. This uniformity enhances the texture and appearance of millet flour, making it more appealing for various culinary applications. Additionally, the gentle processing helps maintain the millet's natural flavor and aroma, which the harsh conditions of conventional milling can compromise.

2.3.3 Drying

Ultrasound exposes millet to high-frequency sound waves during drying, enhancing efficiency, product quality, and nutritional value retention. Compared to traditional drying methods, such as hot air drying, ultrasound-assisted drying promotes faster moisture removal, leading to shorter processing times and increased throughput (Zhang and Abatzoglou 2020). The ultrasonic waves create microscopic cavities within the millet grains, facilitating efficient water vapor diffusion and evaporation

(Mounika et al. 2024). This accelerated drying process helps to preserve the millet's inherent qualities, including its color, flavor, and texture. An investigation revealed that within various drying methodologies, the synergistic application of ultrasonication (20 min) in conjunction with dry heat (8 h) resulted in an augmentation of resistant starch content (14.49–31.69%) and a reduction in glycemic index (58.25–48.49) (Vidhyalakshmi and Meera 2023a). Furthermore, ultrasound technology has a profound impact on the millet's microstructure. The cavitation generated by the sound waves breaks down cellular structures, increasing the surface area available for moisture transfer (Fig. 2.3). This disruption also aids in inactivating enzymes and microorganisms, extending the millet's shelf life and improving its safety (Ahmed et al. 2024a). Another critical advantage of ultrasound-assisted drying is its potential to enhance the nutritional profile of millet. The gentle drying process helps to retain essential vitamins, minerals, and antioxidants. Additionally, ultrasound-induced cavitation can improve the bioavailability of certain nutrients by increasing their solubility and digestibility (Ahmed et al. 2024b).

2.3.4 Hydration

Ultrasound-assisted hydration significantly improves millet processing by enhancing water absorption rates during the soaking phase. Traditional hydration methods typically require prolonged soaking times to achieve adequate moisture levels in millet grains, which can be inefficient and energy-intensive. In contrast, applying

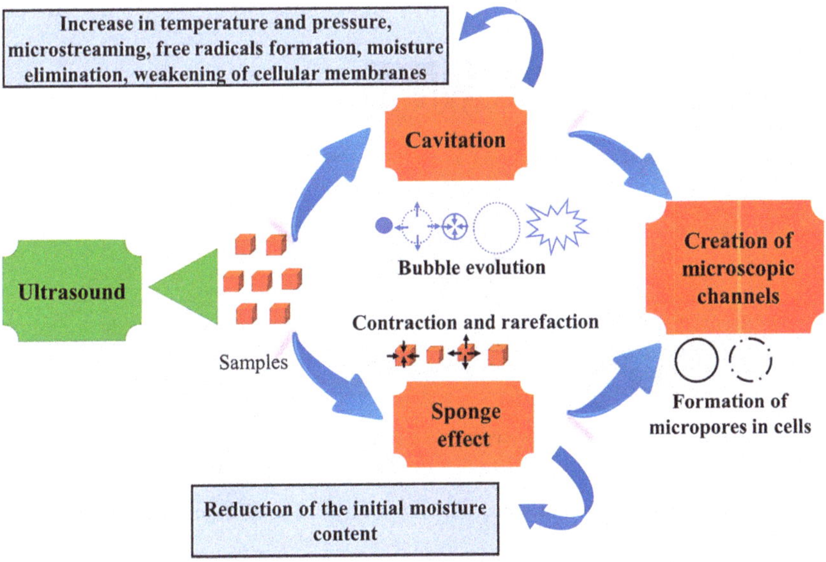

Fig. 2.3 A schematic diagram of the ultrasound-assisted drying process of food samples. (Adapted from Xu et al. (2022))

ultrasonic waves accelerates the hydration process through acoustic cavitation, producing microscopic bubbles imploring and creating shock waves. This phenomenon increases the permeability of the grain's outer layer, allowing water to penetrate more quickly and evenly. This results in a substantial reduction in soaking time, which enhances overall processing efficiency and throughput (Fig. 2.4) (Safwa et al. 2023). Conventional long-duration soaking can lead to the leaching of essential nutrients into the soaking water, diminishing the nutritional value of the final product. However, the rapid and efficient hydration provided by ultrasonic waves minimizes nutrient loss, ensuring that more vitamins and minerals are retained within the grains (Pateiro et al. n.d.). Additionally, this method can improve the texture and consistency of the millet by promoting a more uniform moisture distribution, which is beneficial for subsequent milling and cooking processes. Ultrasonication has been demonstrated to significantly reduce antinutrients in millets. For instance, in finger millet, ultrasound-assisted hydration reduced phytates and tannins by up to 73% and 71%, respectively (Dubey and Tripathy 2024). Overall, ultrasound-assisted hydration presents a more effective and nutrient-preserving alternative to traditional soaking methods in millet processing.

2.3.5 Extraction

Ultrasound-assisted extraction (UAE), on the other hand, leverages high-frequency sound waves to create cavitation bubbles in the extraction medium. These bubbles

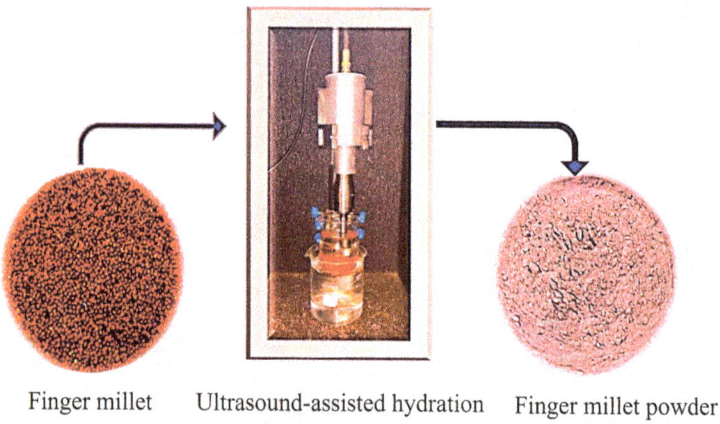

Fig. 2.4 A schematic diagram of ultrasound-assisted hydration of finger millet. (Adapted from Yadav et al. (2021))

implode, generating intense localized pressure and temperature. The intense mechanical effects produced by ultrasound waves enhance the penetration of the solvent into the plant matrix, leading to a more efficient release of bioactive compounds such as phenolics, flavonoids, and antioxidants (Fig. 2.5). This improvement not only boosts the nutritional value of millet-based products but also reduces the quantity of solvent required, making the process more sustainable and cost-effective. Furthermore, the UAE significantly reduces the extraction time. Conventional solvent extraction methods can take several hours to complete, whereas UAE can achieve similar or even superior results in a fraction of the time. For instance, Balasubramaniam et al. (Balasubramaniam et al. 2019) showed that the application of the UAE technique for the finger millet polyphenol extraction has shortened the extraction time threefold, decreased the requirement for solvent threefold, and lowered the operating temperature by 10 °C compared to the conventional heat reflux method. This increased efficiency is particularly beneficial for industrial applications, where time and throughput are critical factors. The shorter processing time also helps preserve the integrity and bioactivity of the extracted compounds, which can be degraded or lost during prolonged extraction periods. Zheng et al. (2022) demonstrated the highest antioxidant activity for phenolic extracts using UAE with deep eutectic solvents (UAE-DES) through three in vitro antioxidant assays. These extracts also showed better antioxidant and acetylcholinesterase inhibitory activities, with IC_{50} values of 715 μg/mL for UAE-DES and 832 μg/mL for conventional solvent extraction. The lower IC_{50} value for UAE-DES suggests increased inhibitory activity against acetylcholinesterase, likely due to higher total phenolic content and hydroxycinnamic acid amides.

2.4 Application of Ultrasound on Millet and Millet-Based Products

2.4.1 Bioactive Compound Extraction from Millet

The UAE effectively extracts intracellular contents such as saponins, phytates, phe-

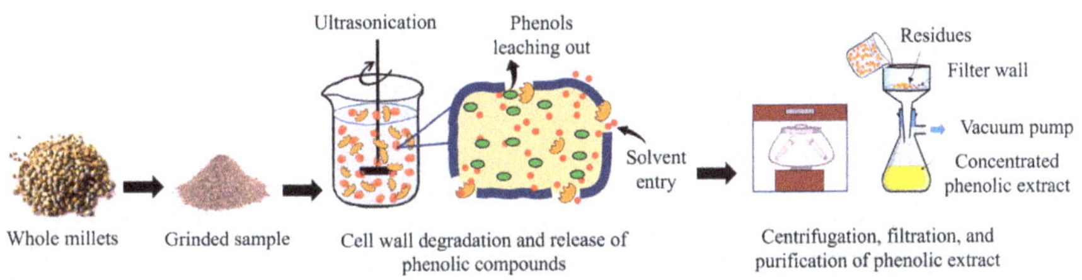

Fig. 2.5 A schematic showing the sequential steps for ultrasound-assisted extraction of bioactive compounds from millet. (Adapted from Nayak et al. (2024))

nolics, flavonoids, and alkaloids from millet into the medium. Studies have shown that UAE can extract free-bound phenolics from sorghum flour, yielding higher total phenolic content (TPC) through continuous ultrasonication (709 mg GAE/100 g d.m.) compared to batch processing (656 mg GAE/100 g d.m.). The constant method's lower specific energy (33%) and 35% reduced processing time contribute to this efficiency (Table 2.1) (Lohani and Muthukumarappan 2021). Additionally, UAE, with aqueous ethanol as the solvent, has extracted 29% more polyphenolic compounds from red sorghum bran than conventional methods (Luo et al. 2018). Ultrasonication accelerates the soaking and germination of stored seeds. For instance, Hassan et al. (2020) found that sonication at 40% amplitude for 5 min increased sorghum sprout germination rates from 80% to 94%. Furthermore, the same treatment decreased the samples' levels of anti-nutritional factors (saponins, phytates, and alkaloids). High-intensity ultrasound pretreatment improves the nutritional quality of millet bran by converting insoluble/bound dietary fibers and antioxidants into soluble/free forms, making them more absorbable. Mustac et al. (2019) used ultrasound pretreatment to increase bioactive compounds in proso millets, achieving a 16% higher TPC and a 17% improvement in antioxidant activity after sonication at 80% amplitude (400 W/cm^2) for 13 min. Raj et al. (2006) suggested that sonication enhances TPC yield by deactivating polyphenol oxidase, reducing polyphenol breakdown, and resulting in higher yields. Also, using a deep eutectic solvent, UAE extracted phenolic compounds from foxtail millet bran. This method achieved a phenolic content of 8 mg FAE/g and a total flavonoid content of

Table 2.1 Application of ultrasound process technology on millet and millet-based product

Millet/ millet-based product	Purpose	Ultrasonic conditions	Key findings	Ref.
Sorghum flour	Phenolic content Antioxidant activity	Batch ultrasonication: 10, 15, and 20% (w/v) 30, 55, and 80 w/cm^2 Continuous ultrasonication: 10, 20, and 30% (w/v) 20, 40, and 60 w/cm^2	Compared with batch ultrasonication, the continuous flow process saved 35% time and 33% energy consumption to obtain the sorghum flour's comparatively higher total phenolic content (66 mg GAE/100 g d.m.) and antioxidant activity (141.0 μmol TE/100 g d.m.)	Lohani and Muthukumarappan (2021)
Sorghum sprouts	Germination rates and phytochemical constituents	40% amplitude, 5 minutes	Germination rates from 80% to 94% and the levels of alkaloids, saponins, tannins, and phytates were significantly ($p \leq 0.05$) decreased, having the amount of 0.04 ± 0.002 mg/100 g, 0.10 ± 0.02 mg/100 g, 0.13 ± 0.004 mg CE/100 g DM, and 143.25 ± 2.32 mg SPE/100 g DM, respectively	Hassan et al. (2020)

(continued)

Table 2.1 (continued)

Millet/ millet-based product	Purpose	Ultrasonic conditions	Key findings	Ref.
Millet bran	Oil extraction	1:10 (solid-to-liquid), 40 kHz, and 300 W	Ultrasonication of millet bran oil demonstrated notable yield (87.68%) under optimal conditions of n-butyl alcohol as extraction solvent, solid-to-liquid ratio of 1:10 (g/mL), two repeated cycles of extraction at 60 °C for 20 min, ultrasonic frequency of 40 kHz, and ultrasonic power of 300 W. The oil revealed 13.08% hexadecanoic acid, 56.73% 9,12-octadecadienoic acid, 19.78% 9-octadecenoic acid, 5.41% octadecanoic acid, and 0.8% β-sitosterol	Chen et al. (2013)
Browntop millet	Functional qualities of millet protein	200 W, 0, 5, 10, 15, and 20 minutes	Protein hydrolysates-15 exhibited significantly enhanced hydrolysis (66%), protein digestibility (85%), free sulfhydryl content (10 μM/g), and emulsifying activity (21 m^2/g)	Rao et al. (2024)
Millet-based probiotic beverage	Fermentation kinetics and beverage quality	23 kHz, 10 μm, 3–5 min	Increased the *lactobacillus brevis* bacterial counts (1.09 log CFU/mL), specific growth rate (58.17–82.59%), cellular membrane permeability (16–36.25%), protein production (31.76%), and gamma-aminobutyric acid production (31.76%)	Meena et al. (2024)
Pearl millet flour	Modification of the starch properties	44 Hz, 10, 20, and 30 min	Increased resistant starch content (15–32%) and decreased glycemic index (58–49)	Vidhyalakshmi and Meera (2023b)
Pearl millet starch-based films	Improvement of stability and gel texture of starches	1:10 (solid-to-liquid), 33 kHz, 45 min	Highest amylose (19.5 ± 0.3%), swelling power (17.53 ± 0.4 g/g), water solubility (35.40 ± 2.2%) and opacity (2.191 ± 00%) were increased significantly by sonification compared to untreated starch	Punia Bangar et al. (2021)
Barnyard millet milk	Process parameters optimization for functional properties	20 kHz, 100 W, 10, 20, and 30 min	Enhanced the viscosity of millet milk from 11.60 to 16.59 cP	Shunmugapriya et al. (2020)

9 mg RE/g, doubling the yield compared to conventional solvent extraction (Zheng et al. 2022). These studies confirm that ultrasonication is a potent green extraction method for millet bioactive compounds, showcasing its efficacy in leaching diverse constituents from cereal flour. Furthermore, the ability of ultrasound to facilitate oil release under mild conditions is particularly advantageous. Unlike conventional extraction techniques requiring high temperatures or harsh solvents, UAE can be performed at lower temperatures. This mild processing environment helps preserve the quality of the extracted oil, maintaining its nutritional and sensory properties and enhancing the overall yield. For example, a study in the UAE of millet bran oil demonstrated a yield of 87.68% under optimal conditions. These conditions included a solid-to-liquid ratio of 1:10, ultrasonic frequency of 40 kHz, and power output of 300 W (Chen et al. 2013).

2.4.2 Millet Protein Concentrate

Ultrasound processing technology significantly improves the protein concentrate of millet by applying high-intensity ultrasound waves typically ranging from 20 to 100 kHz under higher intensities from 10 to 1000 W/cm^2 (Kumari et al. n.d.). This technology relies on cavitation, where rapid gas bubble formation and collapse in the aqueous medium create intense localized pressure differentials over microsecond timescales. This cavitation induces massive shear forces, turbulence, and localized temperature increases around the collapsing bubbles, fundamentally altering the protein structures within the millet. The primary impact of ultrasound processing on millet proteins involves modifications to their native structures, leading to changes in molecular aggregation properties and interfacial activity. These alterations improve techno-functional properties and enhance protein digestibility (Meherunnahar et al. 2023). For example, a study found that ultrasound treatment significantly increased the solubility of millet protein concentrate compared to the native form. While low-intensity ultrasound reduced foaming capacity, high-intensity treatments notably enhanced it and improved emulsifying activity index and emulsion stability. The optimal treatment of 74 W/cm^2 for 13 minutes resulted in reduced molecular weight and a 36% increase in negative surface charge, as confirmed by SDS-PAGE and Zeta potential analysis (Nazari et al. 2018). The cavitation process decreases the size of protein aggregates and partially unfolds their three-dimensional structures, exposing hydrophobic and sulfhydryl groups that are otherwise buried within the molecule. This exposure enhances the protein's solubility and emulsifying capabilities, making the resulting protein concentrate more functional for various food applications. Furthermore, the structural modifications induced by ultrasound improve the interfacial activity of millet proteins, which is crucial for their performance in food systems. Enhanced interfacial activity means the proteins can more effectively stabilize emulsions, foams, and other colloidal systems, improving millet-based products' texture, consistency, and overall quality. These improvements are precious for the food industry, where protein functionality directly impacts the sensory and nutritional qualities of the final product. In addition to improving functionality, ultrasound processing enhances millet protein digestibility. The partial unfolding of protein structures makes them more accessible to digestive enzymes, facilitating better protein absorption in the human body. Akharume and Adedeji (2023) demonstrated that high-power ultrasound treatment significantly enhanced prolamin and glutelin fractions' solubility and in vitro digestibility from two proso millet cultivars (Dawn and Plateau). Notably, there was an average increase of 30–58% in digestibility and up to 235% in solubility, with the foaming capacity increasing significantly only with longer treatment times and the emulsion activity index increasing slightly after 5 minutes but not 10 min of treatment, attributed to the multimodal distribution of particle sizes. Another study investigated browntop millet protein isolates modified by ultrasound pretreatment (200 W, 15 min) and bromelain hydrolysis, revealing that ultrasound-pre-treated millet protein hydrolysates-15 (UBMPH-15) exhibited significantly enhanced hydrolysis (66%), protein digestibility (85%), free sulfhydryl content (10 μM/g), and emulsifying activity (21 m^2/g). Additionally, UBMPH-10 showed higher solubility (89%), hydrophobicity (19 μg/g), and foaming capacity (18.52%). Both UBMPH-15 and UBMPH-10 demonstrated improved antidiabetic activity, with UBMPH-15 also achieving the highest antioxidant and antilipidemic activities, suggesting their potential as bioactive hydrolysates (Rao et al. 2024).

2.4.3 Millet-Based Probiotic Products

Ultrasound technology has significantly enhanced millet-based probiotic beverages by improving their nutritional, functional, and sensory properties. Through ultrasound-assisted fermentation, the growth and activity of essential probiotic bacteria, such as *Lactobacillus* and *Bifidobacterium*, have been markedly increased. This enhancement in probiotic activity is crucial for fermentation, producing bioactive compounds like peptides and exopolysaccharides, which offer various health benefits, including antioxidant, anticancer, and anti-inflammatory properties (Amadou 2019). One of the notable advantages of ultrasound treatment is the reduction in fermentation time. Studies observed that the minimum fermentation time required for ultrasound-treated white finger millet (*Eleusine coracana* L.) variety KMR-340 was 2.6 times lower than untreated samples. The ultrasound treatment postinoculation resulted in a higher lactic acid rate constant, indicating more efficient fermentation. Low-intensity ultrasound treatment effectively promoted the proliferation of probiotic microbes in millet-based beverages, suggesting potential alterations in their physicochemical properties (Meena et al. 2024). Moderate ultrasonic treatments (23 kHz, 10 μm magnitude for 3–5 min) significantly improved the probiotic attributes of millet-based beverages. There was a noticeable increase of 1.09 log CFU/mL in *Lactobacillus brevis* bacterial counts, a 58.17–82.59% rise in specific growth rate, a 16–36.25% increase in cellular membrane permeability, and up to a 36.42% rise in protein breakdown, along with a 31.76% increase in gamma-aminobutyric acid (GABA) production. These findings indicate that low-intensity ultrasonic waves effectively produce GABA-rich fermented products using *L.*

brevis as a probiotic adjunct (Pandey et al. 2021). Furthermore, ultrasound treatment has been shown to enhance the viability of encapsulated probiotics, ensuring their stability and effectiveness during digestion. Studies have demonstrated that ultrasound technology boosts probiotics' growth and fermentative profiles, leading to higher biomass and improved functional properties of the final product (Niamah 2019).

2.4.4 Millet Flour and Starch

Ultrasound technology has shown significant potential in improving the properties of millet flour and its byproducts, making it a valuable tool in the food industry. A study on pearl millet flour treated under various conditions (10, 20, or 30 minutes at 30 °C with a frequency of 44 kHz) revealed several notable changes (Vidhyalakshmi and Meera 2023b). For instance, dry heat-treated samples exhibited a 10% decrease in enthalpy change (ΔH) after 8 hours compared to untreated flour. Additionally, ultrasonicated flour showed significant increases ($p < 0.05$) in peak viscosity, hot paste viscosity, and setback viscosity compared to untreated polished flour. The resistant starch content in the treated flour also increased significantly ($p < 0.05$), while both the hydrolysis index and the predicted glycemic index decreased significantly ($p < 0.05$) relative to the control.

Ultrasound treatment can modify the structural properties of starches in millet, similar to its effects on other grains like tef and rice. It promotes starch fragmentation, increases amylose content, and enhances water absorption and swelling power, improving pasting and rheological properties (Vela et al. 2023). Dual-frequency ultrasound treatment has been particularly effective in modifying starches with large granule sizes, such as potato starch, suggesting that similar benefits could be achieved with millet starch (Hu et al. 2019). Moreover, this technology accelerates grain hydration and germination processes without altering starch properties, which could be beneficial for improving the functional properties of millet flour. Furthermore, ultrasound has been used to enhance the extraction and quality of starch from other cereals like corn, resulting in higher starch yields and improved pasting properties (Elfawakhry et al. 2011). Ultrasound has also significantly affected the properties of pearl millet starch-based films, enhancing their functional characteristics. The study on starch-based films under sonication treatment at 33 kHz for 45 minutes highlights several significant changes in the physicochemical properties of the films (Punia Bangar et al. 2021).

2.4.5 Millet Milk

Ultrasound technology has also been utilized in dairy applications for emulsification, microbial inactivation, and homogenization, showcasing its potential for enhancing millet milk production. This technology shows significant promise in processing millet milk by enhancing its functional properties. Combining ultrasound with other techniques, such as germination, has been shown to reduce antinutrient concentrations and improve the functional properties of millet milk. This makes ultrasound a valuable tool for developing nutritious plant-based milk alternatives. Controlled time and power parameters during ultrasonication have been found to increase the sugar concentration and total phenolic activity of millet milk, elevating these levels from 7.8% to 8.12% and 6.49% to 6.54%, respectively. Conversely, ultrasonication has also demonstrated the capacity to decrease antinutrient and protein contents by up to 32.27% (Saxena et al. 2023). Additionally, ultrasonication has been shown to enhance the whiteness and viscosity of millet milk, with viscosity improvements from 11.60 to 16.59 cP. Ultrasonicated millet milk demonstrated significant improvements in various properties compared to non-ultrasonicated versions. The highest separation value of 72.61 ± 1.72 mL/h was recorded in ultrasonicated barnyard millet milk, indicating enhanced stability (Shunmugapriya et al. 2020). Furthermore, the whiteness of millet milk was notably improved through ultrasonication, contributing to a more visually appealing product. The increased viscosity of millet milk with extended sonication time suggests that ultrasonication effectively modifies the physical properties of millet milk, potentially enhancing its texture and mouthfeel.

2.5 Challenges and Future Directions

Ultrasound processing technology in millet processing faces several notable challenges that hinder its widespread adoption and optimization. One of the primary challenges in processing millet grains is their complex and heterogeneous structure. Millet grains consist of varying proportions of starch, protein, and fiber, each contributing to the overall complexity of the grain. This diversity necessitates precise control over the ultrasound parameters during processing to achieve optimal results (Akhi et al. 2024). For instance, different processing objectives, such as milling, starch, and protein extraction, require optimal ultrasound conditions to target the desired component effectively. Moreover, the presence of anti-nutritional factors in millet can influence the effectiveness of ultrasound treatment. Therefore, achieving the desired improvements in nutritional and physical properties without causing adverse effects is a delicate balance. For instance, high-intensity ultrasound treatment of proso millet bran requires precise optimization to enhance antioxidant activity and dietary fiber content while minimizing enzymatic browning and ensuring water retention (Datta Mazumdar et al. 2022). This delicate optimization process is crucial for maintaining the quality and health benefits of millet-based products. Techniques such as ultrasound-assisted enzymatic treatment, thermal processing, and fermentation can complement ultrasound by enhancing nutrient extraction, reducing anti-nutritional factors, and improving functional properties.

Integrating ultrasound technology into industrial processes is also complex, as it involves overcoming technical hurdles related to equipment design and scalability. The specialized equipment required, such as ultrasonic transducer devices and circulation systems, poses a significant challenge, especially for small enterprises that often lack the resources for such investments. The high initial costs and the need for solutions to fit specific production lines further complicate the widespread adoption of this technology in the millet processing industry (Yao et al. 2020). Furthermore, balancing energy efficiency and product quality is another critical challenge. Ultrasound technology is renowned for its potential to save energy while preserving high product quality. However, consistently achieving these benefits across different food matrices, including millets, can be difficult. The nonthermal nature of ultrasonication, although advantageous for preserving heat-sensitive nutrients, requires careful control to avoid incomplete microbial inactivation or enzyme deactivation. These issues could compromise food safety and shelf life, highlighting the need for meticulous control and monitoring during processing (Aghababaei et al. 2024). Moreover, the variability in the effects of ultrasound on different types of millets and their byproducts necessitates extensive research. Tailoring the technology to specific applications, such as enhancing bioactive compound extraction or improving protein digestibility, requires a deep understanding of the interactions between ultrasound parameters and the unique properties of various millet types. This variability demands comprehensive studies to develop optimized protocols that can be reliably applied across different millet-based products.

In future directions, combining ultrasound technology with other emerging processing methods, such as high-pressure processing (HPP), pulsed electric fields, or enzymatic treatments, can offer a hurdle approach to address existing challenges. This synergistic combination can enhance microbial inactivation, enzyme deactivation, and bioactive compound extraction while preserving product quality and energy efficiency. For instance, integrating ultrasound with HPP could improve microbial safety without compromising nutrient retention, while coupling it with enzymatic processes may optimize protein digestibility and extraction yields. Such multi-technology strategies can help tailor solutions for the unique properties of millet-based products, ensuring scalability and broader applicability in the food industry.

Furthermore, integrated ultrasound technology can modify the structure of plant-based beverages and proteins, thereby improving the bio-accessibility of nutrients, protein solubility, and digestibility while also effectively inactivating microbial contaminants. For instance, when combined with germination, ultrasound processing can significantly enhance millet milk's physicochemical, proximate, adsorptive, antioxidant, and sensory characteristics. This integration increases phenolic and antioxidant activity while reducing antinutrient concentrations, producing superior quality and nutrient-rich products (Rojas et al. 2022).

Millet comprises several varieties, each with unique physical and chemical properties. This diversity poses a challenge in standardizing ultrasound processing parameters. Different millet types may require specific ultrasonic frequencies, intensities, and treatment durations to achieve optimal results, making it challenging to develop a one-size-fits-all solution (Kaur et al. 2024). Future research should focus on characterizing the specific requirements of different millet varieties under ultrasound treatment. Advanced sensors and artificial intelligence-driven algorithms could be employed to automatically adjust ultrasound parameters based on the detected properties of the millet being processed. This approach would enhance the efficiency and effectiveness of ultrasound treatment across various millet types.

Ensuring consistent quality control during ultrasound processing of millet is a significant challenge. Variations in ultrasound treatment can lead to inconsistencies in the final product, affecting its nutritional value, texture, and taste. Maintaining uniformity in ultrasound exposure across different batches is crucial to meet quality standards and consumer expectations (Ahmed et al. 2022). Implementing real-time monitoring and feedback systems within the ultrasound processing units can be beneficial to overcome this challenge. These systems can continuously track and adjust the ultrasound parameters to ensure uniform treatment. Additionally, developing standardized protocols and best practices for the ultrasound processing of millet can help maintain consistency in the final product.

While ultrasound technology can enhance millet processing efficiency, its environmental impact cannot be overlooked. The energy consumption of ultrasound devices and the potential generation of ultrasonic waste pose environmental concerns. Balancing the benefits of ultrasound processing with its ecological footprint is essential. Research into alternative energy sources, such as solar power, to run ultrasound equipment could significantly reduce its environmental impact. Furthermore, designing systems that minimize waste generation and promote sustainable processing practices will be crucial for the long-term viability of ultrasound technology in millet processing.

2.6 Conclusion

In summary, the integration of ultrasound technology into millet processing presents an advancement with the potential to address current challenges in the industry. The traditional methods of millet processing often face limitations such as inefficiency, nutrient loss, and suboptimal quality of the final product. However, the application of ultrasound processing technology offers promising solutions to these issues. Ultrasound processing enhances the efficiency of dehulling, milling, and extraction processes, leading to improved yields and better retention of essential nutrients. This technology also facilitates the modification of millet-based products, improving their functional and sensory properties. The application of ultrasound has shown potential in reducing processing time and energy consumption, thus contributing to more sustainable and cost-effective practices. This chapter highlights several applications of ultrasound technology in various aspects of millet processing. These include improved water absorption during soaking, enhanced starch and bioactive compound extraction, and better structural modification of millet flour. The positive outcomes observed in these applications suggest that ultrasound technology can play a crucial role in modernizing millet processing and making millet-based products more competitive in the market. Looking forward, the future implications of ultrasound technology in millet processing are substantial. Further research and development are needed to optimize processing parameters and explore new applications of ultrasound in this field. There is also a need for scaling up these technologies for industrial applications and assessing their economic viability. By addressing these aspects, ultrasound processing can be fully integrated into the millet processing industry, leading to higher-quality products, increased nutritional value, and more sustainable processing practices. In conclusion, the adoption of ultrasound technology in millet processing offers a pathway to overcome existing challenges and unlock new opportunities. This chapter underscores the importance of continued research and innovation in this area to fully harness the potential of ultrasound in enhancing millet processing and its applications.

Author Contributions Conceptualization, writing-original draft preparation, and reviewing, supervision, T.A.; Writing-original draft preparation and visualization- M.S., K.A., M.H.; all authors have read and agreed to the published version of the manuscript.

Conflicts of Interest The authors declare no conflicts of interest.

References

Abass AB, Ndunguru G, Mamiro P, Alenkhe B, Mlingi N, Bekunda M (2014) Post-harvest food losses in a maize-based farming system of semi-arid Savannah area of Tanzania. J Stored Prod Res 57:49–57

Abubakar A, Ishak MY, Uddin MK, Sulaiman Zangina A, Ahmad MH, Shehu Danhassan S (2023) Impact of climate change and adaptations for cultivation of millets in Central Sahel. Environ Sustain 6:441–454

Aghababaei F, McClement DJ, Hadidi M (2024) Ultrasound processing for enhanced digestibility of plant proteins. Food Hydrocoll 11:110188

Ahmed T, Rana MR, Maisha MR, Sayem ASM, Rahman M, Ara R (2022) Optimization of ultrasound-assisted extraction of phenolic content & antioxidant activity of hog plum (Spondias Pinnata L. f. Kurz) Pulp by response surface methodology. Heliyon 8:e11109

Ahmed T, Suzauddula M, Akter K, Hossen M, Islam MN (2024a) Green technology for fungal protein extraction—a review. Separations 11:186

Ahmed T, Suzauddula M, Safwa SM, Talukder S, Sharma M, Inbaraj BS, Sridhar K (2024b) Biotechnological tools in microbial vitamins and carotenoids production. In: Microbial vitamins and carotenoids in food biotechnology. Elsevier, pp 63–104

Akanbi TO, Timilsena Y, Dhital S (2019) Bioactives from millet: properties and effects of processing on bioavailability. Bioactive Factors and Processing Technology for Cereal Foods:171–183

Akharume F, Adedeji A (2023) Effects of high-power ultrasound on the in vitro digestibility, physicochemical and functional properties of Proso millet prolamin and glutelin proteins. J Food Meas Charact 17:178–186

Akhi A, Ahmed T, Ara R, Rana MR (2024) Response surface optimization of Thermo-sonication conditions and Taro mucilage concentrations for the preparation of soy yogurt. J Agric Food Res 15:100918

Amadou I (2019) Millet based fermented beverages processing. In: Fermented beverages. Elsevier, pp 433–472

Balasubramaniam VG, Ayyappan P, Sathvika S, Antony U (2019) Effect of enzyme pretreatment in the ultrasound assisted extraction of finger millet polyphenols. J Food Sci Technol 56:1583–1594

Bouhallel S, Belhadi B, Souilah R, Djabali B D (2021) Nadjemi isolation of starch from seven pearl millet grain landraces by two processes; wet milling and ultrasound application. Algerian J Env Sc Technol 7:1687–1697

Chang L, Shi R, Dai F, Zhao W, Zhao Y, Chen J (2024) Current flaxseed dehulling technology in China. Agriculture 14:632

Chege, C.; Onyango, K.; Bodjerenou, S.; Schmidt, M.; Heike, H.; Ostermann, H. Adapting green innovation centers to climate change: reducing post-harvest loss and improving processing for target groups. 2024

Chen H, Gu N, Lu Z, Zhao H (2013) Optimization of ultrasonic-assisted extraction of millet bran oil and GC-MS analysis of its fatty acid composition. Food Sci 34:32–36

Datta Mazumdar S, Priyanka D, Akhila Y (2022) Emerging Technologies in Millet Processing. Handbook of Millets-Processing, Quality, and Nutrition Status:231–263

Dey S, Saxena A, Kumar Y, Maity T, Tarafdar A (2024) Synergistic effects of germination and Ultrasonication on nutritional and structural characteristics of Kodo (Paspalum Scrobiculatum) and little (Panicum Sumatrense) millet. J Food Qual 2024:4951196. https://doi.org/10.1155/2024/4951196

Dubey A, Tripathy PP (2024) Ultrasound-mediated hydration of finger millet: effects on Antinutrients, techno-functional and bioactive properties, with evaluation of ANN-PSO and RSM optimization methods. Food Chem 435:137516. https://doi.org/10.1016/j.foodchem.2023.137516

Elfawakhry H, Hussein MA, Becker T (2011) Possibility of using acoustic techniques for dough processing evaluation. Procedia Food Sci 1:2023–2032

Gowda NAN, Siliveru K, Prasad PVV, Bhatt Y, Netravati BP, Gurikar C (2022) Modern processing of Indian millets: a perspective on changes in nutritional properties. Foods 11:499

Gudadhe NN, Raut AA, Bisht K, Bisht K (2023) Assessment of production trends, nutritional benefits, constraints, policy support, processing and value addition of millets for achieving higher productivity and nutritional security. Indian J Fertilis 19:1020–1031

Hassan S, Imran M, Ahmad MH, Khan MI, Xu C, Khan MK, Muhammad N (2020) Phytochemical characterization of ultrasound-processed sorghum sprouts for the use in functional foods. Int J Food Prop 23:853–863

Hassan ZM, Sebola NA, Mabelebele M (2021) The nutritional use of millet grain for food and feed: a review. Agric Food Sec 10:1–14

Hu A, Li Y, Zheng J (2019) Dual-frequency ultrasonic effect on the structure and properties of starch with different size. Lwt 106:254–262

Jacob J, Krishnan V, Antony C, Bhavyasri M, Aruna C, Mishra K, Nepolean T, Satyavathi CT, Visarada KBRS (2024) The nutrition and therapeutic potential of millets: an updated narrative review. Front Nutr 11:1346869. https://doi.org/10.3389/fnut.2024.1346869

Jadaun KK, Khan AA (2023) Impact of climate change on millet yield in India since 1991: an econometric analysis. J Res ANGRAU 51:129–139

Jeyalakshmi S (2024) Consumer perception towards millets among households in Coimbatore City

Joshi TJ, Singh SM, Rao PS (2023) Novel thermal and non-thermal millet processing technologies: advances and research trends. Eur Food Res Technol 249:1149–1160

Joshi RK, Ghimire KH, KC, H.B. GC A (n.d.) Insect Pests of Millets and Their Management in Nepal

Kane-Potaka J, Anitha S, Tsusaka TW, Botha R, Budumuru M, Upadhyay S, Kumar P, Mallesh K, Hunasgi R, Jalagam AK et al (2021) Assessing millets and sorghum consumption behavior in urban India: a large-scale survey. Front Sustain Food Syst 5:680777

Kaur G, Ahmadzadeh-Hashemi S, Amir S, Khan ZS, Gulsunoglu-Konuskan Z, Karimidastjerd A, Fayaz S, Bhat MS, Rustagi S, Bekhit AE-DA et al (2024) Exploring sustainable novel millet protein: a look at the future foods through innovative processing. Future Foods 9:100367

Kumari P, Singh A, Chauhan K (n.d.) Proso millet (Panicum Miliaceum L.)

Lazaro E, Favier J (2005) Improvement of the Dehulling efficiency of sorghum and millet using hydrothermal treatments. J Agric Sci Technol 3:15–30. https://doi.org/10.4314/jagst.v3i2.31677

Lohani UC, Muthukumarappan K (2021) Study of continuous flow Ultrasonication to improve Total phenolic content and antioxidant activity in sorghum flour and its comparison with batch Ultrasonication. Ultrason Sonochem 71:105402

Luo X, Cui J, Zhang H, Duan Y, Zhang D, Cai M, Chen G (2018) Ultrasound assisted extraction of polyphenolic compounds from red sorghum (sorghum bicolor L.) bran and their biological activities and polyphenolic compositions. Ind Crop Prod 112:296–304

Ma K, Zhao X, Lu B, Wang Y, Yue Z, Zhang L, Diao X, Yuan X (2024) Effect of ecological factors on nutritional quality of foxtail millet (Setaria Italica L.). Agronomy 14:387

Meena L, Malini B, Byresh TS, Sunil CK, Rawson A, Venkatachalapathy N (2024) Ultrasound as a pre-treatment in millet-based probiotic beverage: it's effect on fermentation kinetics and beverage quality. Food Chem Adv 4:100631

Meherunnahar M, Ahmed T, Chowdhury RS, Miah MAS, Sridhar K, Inbaraj BS, Hoque MM, Sharma M (2023) Development of novel foxtail millet-based nutri-rich instant noodles: chemical and quality characteristics. Foods 12:819

Meherunnahar M, Hoque MM, Satter MA, Ahmed T, Chowdhury RS, Aziz S (2024) Effect of cooking characteristics, amino acid consistency, and functional properties of composite noodles made from foxtail millet. Measurement: Food 13:100138

Mounika A, Shanmugam A, Rao TJ, Sinija VR, Rawson A, Baskaran N (2024) Evaluation of quality characteristics of ultrasound–treated browntop millet grains. Chem Eng Proc—Process Intensific 203:109919. https://doi.org/10.1016/j.cep.2024.109919

Mustač NČ, Voučko B, Novotni D, Drakula S, Gudelj A, Dujmić F, D.Ć (2019) Optimization of high intensity ultrasound treatment of proso millet bran to improve physical and nutritional quality. Food Technol Biotechnol 57:183

Nayak N, Bhujle RR, Nanje-Gowda NA, Chakraborty S, Siliviru K, Subbiah J, Brennan C (2024) Advances in the novel and green-assisted techniques for extraction of bioactive compounds from millets: a comprehensive review. Heliyon 10:e30921

Nazari B, Mohammadifar MA, Shojaee-Aliabadi S, Feizollahi E, Mirmoghtadaie L (2018) Effect of ultrasound treatments on functional properties and structure of millet protein concentrate. Ultrason Sonochem 41:382–388

Niamah AK 2019 Ultrasound treatment (low frequency) effects on probiotic bacteria growth in fermented Milk

Pandey P, Mettu S, Mishra HN, Ashokkumar M, Martin GJO (2021) Multilayer co-encapsulation of probiotics and γ-amino butyric acid (GABA) using ultrasound for functional food applications. Lwt 146:111432

Pateiro M, Vargas-Ramella M, Franco D, da Cruz AG, Zengin G, Kumar M, Dhama K, Lorenzo JM (n.d.) The role of emerging technologies in the dehydration of berries: quality, bioactive compounds, and shelf life. Food Chem: X2022 16:100465

Pingret D, Durand G, Fabiano-Tixier A-S, Rockenbauer A, Ginies C, Chemat F (2012) Degradation of edible oil during food processing by ultrasound: electron paramagnetic resonance, physicochemical, and sensory appreciation. J Agric Food Chem 60:7761–7768. https://doi.org/10.1021/jf301286f

Punia Bangar S, Nehra M, Siroha AK, Petruu M, Ilyas RA, Devi U, Devi P (2021) Development and characterization of physical modified pearl millet starch-based films. Foods 10:1609

Raj SN, Sarosh BR, Shetty HS (2006) Induction and accumulation of polyphenol oxidase activities as implicated in development of resistance against pearl millet downy mildew disease. Funct Plant Biol 33:563–571

Rao TJ, Mounika A, Shanmugam A (2024) Improved bioactivity and functional qualities of millet protein enzyme hydrolysates by Ultrasonication. Indus Eng Chem Res 63:5380–5396

Rasika A, Chavan SM, Ghatge J, Sravya G, Prathapan K (2024) Review paper on millet: production, nutrients, processing, and food products for health and sustainability. Int J Environ Agric Biotechnol 9:9

Rojas ML, Kubo MTK, Miano AC, Augusto PED (2022) Ultrasound processing to enhance the functionality of plant-based beverages and proteins. Curr Opin Food Sci 48:100939

Sabuz AA, Rana MR, Ahmed T, Molla MM, Islam N, Khan HH, Chowdhury GF, Zhao Q, Shen Q (2023) Health-promoting potential of millet: a review. Separations 10:80

Safwa SM, Rana MR, Ahmed T, Rahman S, Kabir MA (2023) Bin maximization and characterization of ultrasonic-assisted extraction of Taro corms mucilage using response surface optimization and comparison with conventional methods. Food Anal Methods 16:1724–1737

Saxena S, Vasudevan H, Saini S, Sasmal S (2023) Comparative nutritional assessment of millet-based Milk produced by ultrasound, germination, and a combined approach. ACS Food Sci Technol 3:600–607

Sharma S, Jan R, Riar CS, Bansal V (2021) Analyzing the effect of germination on the pasting, rheological, morphological and in- vitro antioxidant characteristics of Kodo millet flour and extracts. Food Chem 361:130073. https://doi.org/10.1016/j.foodchem.2021.130073

Shunmugapriya K, Kanchana S, Maheswari TU, Saravanakumar R, Vanniarajan C (2020) Optimization of the process parameters for extraction of millet Milk. Int J Biochem Res Rev:39

Singh OP, Singh PK (2024) Trends in millets production, consumption and export from India. J Res ANGRAU 52:121–133

Singh S, Vemireddy V (2003) Transitioning diets: a mixed methods study on factors affecting inclusion of millets in the urban population. BMC Public Health 2023:23

Singh SM, Joshi TJ, Rao PS (2024) Technological advancements in millet Dehusking and polishing process; an insight into pretreatment methods, machineries and impact on nutritional quality. Grain Oil Sci Technol 7:186

Sivarajasekar N (2023) 2. millet seeds-production technology, harvesting and post harvesting harvest management, processing, and storage

Srinivasa Reddy M, Linga Reddy D (2015) Investigation of ultrasonic parameters—effect of high frequency ultrasound on disruption of cell density of a. Niger and P. Chrysogenum, India

Sun Y, Ritchey M, Zhong H, Tang L, Sergusheva E, Shi T, Song J, Li H, Dong G, Liu X (2024) Variation of millet grain size and cooking techniques across Asia between the late fourth and first millennia BC. Antiquity 98:401–416

Tan X, Shu X, Guo J, Parajuli K, Zhang X, Duan Z (2018) Effects of low-frequency ultrasound on Microcystis Aeruginosa from cell inactivation to disruption. Bull Environ Contam Toxicol 101:117–123. https://doi.org/10.1007/s00128-018-2348-y

Vela AJ, Villanueva M, Ozturk OK, Hamaker B, Ronda F (2023) Modification of the microstructure of Tef [Eragrostis Tef (Zucc.) trotter] flour ultrasonicated at different temperatures. Impact on its techno-functional and rheological properties. Curr Res Food Sci 6:100456

Verma A, Lata K, Verma S, Rani S, Kumar A, Kumar PN (2024) Attitude and behavior regarding millet consumption in children with type 1 diabetes mellitus. Asian J Med Sci 15:218–222

Vidhyalakshmi R, Meera MS (2023a) Dry heat and ultrasonication treatment of pearl millet flour: effect on thermal, structural, and in-vitro digestibility properties of starch. Food Measure 17:2858–2868. https://doi.org/10.1007/s11694-023-01832-9

Vidhyalakshmi R, Meera MS (2023b) Dry heat and ultrasonication treatment of pearl millet flour: effect on thermal, structural, and in-vitro digestibility properties of starch. J Food Meas Charact 17:2858–2868

Xu B, Tiliwa ES, Yan W, Azam SMR, Wei B, Zhou C, Ma H, Bhandari B (2022) Recent development in high quality drying of fruits and vegetables assisted by ultrasound: a review. Food Res Int 152:110744

Yadav S, Mishra S, Pradhan RC (2021) Ultrasound-assisted hydration of finger millet (Eleusine Coracana) and its effects on starch isolates and antinutrients. Ultrason Sonochem 73:105542

Yang L, Zhibing L, Xibin W, Tao H (2021) Experimental study on cutting force and surface quality in ultrasonic vibration-assisted milling of C/SiC composites. Int J Adv Manuf Technol 112:2003–2014

Yao Y, Pan Y, Liu S (2020) Power ultrasound and its applications: a state-of-the-art review. Ultrason Sonochem 62:104722

Yousaf L, Hou D, Liaqat H, Shen Q (2021) Millet: a review of its nutritional and functional changes during processing. Food Res Int 142:110197

Zhang Y, Abatzoglou N (2020) Fundamentals, applications and potentials of ultrasound-assisted drying. Chem Eng Res Des 154:21–46

Zheng B, Yuan Y, Xiang J, Jin W, Johnson JB, Li Z, Wang C, Luo D (2022) Green extraction of phenolic compounds from foxtail millet bran by ultrasonic-assisted deep eutectic solvent extraction: optimization, comparison and bioactivities. Lwt 154:112740

3 Advancing Millet Processing with Cold Plasma: Practical Applications and Future Opportunities

Saima Nigar, Kadavakollu Subrahmanyam, Sibasish Sahoo, Khalid Gul, and Rachna Sehrawat

3.1 Introduction

Millets are a group of small-seeded grasses belonging to the *Poaceae* family that have been cultivated for thousands of years, particularly in Africa and Asia (Maharajan et al. 2024). Their exceptional resilience and adaptability to arid, high-temperature environments make them essential crops for sustaining food security in regions with limited water resources (Rotela et al. 2021; Singh et al. 2020; Niranjan et al. 2024). Nutritionally, millets primarily consist of carbohydrates (65–75%), protein (7–12%), dietary fiber (8–15%), and a low-fat content (2–5%) (Rotela et al. 2021). They are also rich in essential vitamins and minerals, including iron, magnesium, phosphorus, and calcium (Chandrakar and Shahi 2023). A detailed overview of the nutritional composition of various millet types is presented in Table 3.1 (Dey et al. 2022). Millets offer numerous health benefits. High dietary fiber content in millets aids digestion, supports gut health, and assists in weight management by enhancing satiety (Maurya et al. 2023). With a low glycemic index, millets are well-suited for blood sugar management, contributing to diabetes prevention. Additionally, the presence of antioxidants, such as phenols and flavonoids, helps in lowering the risk of chronic diseases, including diabetes, cardiovascular conditions, and certain types of cancer (Bunkar 2021; Maurya et al. 2023). These antioxidants also promote cardiac health by contributing to cholesterol reduction. Furthermore, the gluten-free nature of millets makes them an excellent option for individuals with celiac disease or gluten sensitivity (Abah et al. 2020; Bunkar 2021).

Despite their environmental and nutritional benefits, millets and their derivatives, such as flour, starch, and protein, face significant challenges that limit their adoption in the food industry (Latha Ravi and Rana 2024). These challenges include

S. Nigar · K. Subrahmanyam · S. Sahoo · K. Gul (✉) · R. Sehrawat
Department of Food Process Engineering, National Institute of Technology, Rourkela, Odisha, India
e-mail: gulmk@nitrkl.ac.in

S. Jaddu, R. Chandra Pradhan (eds.), *Innovative Millet Processing*,
https://doi.org/10.1007/978-3-031-99864-5_3

Table 3.1 Proximate composition and vitamin and mineral contents of millets

Millets	Moisture (g)	Protein (g)	Fat (g)	Ash (g)	Carbohydrate (g)	Energy (kcal)	Total dietary fiber (g)	Thiamine B1 (mg)	Riboflavin B2 (mg)	Niacin B3 (mg)	Total folates B9 (µg)	Magnesium (mg)	Phosphorus (mg)	Iron (mg)	Zinc (mg)	Sodium (mg)	Calcium (mg)
Kodo millet	14.23	8.92	2.55	1.72	66.19	331.74	6.39	0.29	0.20	1.49	39.49	122.00	101.00	2.34	1.65	3.35	15.27
Little millet	11.36	10.13	3.89	1.34	65.55	346.32	7.72	0.26	0.05	1.29	36.20	91.41	130.00	1.26	1.82	4.77	16.06
Proso millet	11.90	12.50	1.10	1.90	70.40	341	0.00	0.20	0.18	2.30	0.00	153.00	206.00	0.80	1.40	8.20	14.00
Finger millet	13.10	7.30	1.30	2.70	72.00	328	11.50	0.42	0.19	1.10	18.30	137.00	283.00	3.90	2.30	11.00	344.00
Foxtail millet	11.20	12.30	4.30	3.30	60.90	331	2.40	0.59	0.11	3.20	15.00	81.00	290.00	2.80	2.40	4.60	31.00
Barnyard millet	11.90	6.20	2.20	4.40	65.50	307	1.98	0.33	0.10	4.20	0.00	82.00	280.00	5.00	3.00	0.00	20.00
Pearl millet	12.40	11.60	5.00	2.30	67.50	361	11.30	0.33	0.25	2.30	45.50	137.00	296.00	8.00	3.10	10.90	42.00

suboptimal functional properties such as limited solubility, reduced swelling capacity, weak pasting and gelling characteristics, short shelf life, coarse or gritty textures, microbial contamination, and the presence of antinutritional factors (Takhellambam et al. 2016). For instance, the absence of gluten in millet flour results in doughs with reduced elasticity and weaker structural integrity, impacting the quality of bakery products (Chakraborty et al. 2018). Others such as phytic acid and tannins limit protein bioavailability (Samtiya et al. 2021). The high-fat content of certain millets, particularly pearl millet, makes them susceptible to oxidative rancidity, which significantly reduces their shelf life. Millets are also prone to microbial and mycotoxin contamination, which impacts food safety and nutritional value (Yogendra et al. 2024). Pathogens such as *Escherichia coli* and *Salmonella* have been detected, particularly in inadequately handled or stored products. Additionally, mycotoxins like aflatoxins, fumonisins, and ochratoxins, produced by *Aspergillus*, *Penicillium*, and *Fusarium molds*, can contaminate grains during pre- and postharvest periods. These toxins pose serious health risks, including carcinogenic and immunosuppressive effects (Harikrishna et al. 2023). Such contamination not only degrades the nutritional quality of millets but also poses significant public health concerns. Moreover, the food industry faces limitations due to the lack of specialized technological and processing infrastructure for millets, hindering large-scale adoption. However, adopting advanced processing techniques such as cold plasma (CP) could mitigate these issues and significantly expand the use of millets in the food industry.

Recently, CP technology has emerged as a promising nonthermal technique that holds potential for enhancing the quality and functional properties of millets and their derivatives (Ucar et al. 2021). CP is an ionized gas containing reactive species such as electrons, ions, and free radicals, and it operates at or near room temperature, making it suitable for the treatment of heat-sensitive food matrices (Subrahmanyam et al. 2023). CP has demonstrated major potential in microbial inactivation and mitigating mycotoxin levels while preserving the quality of millets. A key advantage of CP is its ability to effectively eliminate microorganisms without inducing resistance (Ismail and Tannous 2024; Subrahmanyam et al. 2024b). CP has shown promising results in modifying the surface characteristics of millets, enhancing hydration properties and germination (Ling et al. 2015; Yodpitak et al. 2019), and improving the functional characteristics of millet derivatives, such as water absorption capacity, solubility, viscosity, and digestibility (Chen et al. 2024; Ge et al. 2022; Balendran et al. 2024). Furthermore, CP can effectively reduce antinutritional factors, potentially enhancing the nutritional quality and bioavailability of essential nutrients in millet-based products (Sarkar et al. 2023). In light of these promising results, this chapter presents a comprehensive review of CP technology, covering its fundamental principles and the different systems used, along with the reactive species produced. We also delve into the effects of CP on major fractions of millets, such as starch, proteins, lipids, and bioactive compounds, highlighting both the benefits and practical limitations of this technology. The impact of cold plasma treatment under varying operating parameters on different millet fractions is presented in Table 3.2. Additionally, we discussed how CP treatment can improve the overall quality of millets and their derivatives, promoting wider acceptance and utilization within the food industry.

Table 3.2 Impact of cold plasma treatment under varying operating parameters on different millet fractions

Millet fraction	Millet type	Type of reactor	Operating parameters		Effects	References
			Voltage	Duration		
Starch	Kodo millet	Pin type cold plasma reactor	10, 20, and 30 kV	10, 20, and 30 min	Increase in water and oil absorption capacities Decrease in amylose content by 16% pH decreased from 6.9 to 7.5. Decrease in relative crystallinity by 6.2% Increase in Whiteness index value	Sonkar et al. (2023)
	Proso millet	Dielectric barrier discharge	40 V	1, 3, and 5 min	Holes, cavities, and fissures were observed. Fluorescent intensity became weaker. Decrease in % relative crystallinity. Decrease in RS content in combination with UHP treatment.	Sun et al. (2022a)
	Finger millet	Pin to plate system	170, 200, and 230 V	10, 20, and 30 min	Water solubility increased by 6.7 times pH continuously decreased Decrease in zeta potential Consistent rise in the reducing sugar content Reduction in amylose content	Kaur and Annapure (2024a, b)
	Proso millet	Pin to plate type	170–230 V	5–15 min	Water absorption capacity increased from 1.54 to 1.93 g/g. Oil absorption capacity increased from 2.22 to 2.79 g/g. Increase in solubility and swelling power by 76.39% and 42.76% respectively. Decrease in amylose content by 9.88%. Relative crystallinity increased significantly.	Chauhan et al. (2023)
Flour	Little millet	Multipin cold plasma discharge	13–24 W	10, 20, and 30 min	Increase in water and oil absorption capacity, swelling capacity and solubility index. Decrease in viscosity of cooked paste. Decreased trend in peak gelatinization temperature and crystallinity.	Jaddu et al. (2022, b)

	Pearl millet	Atmospheric pressure jet plasma system	40–45 kV	5, 10, and 15 min	Crude fat content increased by 36.13% Crude fiber content decreased by 40.27% Protein content decreased by 11.67% Decrease in moisture content Water and oil absorption capacity along with solubility and swelling power increased significantly. Improved pasting properties of the treated millet.	Lokeswari et al. (2021a)
	Little millet	Multipin reactor	10–20 kV	10–30 min	Foaming capacity increased up to 51.47%. Increase in Foaming stability, emulsification ability, and emulsion stability up to 47.02%. Saponin content decreased by 58% after exposure to 20 kV for 20 min. Total phenolic content (TPC) rose from 527.54 ± 8.94 to 575.82 ± 3.58 mg GAE/100 g. Antioxidant activity increased by 59%.	Jaddu et al. (2024a)
	Kodo millet	Multipin cold plasma electric reactor	10–20 kV	10–30 min	Increase in water absorption capacity after 20 min. Significant increase in oil absorption capacity. Solubility index decreased when voltage exceeded 16 kV. Swelling power increased.	Jaddu et al. (2022a)
Protein	Foxtail millet	Dielectric barrier discharge	1–2 kV	5, 15, and 25 min.	Sample treated for 25 min at 1 kV and 2 kV exhibited the highest solubility. Increase in water and oil absorption capacities. Protein characteristics significantly changed. CP treatment induced oxidation in treated sample, leading to a drastic increase in carbonyl content.	Monica et al. (2024)
	Pearl millet	Dielectric barrier discharge	2 kV	5 min	Decrease in the protein content by 0.7%	Charu et al. (2024)

(continued)

Table 3.2 (continued)

Millet fraction	Millet type	Type of reactor	Operating parameters		Effects	References
			Voltage	Duration		
	Barnyard millet	Dielectric barrier discharge	2 kV	5 min	Protein content increased by 0.06%	Charu et al. (2024)
Lipids	Foxtail millet	Dielectric barrier discharge	1 kV and 2 kV	1, 3, and 5 min	10.3% reduction in total fat content.	Monica et al. (2023)
	Common millet seeds	Radio frequency glow discharge plasma	40–250 W	–	High doses of plasma led to degradation of lipids.	Perner et al. (2024)
	Foxtail millet	Dielectric barrier discharge	0, 15, 25, 35 kV	0, 2, 4, 8, and 12 min.	Drastic increase in MDA[a] (malondialdehyde) content at 25 kV. Significant decrease in MDA content than the control when treated at 15 kV.	Wang et al. (2022b)
	Pearl millet	Atmospheric system jet plasma system	40–45 kV	5, 10, and 15 min	Significant increase in the extracted fat content with treatment time and voltage.	Lokeswari et al. (2021a)
Enzymes	Pearl millet and barnyard millet	Dielectric barrier discharge	2 kV	5 min	Reduction in phytase concentration was observed that caused the activation of enzymes involved in vitamin synthesis. Improvement in the digestibility of proteins was inferred by enhancement in the function of gastrointestinal enzymes such as trypsin, amylase, lipase, and chymotrypsin within the human system.	Charu et al. (2024
	Foxtail millet	Dielectric barrier discharge	15 kV	2–12 min	Lipoxygenase activity decreased by 57.5% Lipase activity decreased by 73.04%	Wang et al. (2022b)

Bioactive compounds	Pearl millet and barnyard millet	Dielectric barrier discharge	2 kV	5 min	Gradual decrease in the total phenolic content of 2% in cold plasma treated pearl millet and 19% decrease in cold plasma treated barnyard millet. In the case of cold plasma treated pearl millet there was 10% decrease and in cold plasma treated barnyard millet there was 7% decrease. Increment in the phytochemical content was observed. Improvement in the extraction of bioactive compounds.	Charu et al. (2024)
	Pearl millet	Multipin discharge cold plasma reactor	20–30 kV	10–20 min	Decrease in the total phenolic content with higher voltage and treatment time.	Sarkar et al. (2023)
	Barnyard millet	Multipin discharge cold plasma reactor	30 kV	30 min	Increase in the total phenolic and flavonoid content of the flour with the treatment duration and voltage by 22.01% and 40.84% respectively.	Sharma et al. (2024)
Antinutritional compounds	Pearl millet and barnyard millet	Multipin discharge plasma reactor	2 kV	5 min	There was 57% and 61% decrease in phytic acid content of cold plasma treated pearl and barnyard millet, respectively. The initial tannin concentration of native pearl and barnyard millet was 236 ± 12.53 mg/100 g and 232.33 ± 9.29 mg/100 g, respectively. After cold plasma treatment, there was 18% and 24% reduction in tannin content of cold plasma treated pearl and barnyard, respectively. Reduction in the tannin content of cold plasma treated pearl and barnyard millet by 18% and 24%, respectively.	Charu et al. (2024)

(continued)

Table 3.2 (continued)

Millet fraction	Millet type	Type of reactor	Operating parameters		Effects	References
			Voltage	Duration		
	Pearl millet	Dielectric barrier discharge system	180 V	60–120 min	Reduction in the phytic acid content for 1 h and 2 h of exposure duration by 60.66% and 39.27% respectively.	Lokeswari et al. (2021a)
	Foxtail millet	Dielectric barrier discharge system	1–2 kV	1, 3, and 5 min	Significant reduction in the phytic acid content to 1.44 and 1.10 mg/g was observed at 1 kV for 5 min and 2 kV for 3 min treatment conditions, respectively. Decrease in the tannin content to 0.86 and 0.80 mg/100 g was observed at 1 kV and 2 kV for 5 min and 3 min, respectively.	Monica et al. (2023)
	Little millet	Multipin reactor	20 kV	20 min	Reduction in the Saponin content by 58% after an exposure to 20 kV for 20 min treatment conditions. Reduction in the phytic acid content by 28% at 10 kV and 10 min treatment conditions.	Jaddu et al. (2024a)
	Pearl millet	Plasma bubbling system	180 V	1–2 h	Phytic acid content decreased by 60.66% and 39.27% after 1 and 2 h of treatment respectively.	Lokeswari et al. (2021a)
	Pearl millet	Multipin discharge reactor	30 kV	20 min	Tannin content decreased significantly by 17.35%. Phytate content significantly reduced by 27.46%	Sarkar et al. (2023)
	Little millet	Multipin discharge reactor	10–20 kV	10–30 min	40% reduction in the tannin content. 28% reduction in phytic acid content. 58% decrease in saponin content.	Jaddu et al. (2024a)

[a]MDA is the byproduct of lipid peroxidation

3.2 Plasma Basics

3.2.1 Cold Plasma

Plasma is also known as the fourth state of matter (Fig. 3.1a). It is an ionized gas consisting of various reactive species, such as electrons, free radicals, and ions, in either their fundamental or excited states, while maintaining an overall neutral charge. It is generated by energizing a neutral gas under different temperatures and pressures (Niemira 2012). Depending on thermal equilibrium, plasma is classified into two types: nonthermal plasma (nonequilibrium plasma) and thermal plasma (quasi-equilibrium plasma), as shown in Fig. 3.1b. In thermal plasma, all species are in local thermal equilibrium, meaning their temperatures are approximately equal (e.g., arc plasma, where the temperature of electron is about 10,000 K, similar to the temperature of heavier species) (Mehta and Yadav 2022; Pankaj et al. 2018). Thermal plasma necessitates extremely high pressure (105 Pa) and power (up to 50 MW) for its generation. It is characterized by a thermodynamic equilibrium between electrons and heavier species, resulting from a uniform gas temperature among all constituents (Chizoba Ekezie et al. 2017), whereas, in nonthermal plasma, species are in thermal nonequilibrium, with a significant temperature difference, and the temperature of all species is uniform in localized areas in the plasma (e.g.,

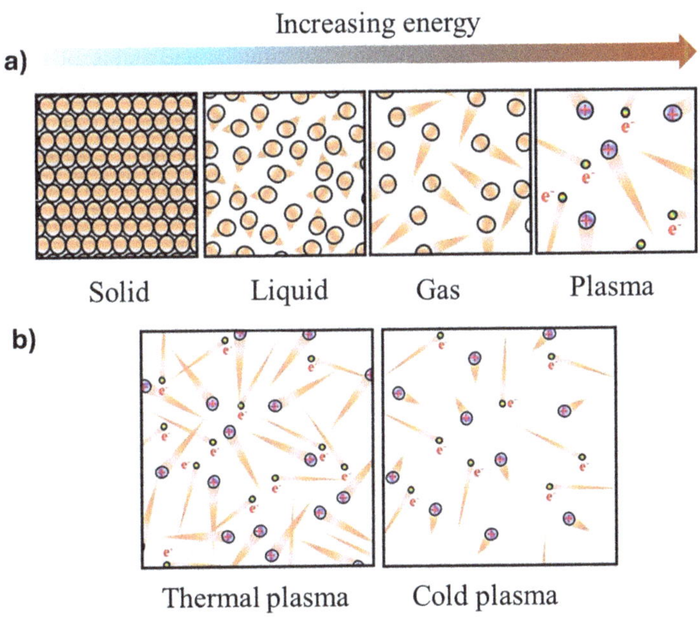

Fig. 3.1 Schematic representation of (**a**) the four fundamental states of matter, with triangular tails indicating the intensity of particle thermal motion, (**b**) thermal plasma and cold plasma, where brown spheres represent neutral atoms, violet spheres indicate positively charged ions, and iridescent spheres depict electrons (Yan et al. 2017)

glow discharges, where electron temperature ranges from 10,000 to 100,000 K, while heavier species are around 300–1000 K) (Pankaj et al. 2018). This type of plasma is generated at low pressures and power levels, and it does not exhibit localized thermodynamic equilibrium. These unique characteristics make CP highly valuable in the food industry (Chizoba Ekezie et al. 2017; Mehta and Yadav 2022).

3.2.2 Cold Plasma Sources

A range of nonthermal plasma sources has been studied to meet the specific demands of diverse food processing applications. Each plasma source offers unique attributes that make it suitable for various processes, such as microbial decontamination, toxin degradation, and functional property enhancement. Their efficacy depends on operating parameters, such as gas composition, power input, treatment duration, electrode distance, and the nature of reactive species generated. The most commonly used setups for plasma generation in food processing include dielectric barrier discharge, plasma jet, radiofrequency, and microwave plasma. Each of these setups is detailed in the following sections.

3.2.2.1 Dielectric Barrier Discharge

Dielectric barrier discharge (DBD) plasma is generated by applying a high voltage between two metal electrodes (Fig. 3.2a) where one or both are coated with a dielectric material such as polymers, glass, quartz, or ceramics (Ollegott et al. 2020). The dielectric barrier prevents arcing, enabling continuous and stable plasma generation. When the voltage across the electrodes reaches a critical threshold, it breaks down the gas molecules, ionizing them into plasma (Chizoba Ekezie et al. 2017). The electrode spacing can vary from 0.1 mm to several centimeters, depending on operational needs. Typical DBD operating conditions include gas pressures ranging from 10,000 to 1,000,000 Pa, frequencies between 10 and 50 MHz, and voltage amplitudes of 1–100 kV_{rms} (Laroque et al. 2022). To maintain charge neutrality, surface charges on the dielectric barriers are often neutralized using bipolar pulsed DC power (Bárdos and Baránková 2010). DBD systems are available in various configurations with different electrode shapes and dielectric barriers to suit specific needs in food processing applications. Despite their simple design and widespread use, DBD systems face challenges such as electrode wear and dielectric degradation, which can affect system longevity and stability over extended use (Subrahmanyam et al. 2024b).

3.2.2.2 Plasma Jet

A plasma jet (PJ) system (Fig. 3.2b) is a versatile plasma generation technique characterized by various electrode configurations, where plasma is generated using an auxiliary plasma source. Common designs include configurations with coaxial electrodes or dual rings, where gas flows between the electrodes (Bárdos and Baránková 2010). In these systems, the outer electrode is usually grounded, while the central electrode is excited by RF at 13.56 MHz (Sahu and Han 2016). This excitation

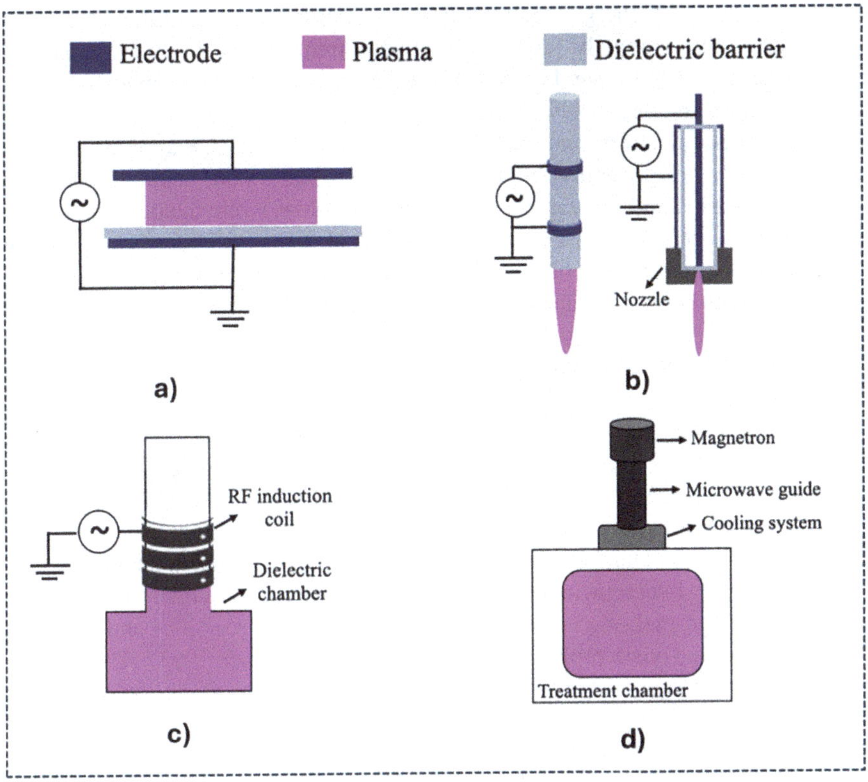

Fig. 3.2 Primary configurations of cold plasma systems: (**a**) dielectric barrier discharge, (**b**) plasma jet, (**c**) corona discharge, and (**d**) microwave plasma (Laroque et al. 2022)

accelerates free electrons, which then collide with gas molecules to generate reactive species. A noble gas or a mixture with a reactive gas supplied at a high rate, projecting the plasma beyond the electrode region as a jet, thus releasing plasma species into the open environment. This design allows for a stable, homogeneous, and uniform discharge at atmospheric pressure (Laroque et al. 2022). However, due to its limited surface coverage, multiple jets are often required for applications over larger areas (Winter et al. 2015).

3.2.2.3 Radiofrequency Plasma

Radiofrequency (RF) plasma generation uses alternating electromagnetic fields to ionize gas molecules. The oscillating electric fields energize the gas, splitting molecules into ions and electrons to form a controlled, reactive plasma as shown in Fig. 3.2c (Niemira 2012). The primary configurations for RF plasma generation include capacitively coupled plasmas (CCP), inductively coupled plasmas (ICP), and helicon wave sources. Of these, CCP and ICP systems are the most widely adopted in industrial settings. In the CCP configuration, two parallel electrodes are positioned a few centimeters apart within a vacuum chamber, where an alternating voltage generates RF energy (Mehta and Yadav 2022). Typically, this setup uses an RF source of approximately 1 kW, operating at a frequency of 13.56 MHz, though frequencies can range from 1 to 500 MHz. The oscillating electromagnetic field ionizes the gas, creating a plasma with a typical density between 10^{15} and 10^{16} m^{-3}, comprising both electrons and ions (Moisan et al. 1991). In ICP systems, two RF power supplies are generally employed: one energizes a coil, usually positioned outside the plasma and separated by a dielectric window, while the other biases the substrate holder to control ion energy (Laroque et al. 2022). RF plasma systems are complex and usually need higher voltage levels to produce the desired effects, which increases energy use and operating costs (Conrads and Schmidt 2000).

3.2.2.4 Microwave Plasma

In microwave (MW) plasma generators (Fig. 3.2d), electromagnetic waves, typically at 2.45 GHz, are emitted by a magnetron to create microwave discharges. These waves accelerate the electrons in gas molecules, generating plasma without the need for electrodes (Lebedev 2010). The closed-chamber design allows the MW plasma system to operate effectively at both low and atmospheric pressures (Chizoba Ekezie et al. 2017). The waves are directed into a treatment chamber via a waveguide, where they interact with gas electrons. Upon absorption of microwave energy, electrons undergo inelastic collisions, initiating ionization reactions and releasing energy as UV and visible photons (Moisan and Wertheimer 1993). This system offers advantages such as high electron density and efficient generation of reactive species, making it particularly suitable for applications like surface sterilization and decontamination (Ponraj et al. 2017; Tu Chau et al. 1996). Key advantages of this system include the generation of high electron densities and an efficient production of reactive species, which makes it particularly suited for applications such as surface sterilization and microbial decontamination. However, the high operational costs and specialized maintenance requirements limit the widespread adoption of this technology (Laroque et al. 2022). Microwave power at frequencies such as 896 MHz, 915 MHz, 922 MHz, 2.45 GHz, and 5.6 GHz are commonly used to create nonequilibrium plasma columns and jets. Nonetheless, significant power requirements and the dimensional constraints of waveguides present limitations in scaling microwave plasma systems, though plasma size can be increased as needed (Bárdos and Baránková 2010).

3.2.3 Plasma Chemistry

Plasma chemical reactions are intrinsically complex, involving a range of reactive plasma species (RPS) with lifetimes that vary across multiple orders of magnitude, from nanoseconds to hours (Khlyustova et al. 2019). The complexity arises from the mixture of single and multi-atomic molecules in the gas injected into the plasma, which is further increased by the presence of water molecules, either from the gas source or from food materials added to the discharge. The concentration and variety of RPS produced during plasma processing are influenced by factors such as the plasma source, processing conditions, system setup, power level, treatment mode, feed gas composition, electrode spacing, treatment time, and gas flow rate (Zhang et al. 2019). Various gases can be employed to generate CP, each exhibiting distinct properties. Inert gases such as argon, neon, and helium are often favored for their chemical stability and low ionization potential, which enhances the efficiency of ionization (Vagin et al. 2020). However, these gases can be quite costly. On the other hand, using ambient air offers a more economical option. It is readily accessible and contains both oxygen and nitrogen, which leads to the generation of plasma rich in active species, such as reactive oxygen-nitrogen species (RONS) (Rehman et al. 2016). This makes air an ideal choice for many applications, as it is cost-effective and contains a wide variety of chemically reactive components. When CP is applied to a food substrate using atmospheric air as the feed gas, the system can be divided into three distinct phases, (i) the gas phase, (ii) the plasma-liquid interface, and (iii) the liquid phase, with the liquid phase specifically referring to the moisture present in the food. A schematic representation illustrating the generation of various RPS within these regions is depicted in Fig. 3.3. Primary reactive species, such as nitric oxide radicals (NO•), hydroxyl radicals (•OH), singlet oxygen ($^1O_2$), superoxide radicals (O_2^-), atomic oxygen (O), and excited nitrogen (N^*), are generated directly within the discharge region (gas phase). These species form as a result of collisions between fast-moving electrons and neutral particles (Lim et al. 2020; Lu et al. 2019). They have relatively short lifetimes and exist in high concentrations within the discharge region. For example, the primary reactive species—NO•, •OH radicals, and O_2^-—have lifetimes of about 1.2, 2.7, and 1.4 μs, respectively (Attri et al. 2015). Some of these reactive species undergo rapid radiative decay, while others react with neutral particles, RPS, and water molecules. As a result, the RPS are transformed into secondary species such as hydrogen peroxide (H_2O_2), ozone (O_3), nitrogen dioxide (NO_2), and nitrate (NO_3) in the surrounding environment (Ghimire et al. 2018). The reactive oxygen and nitrogen species (RONS) produced

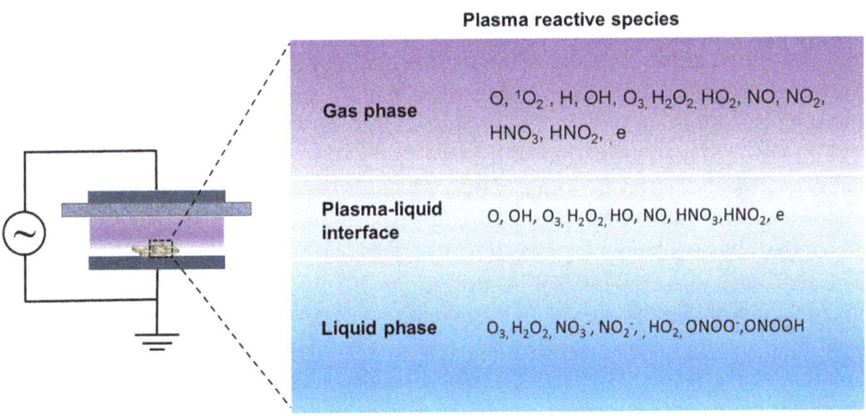

Fig. 3.3. A schematic illustration detailing reactive species generated at the interface between plasma and the product or liquid medium

in the gas phase penetrate into the liquid phase or the moisture within the food sample, where they dissolve and form tertiary reactive species (Lu et al. 2019). These tertiary reactive species, including H_2O_2, O_3, NO_2^-, and NO_3^-, are considered long-lived, with lifetimes ranging from several milliseconds to days (Kondeti et al. 2018). NO_2, NO_3, and H_2O_2 readily dissolve in water, where NO_2 becomes NO_2^- and NO_3 transforms into NO_3^- upon dissolution (Lamichhane et al. 2020). As these RPS dissolve in the liquid, the pH of the liquid may decrease. However, the formation of reactive oxygen species (ROS) and reactive nitrogen species (RNS) within a material depends on whether the material is in a dry or liquid state (Judée et al. 2018). RPS produced in different phases can be characterized using a range of diagnostic techniques. For analyzing the gas-phase plasma, methods such as optical absorbance spectroscopy, optical emission spectroscopy, mass spectrometry, microwave diagnostics, and heterodyne interferometry are employed. In contrast, for the liquid phase, techniques like UV-visible absorption spectroscopy, electron paramagnetic resonance spectroscopy, mass spectrometry, and direct chemical analysis are used for identification (Zhang et al. 2019).

3.3 Effect of Cold Plasma on Millet Components and Their Functionalities

3.3.1 Starch

Starch, a complex carbohydrate and one of the most abundant biopolymers, serves as a critical energy reserve and functional component in numerous food applications (Subrahmanyam et al. 2024a). Millets, known for their high starch content, comprise approximately 50–70% starch by weight. Structurally, starch is organized as granules primarily made up of two glucose polymers: amylose and amylopectin. Amylose is a relatively linear polymer composed of α (1→4) glycosidic bonds, while amylopectin is a highly branched polymer with additional α (1→6) linkages (Apriyanto et al. 2022). The amylose-to-amylopectin ratio varies across starch sources, which in turn influences physicochemical properties and functional applications of starch. Starch granules generally display semicrystalline structures with alternating crystalline and amorphous lamellae, as shown in Fig. 3.4. The crystalline regions are attributed to amylopectin clusters, organized in a double-helix formation stabilized by hydrogen bonding, while the amorphous regions contain amylose chains and parts of amylopectin branches. The arrangement of these regions determines the physicochemical properties of starch, such as swelling, solubility, and digestibility (Nigar et al. 2024). Modifications of starch can alter these properties, enhancing its suitability for specific applications. Studies have shown that CP treatment induces significant structural transformations in starch at both molecular and supramolecular levels (Gupta et al. 2024; Leandro et al. 2024; Kopuk et al. 2022). The reactive species, generated within the plasma field, effectively interact with the surface of starch granules, leading to targeted modifications (Subrahmanyam et al. 2024b).

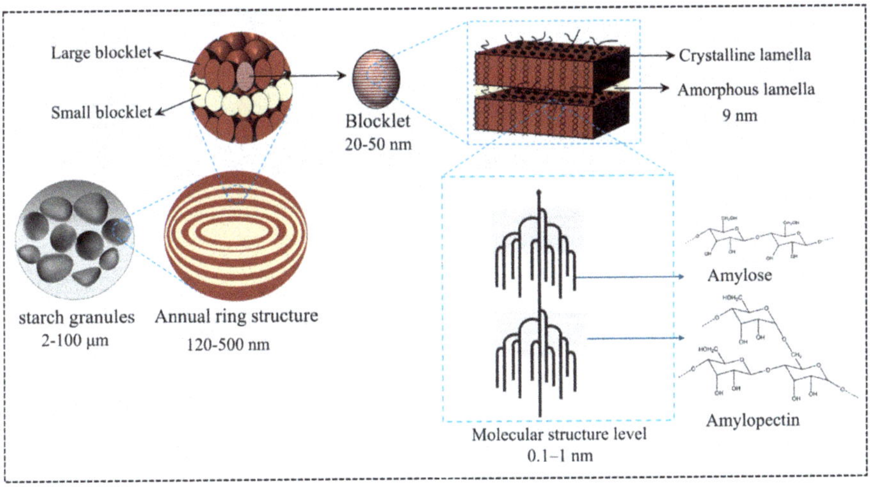

Fig. 3.4. A schematic representation of the hierarchical structure of starch (Lin et al. 2022)

Primary mechanisms of CP modification of starch are cross-linking, depolymerization, and plasma etching (Thirumdas et al. 2017a, b). ROS interacts with hydroxyl (–OH) groups on starch molecules, leading to the formation of new functional groups such as carbonyl (C=O) and carboxyl (COOH) (Gupta et al. 2024). Additionally, CP exposure selectively depolymerizes starch by cleaving glycosidic bonds within the amorphous regions due to the energy imparted by radicals and UV photons (Fig. 3.5). This leads to the breakdown of amylose and amylopectin chains, reducing molecular weight, enhancing solubility, and improving water-binding capacity (Sun et al. 2022a, b). Crystalline regions in starch, though more resistant to alteration, can undergo mild disruptions in amylopectin's ordered helical structures (Thirumdas et al. 2017a). Plasma-generated radicals interact with hydrogen bonds within crystalline lamellae, causing partial disorganization, often evidenced by broadening or weakening in X-ray diffraction peaks, which reveals less-ordered regions in the granules (Ma and Jiang 2024). Moreover, the high-energy environment of CP also enables radical-induced cross-linking on the starch surface, where radicals form covalent bonds between starch molecules. However, the specific mechanism and extent of modification depend on plasma parameters like exposure time, power, gas composition, and the starch structure itself (Zhu et al. 2023). Shorter exposure times or lower power levels typically produce mild surface oxidation, adding functional groups like carbonyls and carboxyls without significantly altering internal structures. In contrast, higher power and extended exposure times provide sufficient energy to penetrate deeper, promoting depolymerization, especially within the amorphous regions where glycosidic bonds are more easily disrupted (Du et al. 2024). The type of gas used also plays a crucial role, oxygen plasmas tend to enhance oxidation, introducing hydroxyl and carboxyl groups, while nitrogen-based plasmas can facilitate grafting by introducing nitrogen-containing functionalities (Taslikh et al. 2022). These CP-induced modifications

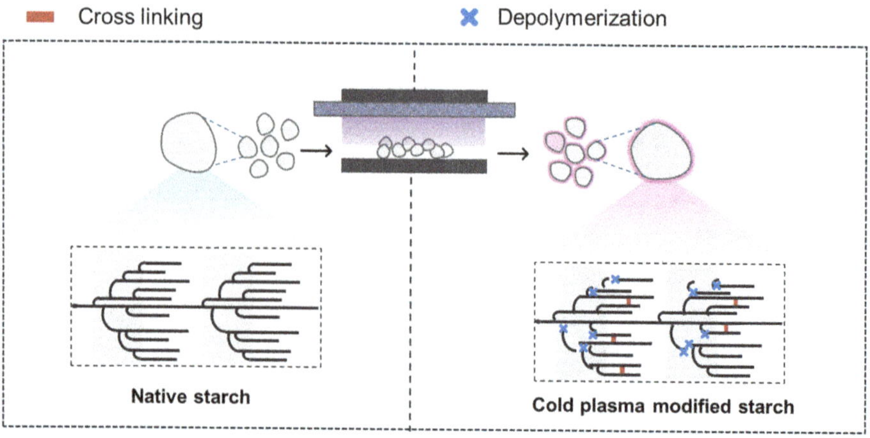

Fig. 3.5 Schematic illustration of cold plasma-induced structural and molecular modifications in starch

can significantly enhance the physicochemical and functional properties of millet starch, making it well suitable for food applications. For instance, Sonkar et al. (2023) reported that CP treatment significantly altered the starch structure and enhanced the techno-functional properties of kodo millet starch. Their results showed that plasma etching enhanced water solubility, water absorption capacity, and swelling power of kodo millet starch due to depolymerization of starch chains, confirmed by a reduction in peak intensity of X-ray diffractograms. Despite a slight drop in pH, the authors observed improvements in color and transparency. In a similar study, Kaur and Annapure (2024a, b) treated finger millet starch with CP at varying voltages and durations. They reported significant depolymerization with reductions in molecular weight and increase in water solubility up to a 6.7-fold at 230 V for 30 min. Moreover, the authors observed oxidation at the C6 position in nuclear magnetic resonance (NMR) analysis, forming carbonyl and carboxyl groups, thereby conforming crosslinking, particularly at lower treatment times and voltages. Similar changes during the CP treatment in the millet starch at varying conditions were also observed by Chauhan et al. (2023), Jaddu et al. (2022a, b), Kaur and Annapure (2023), Amarnath et al. (2024), Kaur and Annapure (2024a, b), and Lokeswari et al. (2021a, b). These findings highlight the potential of CP as an effective, environmentally sustainable approach to starch modification, with promising implications for a range of food industry applications.

3.3.2 Proteins and Enzymes

Proteins are essential macromolecules that play critical roles in nearly all biological processes, from cellular architecture to metabolic regulation (Ma et al. 2022). These biomolecules, composed of amino acid chains linked by peptide bonds, are fundamental to functions such as catalysis, structural support, and cellular signaling

(Subrahmanyam et al. 2024b). Protein structure is organized into four hierarchical levels: the primary structure, which is the linear sequence of amino acids; the secondary structure, comprising alpha-helices and beta-sheets; the tertiary structure, which is the three-dimensional conformation of a single polypeptide chain; and the quaternary structure, involving the assembly of multiple polypeptide subunits (Mehta and Yadav 2022). Millets contain 7–12% protein content, positioning them as a promising source of alternative plant-based proteins (Yousaf et al. 2021). Plant-derived proteins are gaining recognition as sustainable alternatives to animal-based proteins, especially in addressing global food security and environmental sustainability concerns (Poore and Nemecek 2018).

Recent studies have highlighted that CP technology can significantly improve the functional properties of plant proteins (Pankaj et al. 2014). The underlying mechanisms of CP on plant proteins involve complex interactions with reactive species generated during plasma exposure as illustrated in Fig. 3.6. These species interact with amino acid residues, inducing structural modifications from the primary through quaternary levels. At the primary structure level, CP can oxidize or nitrosylate specific amino acid side chains, particularly electron-rich residues such as sulfur-containing amino acids (Sharma and Singh 2020). This process can involve forming new covalent bonds or disrupting existing ones, either altering the integrity of the primary sequence or introducing new functional groups. Such oxidative and nitrosative modifications also impact the higher-order structures of proteins. For instance, CP-induced disruption of hydrogen bonding and disulfide bridges can lead to the destabilization of secondary and tertiary structures, resulting in protein

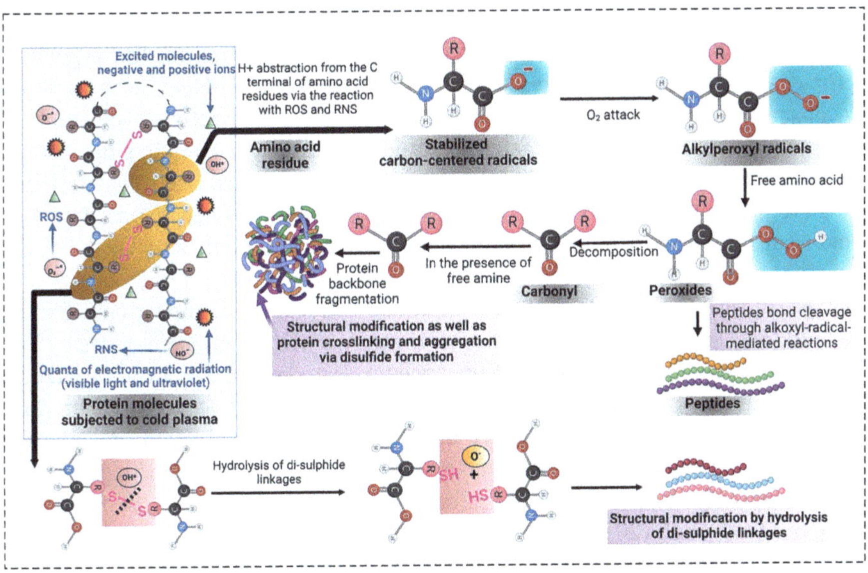

Fig. 3.6 Mechanistic pathways of protein structural modifications induced by cold plasma treatment (Olatunde et al. 2023).

unfolding, changes in surface charges, and, in some cases, aggregation (Basak and Annapure 2022). In the context of quaternary structures, CP exposure can break non-covalent interactions, potentially dissociating subunits or promoting aggregation through cross-linking reactions (Mehta and Yadav 2022). These structural modifications can enhance the solubility, emulsification capacity, and foaming properties of plant proteins, making them more suitable for various food applications (Kumar et al. 2024). For instance, Monica et al. (2024) examined the effects of CP treatment on the structural and functional attributes of protein concentrate (72.88 ± 1.21 g/100 g (dry)) derived from sprouted foxtail millet (SFMPC) under various conditions (1 and 2 kV, with exposure times of 5, 15, and 25 min). The authors found that the CP exposure for 25 min at both 1 kV and 2 kV produced significant impacts on the protein characteristics. The treatment introduced substantial modifications in the physicochemical properties of SFMPC, including reductions in particle size and alterations in zeta potential, carbonyl content, free sulfhydryl content, and surface hydrophobicity. Reactive species generated during CP exposure led to surface etching effects, resulting in increased surface roughness, while a reduction in intrinsic fluorescence intensity suggested tertiary structural changes. Functional properties were also enhanced, with marked improvements in protein solubility (up to 82.08%) and emulsification activity index (up to 80.96 m^2/g) observed following a 25-min CP treatment at 2 kV. In related studies, Monica et al. (2023) and Sarkar et al. (2023) observed an increase in the total soluble protein content of foxtail millet and pearl millet flour, respectively, following CP treatment. This enhancement is attributed to the corrosive and etching actions of free radicals, electrons, and photons, which increase the affinity between protein active sites and water molecules, thereby enhancing protein solubility.

Enzymes play crucial roles in determining the nutritional stability and functional properties of millets. Millet flour contains enzymes such as lipase, lipoxygenase, peroxidase, and polyphenol oxidase, which can lead to deterioration during storage by causing rancidity and off-flavors (Bhargavi et al. 2024). Effective inactivation of these enzymes is essential for maintaining product quality and extending the storage life of millet-derived products (Monica et al. 2023). Enzymes, like other proteins, are susceptible to structural and functional modifications when exposed to CP, which can significantly influence their catalytic efficiency, stability, and activity. CP generates reactive species, such as ROS and RNS, which interact with amino acid residues in enzymes, leading to oxidative and nitrosative modifications (Sarkar et al. 2023). These modifications can alter the primary structure of enzymes by targeting specific amino acids, where new covalent bonds may form or break, leading to changes in folding and stability. Conformational shifts impact substrate binding affinity and enzyme stability, as shown in Fig. 3.7 (Monica et al. 2023). As a result, CP-modified enzymes may exhibit reduced catalytic efficiency, effectively inactivating enzymatic activities without affecting the nutritional quality of millets. A study by Wang et al. (2022a) investigating the effects of DBD-CP on foxtail millet revealed substantial reductions in the activities of lipase and lipoxygenase enzymes. Lipoxygenase activity dropped from 44.0 to 18.7 U g^{-1} min^{-1}, while lipase activity decreased from 56.0 to 15.1 U/mg protein following exposure to 25 kV for 2–12 min.

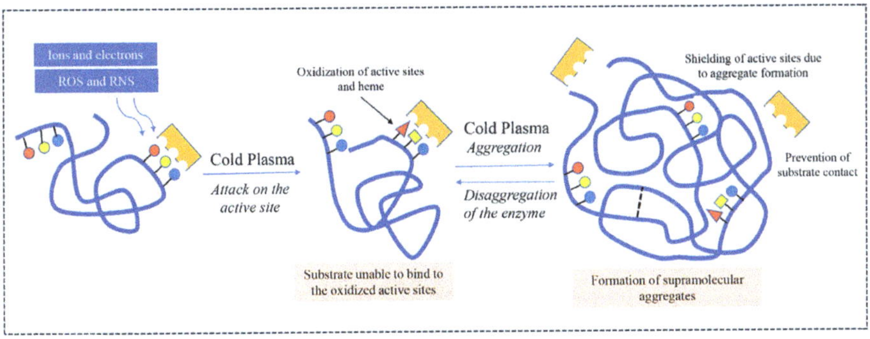

Fig. 3.7 Mechanism of cold plasma-induced modifications on enzyme structure and function (Basak and Annapure 2022)

Importantly, the DBD-CP treatment preserved key nutritional attributes, including the color and total carotenoid content of millets. Treatment at 15 kV for 4 min effectively maintained color and carotenoid stability during 15 days of accelerated storage at 40 ± 2 °C and 75% relative humidity. Although data on specific enzyme activity in millets subjected to CP treatment is limited, evidence suggests that this innovative technique effectively inhibits enzyme activity in millets.

3.3.3 Lipids

Lipid oxidation, a major factor in food quality deterioration, is influenced by ROS and RNS generated during CP treatment. Studies have shown that CP can increase lipid peroxidation, particularly due to ROS generated in oxygen-based plasmas, which accelerate oxidative reactions in lipid molecules (Gavahian et al. 2018). These ROS, including singlet oxygen, superoxide, and hydroxyl radicals, initiate and propagate lipid oxidation, leading to lipid peroxides and secondary oxidation products (Fig. 3.8) (Ke et al. 2022). For instance, Sruthi and Rao (2021) observed that atmospheric CP treatment on millet flour significantly increased lipid peroxidation in the germ and pericarp, resulting in altered sensory properties due to lipolysis and oxidation of de-esterified unsaturated fatty acids. Lipid peroxidation is also influenced by CP operating parameters and can be controlled through optimization. Zhu et al. (2024) reported increased fat oxidation in wheat flour treated with DBD plasma, though oxidative stress was mitigated at lower power settings and shorter exposure times, reducing rancidity over time. Furthermore, nitrogen-based CP has been shown to reduce stress by scavenging free radicals and minimizing lipid peroxidation. For example, Liu et al. (2021) found that nitrogen-based CP treatment on rice flour preserved nutritional and sensory properties more effectively than oxygen-based treatments. While nitrogen-rich plasma is effective against lipid oxidation, it presents challenges for low microbial inactivation efficiency. High voltage and prolonged treatment times can exacerbate lipid oxidation, while lower settings may be less effective against microbial contamination. Therefore, CP parameters must be

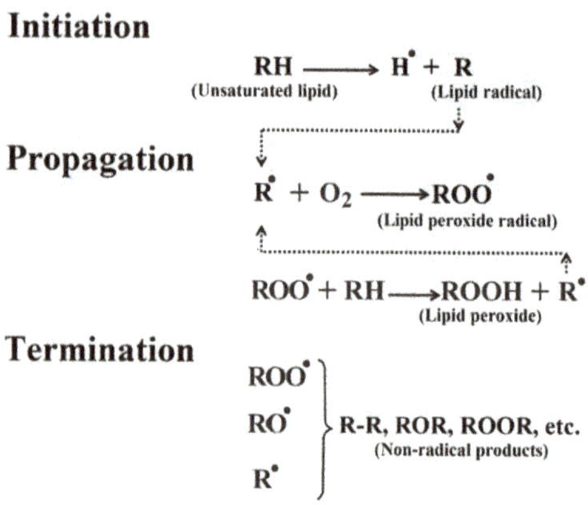

Fig. 3.8 Mechanism of lipid peroxidation (Gavahian et al. 2018)

optimized based on specific application requirements. A few studies have reported an increase in lipid content in millet flour following CP treatment, which may be attributed to the release of bound lipids due to structural modifications induced by plasma treatment (Charu et al. 2024; Kaur and Kaur 2024).

3.3.4 Nutritional and Antinutritional Compounds

Millets are renowned for their rich bioactive compound profile, which offers numerous health benefits, including antioxidant, anti-inflammatory, and protective effects. Key bioactive compounds in millets include polyphenols and flavonoids (Majid and Priyadarshini 2020; Udeh et al. 2017; Wang et al. 2022a). The impact of CP treatment on bioactive and antinutritional compounds shows considerable variability. Some studies reported an increase in bioactive content (Jaddu et al. 2024a, b; Loureiro et al. 2021; Yodpitak et al. 2019), while others noted reductions (Charu et al. 2024; Gnananethri et al. 2024). For instance, Sharma et al. (2024) reported that multi-pin atmospheric CP treatment significantly enhanced the TPC and TFC in barnyard millet flour treated at 30 kV for 30 min. This increase was attributed to the disruption of cell membranes by reactive plasma species, which facilitated the extraction of phenolic compounds. In contrast, Charu et al. (2024) found that CP treatment led to a gradual reduction in bioactive compounds, with a 20% decrease in TPC and a 10% decrease in TFC, likely due to oxidative degradation of flavonoids by ROS produced during the treatment. These contrasting findings underscore the need for careful optimization of CP parameters to achieve desired outcomes.

While millets are rich in bioactive compounds, they also contain antinutritional factors like phytates, tannins, protease inhibitors, trypsin inhibitory factors, oxalates, and enzyme inhibitors, which can affect their overall nutritional value. Although these compounds offer antioxidant benefits, they are often considered antinutrients due to their mineral-chelating and enzyme-inhibiting properties (Dey et al. 2022). These antinutritional factors are secondary metabolites produced by plants as natural defense mechanisms, which can hinder the absorption of essential nutrients like proteins, vitamins, and minerals (Samtiya et al. 2020). Tannins are heat-stable antinutrients that impair protein digestion by reducing protein availability or inhibiting digestive enzymes such as amylase, trypsin, chymotrypsin, and lipase (Monica et al. 2023). This interference can disrupt protein digestion and induce astringency, impacting food palatability. Phytate, a key antinutritional factor in little millet, chelates essential minerals like calcium, iron, zinc, and magnesium, reducing their bioavailability. Similarly, saponins hinder nutrient absorption by compromising cell membrane integrity in the digestive tract (Jaddu et al. 2024a, b). For example, Lokeswari et al. (2021a, b) demonstrated that plasma-processed air bubbling at 180 V with an airflow of 10 liters/hour led to a substantial reduction in phytic acid content in pearl millet, achieving reductions of 60.66% and 39.27% after 1 and 2 h of treatment, respectively. This effect is likely due to the continuous formation and collapse of large bubbles in the plasma treatment, which enhances phytate leaching. Similarly, Sarkar et al. (2023) reported notable reductions in tannins and phytic acid in CP-treated pearl millet flour, with maximum reductions achieved at 30 kV for a 20-min treatment compared to control samples.

3.4 Application of Cold Plasma for Enhancing the Millet Quality

3.4.1 Microbial Inactivation and Shelf-Life Extension

Microbial contamination is a critical challenge in food safety, particularly in minimally processed or heat-sensitive foods. CP has emerged as a novel and effective approach for addressing this concern. Unlike thermal methods that rely on high temperatures to denature microbial proteins and disrupt cellular structures, CP achieves inactivation through the generation of ROS and RNS (Pankaj et al. 2014). These reactive species, O_3, •OH, nitric oxide NO•, and $ONOO^-$, interact with microbial cells to induce oxidative stress, disrupt membrane integrity, and damage proteins and genetic material as shown in Fig. 3.9 (Mehta and Yadav 2022). CP demonstrates efficacy against both Gram-positive and Gram-negative bacteria, though the degree of inactivation may vary. Gram-negative bacteria are generally more susceptible to CP due to their thinner peptidoglycan layer, which makes their outer membrane more vulnerable to reactive species. In contrast, Gram-positive bacteria, with their thicker and more robust peptidoglycan structure, exhibit slightly higher resistance but can still be effectively inactivated with optimized plasma exposure (Laroussi 2005). Additionally, CP has shown promising results against bacterial and fungal spores, which are inherently more resistant due to their protective structures and metabolic dormancy. Prolonged plasma treatment or increased concentrations of reactive species can effectively compromise the spore coat, leading to successful inactivation (Liao et al. 2019). One of the significant advantages of CP technology over conventional thermal methods is its reduced potential to induce microbial resistance. This multifaceted mechanism of action minimizes the likelihood of microbes developing resistance, as they cannot easily adapt to the broad and

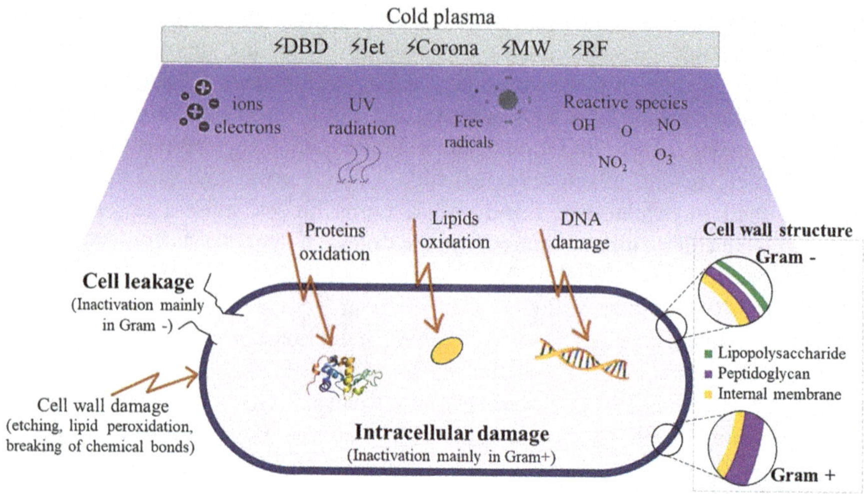

Fig. 3.9 Mechanism of cold plasma inactivation of microorganisms (Laroque et al. 2022)

unpredictable range of oxidative and nitrosative stresses imposed by CP (Scholtz et al. 2015). For instance, antioxidative enzymes such as catalase, superoxide dismutase, and peroxidases can neutralize ROS, thereby reducing the oxidative stress induced by plasma (Cheng et al. 2020). Furthermore, biofilms, which are extracellular matrices produced by many bacteria, provide a physical barrier that limits the penetration of reactive species and shields microbial cells from direct plasma exposure. Such protective mechanisms may necessitate higher plasma intensities, prolonged treatment durations, or synergistic use with other technologies to achieve complete inactivation (Gilmore et al. 2018).

The application of CP technology holds great potential in millet processing, where it can be integrated at various stages to ensure microbial safety while maintaining the grain's nutritional value. For instance, during preprocessing of millets, CP can be employed for surface decontamination, including bacteria, fungi, and spores, without compromising the grain's structure. In milling, it can be used to treat flour, preventing microbial growth and ensuring extended shelf life (Rao et al. 2023). During packaging, CP can act as a final decontamination step to reduce microbial load and maintain food safety standards (Katsigiannis et al. 2022). A recent study by Charu et al. (2024) demonstrated the efficacy of CP in microbial inactivation for millets. The findings revealed that CP treatment at 2 kV for 5 min effectively reduced microbial populations across various millet varieties. Specifically, CP-treated pearl millet exhibited a bacterial reduction from 4.1×10^7 to 6.1×10^1 CFU/mL, while yeast counts decreased from 2.7×10^6 to 9×10^1 CFU/mL. For barnyard millet, bacterial counts declined from 5.0×10^6 to 5.4×10^1 CFU/mL, and yeast counts reduced from 1.4×10^6 to 6.7×10^1 CFU/mL.

3.4.2 Toxin Inactivation

Bacterial and fungal toxins, prevalent in contaminated food and agricultural products, pose serious risks to human and animal health. Bacterial toxins, including exotoxins such as botulinum toxin and enterotoxins, and endotoxins like lipopolysaccharides (LPS) from Gram-negative bacteria, are responsible for conditions like food poisoning, sepsis, and systemic inflammation (Subrahmanyam et al. 2024b). Similarly, fungal toxins, or mycotoxins, such as aflatoxins, fumonisins, ochratoxins, and zearalenone, are secondary metabolites produced by *Aspergillus*, *Fusarium*, and *Penicillium* species (Bosch et al. 2017). These compounds exhibit carcinogenic, hepatotoxic, nephrotoxic, and immunosuppressive properties, with significant health impacts even at low exposure levels (Livingstone et al. 2018). Contaminated cereals, grains, and nuts act as major reservoirs, particularly under improper storage conditions, leading to annual economic losses exceeding USD 1.5 billion in the USA alone (Pal et al. 2015). CP has shown significant potential for neutralizing these toxins while preserving food quality. CP reactive species, including ozone, hydroxyl radicals, and UV photons, interact with bacterial toxins, inducing oxidative damage, structural denaturation, and functional inactivation (Shi et al. 2017). Fungal mycotoxins, such as aflatoxins, are similarly susceptible to CP-induced chemical degradation, which alters their molecular structure and reduces toxicity. Studies have demonstrated that CP is more effective at inactivating mycotoxins than ozone or UV treatments alone, due to its ability to produce a broader spectrum of reactive species (Wang et al. 2018).

Although literature on the effects of CP on toxins in millets is limited, studies in other food matrices highlight its efficacy. For example, Park et al. (2007) reported that microwave-argon plasma completely degraded deoxynivalenol (DON), aflatoxin B1 (AFB1), and nivalenol (NIV) within 5 s of treatment. Similarly, Dasan et al. (2016) showed that atmospheric plasma reduced *Aspergillus parasiticus* and *A. flavus* spore populations on groundnuts by 98% and 99%, respectively, after 24 min at 60 W. However, the degradation rates of mycotoxins vary depending on their chemical structure, food matrix, and treatment conditions. Bosch et al. (2017) observed complete degradation of fumonisin B1 (FuB1) and *Alternaria alternata* (AAL) toxins within 60 seconds of CP exposure, while sterigmatocystin, with its aromatic structure, exhibited greater resistance. Similarly, low-pressure plasma reduced DON levels in barley by 50% in 4 min, while gliding arc discharge plasma achieved an 80% reduction under similar conditions (Kříž et al. 2015).

3.4.3 Physical and Functional Properties

The color of millet is a key quality attribute that significantly influences consumer perception, product appeal, and marketability. Maintaining a stable, vibrant color in millet products enhances their visual appeal and signals quality. CP treatment has demonstrated potential in enhancing the functional properties of millets while also preserving their natural color. However, excessive CP treatment intensity or duration can adversely affect millet color, leading to discoloration due to the oxidation of pigments by ROS in plasma (Lokeswari et al. 2021a, b; Mishra et al. 2024; Sharma et al. 2024; Sonkar et al. 2023). For instance, Wang et al. (2022b) examined the impact of DBD plasma at 15–35 kV for 2–12 min on foxtail millet quality. No significant color changes were observed at lower voltages (15 kV); however, at higher intensities (25 and 35 kV), notable changes occurred regardless of treatment duration due to oxidation of carotenoids. Similarly, Lokeswari et al. (2021a, b) investigated the effects of atmospheric pressure CP on pearl millet, using air as the feed gas at input voltages of 40 and 45 kV for 5, 10, and 15 min. No significant color differences were observed at shorter exposures (40 kV for 5 min and 45 kV for 4 min), but considerable changes were noted at higher voltages and extended durations. These variations in color values likely result from pigment oxidation on the grain surface in the presence of ozone. Another possible explanation is the hydrolysis of starch molecules into smaller sugar and sugar acid fragments under high-energy reactive species. Comparable findings have been reported by Sonkar et al. (2023) where the water absorption capacity, oil absorption capacity, and amylose content, along with other functional characteristics of kodo millet starch, exhibited a rising trend with increasing treatment duration and voltage supplied. Jaddu et al. (2022b) also reported increased functional properties of little millet flour at 24 W and 30-min treatment condition than other lower power supply and treatment duration.

The functional attributes of millets, such as water absorption capacity, oil absorption capacity, solubility, and swelling power, are crucial in defining the texture, stability, and overall quality of food products (Jaddu et al. 2024b). Therefore, understanding the effects of CP treatment on these properties is essential for achieving specific functional outcomes in product development. In a recent study, Sonkar et al. (2023) explored the impact of atmospheric multi-pin CP treatment on the functional properties of kodo millet starch under conditions of 10, 20, and 30 kV at exposure durations of 10, 20, and 30 min. The authors observed significant increases in water absorption capacity (from 1.43 to 1.55 g/g) and oil absorption capacity (from 1.01 to 1.34 g/g), which they attributed to plasma etching that modified the starch surface, thereby facilitating enhanced penetration of water, oil, and heat into the granules. Additionally, CP treatment improved the swelling power (from 3.03 to 4.15 g/g) and solubility index (from 0.102 to 0.130 g/g) of the starch, likely due to amylose depolymerization, which promoted both solubility and swelling capacity. In a related study, Jaddu and Abdullah et al. (2022a) investigated the effects of multi-pin CP on kodo millet flour, observing a notable increase in water absorption capacity and swelling power posttreatment. This enhancement was ascribed to CP-induced depolymerization of starch components, leading to the introduction of hydrophilic carboxyl groups that increased water uptake. Consistent findings reported by Jaddu and Pradhan et al. (2022b) further corroborate the influence of CP treatment on the functional properties of millet flour.

3.4.4 Germination

Germination is a complex physiological process involving the enzymatic breakdown of macromolecules such as starch and proteins into smaller units, providing essential energy and building blocks for seedling development (Sharma et al. 2018). This process enhances nutrient levels, including proteins, starch, and vitamins, while simultaneously reducing antinutritional factors (Ratish Ramanan et al. 2018). Recent research highlights CP as an effective method to enhance germination (de Groot et al. 2018; Li et al. 2014, 2021; Monica et al. 2023). The high kinetic energy of plasma components induces rapid effects on treated seeds, leading to surface etching and deformations, commonly referred to as the "plasma physical effect." These surface modifications increase the hydrophilicity of the seed coat, improving water permeability and facilitating the rapid germination of seeds as shown in Fig. 3.10 (Balendran et al. 2024). In a study by Monica et al. (2023), foxtail millet was exposed to low-pressure air plasma at voltages of 1 kV and 2 kV for varying durations (1, 3, and 5 min). The optimal treatment of 2 kV for 3 min yielded the highest germination rate, reaching 84.00%, followed by 1 kV for 5 min, which achieved 77.33% germination after 48 h. CP exposure significantly improved germination characteristics, including speed, potential, seed length, and vigor indices. Scanning electron microscopy images revealed that the reactive species generated during CP exposure facilitate germination by modifying the seed surface. Additionally, CP-treated foxtail millet exhibited elevated levels of soluble protein (up to 19.00 g/100 g) and γ-amino butyric acid (up to 62.27 mg/100 g), along with reductions in antinutritional factors like phytic acid and tannins, which decreased to 1.1 mg/g and 0.8 mg/100 g, respectively. Additionally, plasma-activated water (PAW) demonstrated a notably greater effect on germination and plant growth compared to tap water (TW) and distilled water (DW). Combining PAW with plasma-treated seeds (20 and 30 min) showed a synergistic effect, with germination rates

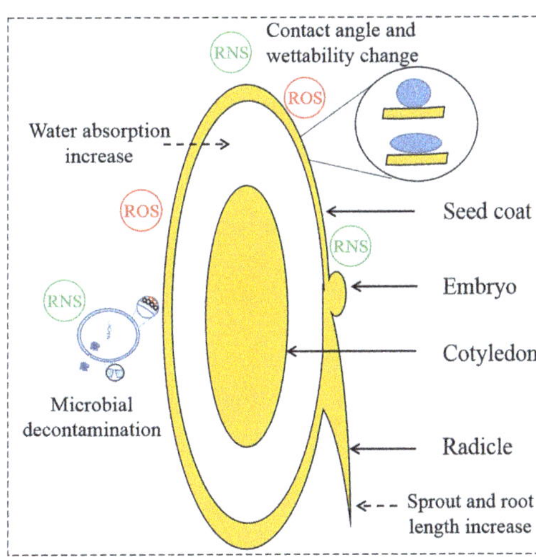

Fig. 3.10 Effects of cold plasma treatment on seed germination and growth parameters (Rajan et al. 2023)

reaching 95 ± 2% by the fifth day and enhanced seedling growth. This suggests that the proposed CP reactors could be effectively applied to promote seed germination and plant development. Among the active species generated by plasma discharge in water, such as NO^{3-}, NO^{2-}, and H_2O_2, was particularly effective in breaking seed dormancy and promoting seedling growth. Even at a concentration of 0.5 ppm, NO^{3-} achieved better germination and healthier plant growth than TW and DW (Mohandoss et al. 2024).

3.5 Challenges and Future Prospects

The application of CP technology in millet processing faces several significant challenges. One primary limitation is the high initial and operational costs associated with CP systems, which arise from the need for specialized plasma generation equipment and rigorous maintenance requirements. Scaling CP for commercial operations presents another hurdle, as the treatment often demands expanding the plasma-generating surface area, necessitating advanced electrode designs and increasing costs. Furthermore, optimizing CP parameters remains complex due to the diverse range of factors influencing treatment efficiency. Key variables, such as gas composition, power input, exposure duration, sample type, and ambient conditions, interact in intricate ways, making it challenging to establish universally effective protocols. Additionally, the reactive plasma species generated during CP treatments exhibit complex chemistries, with interactions that vary significantly based on food matrix components, creating a need for tailored approaches. The safety and regulatory acceptance of CP-treated millet products poses another critical challenge. Despite promising results, limited data on the long-term safety and potential toxicity of plasma-treated food products hampers regulatory approval processes. The absence of clear criteria for assessing the safety of RPS interactions with food constituents further complicates commercialization efforts.

To address these challenges, future research should focus on developing cost-effective CP systems with scalable designs suitable for large-scale millet processing. Innovations in electrode materials and configurations can enhance energy efficiency while reducing operational costs. Additionally, computational modeling and real-time monitoring tools can facilitate the optimization of CP parameters, ensuring consistent treatment outcomes across diverse millet varieties and derivatives. Comprehensive studies are needed to elucidate the effects of RPS on the nutritional, functional, and structural properties of millets for specific applications. Collaborative efforts between researchers, industry stakeholders, and regulatory bodies can establish standardized protocols and safety criteria, expediting the approval process for CP-treated products. The integration of CP with other green technologies, such as ultrasound, superheated steam, or plasma-activated water, holds promise for synergistic benefits. These combinations can potentially enhance the efficacy of treatments, reduce processing time, and minimize oxidative damage to bioactive compounds. Exploring these synergies will be pivotal in maximizing the advantages of CP technology.

3.6 Conclusions

Cold plasma technology has emerged as a transformative nonthermal processing method with immense potential to address the limitations associated with millets and their derivatives. By enhancing physicochemical, functional, and nutritional properties, CP provides innovative solutions to challenges like poor solubility, antinutritional factors, and microbial contamination. Its ability to selectively modify starch, proteins, lipids, and bioactive compounds underpins its versatility in improving the industrial applicability of millets. Additionally, CP-induced structural modifications extend shelf life, enhance safety, and unlock novel functionalities in millet-based products. The adaptability of CP across various processing stages, from seed germination to packaging, illustrates its scalability and relevance for sustainable food systems. While promising results have been achieved, the variability in outcomes emphasizes the need for precise optimization of CP parameters tailored to specific applications.

References

Abah CR, Ishiwu CN, Obiegbuna JE, Oladejo AA (2020) Nutritional composition, functional properties and food applications of millet grains. Asian Food Sci J:9–19. https://doi.org/10.9734/afsj/2020/v14i230124

Amarnath MS, Muhammed A, Antony AK, Yashini M, Sherin AJ, Sunil CK (2024) Millet starch modification: effect of thermal and non-thermal methods on its properties and their application-a review. Food Humanity 3:100356. https://doi.org/10.1016/j.foohum.2024.100356

Apriyanto A, Compart J, Fettke J (2022) A review of starch, a unique biopolymer—structure, metabolism and in planta modifications. Plant Sci 318. Elsevier Ireland Ltd. https://doi.org/10.1016/j.plantsci.2022.111223

Attri P, Kim YH, Park DH, Park JH, Hong YJ, Uhm HS, Kim KN, Fridman A, Choi EH (2015) Generation mechanism of hydroxyl radical species and its lifetime prediction during the plasma-initiated ultraviolet (UV) photolysis. Sci Rep 5:1–8. https://doi.org/10.1038/srep09332

Balendran SD, Kalaivendan RGT, Theagarajan R (2024) Recent advances with cold plasma technology for millet processing: a brief review. J Food Process Eng 47(1). Wiley. https://doi.org/10.1111/jfpe.14409

Bárdos L, Baránková H (2010) Cold atmospheric plasma: sources, processes, and applications. Thin Solid Films 518(23):6705–6713. https://doi.org/10.1016/j.tsf.2010.07.044

Basak S, Annapure US (2022) Recent trends in the application of cold plasma for the modification of plant proteins - a review. Future Foods 5. https://doi.org/10.1016/j.fufo.2022.100119

Bhargavi HA, Singh SP, Goswami S, Yadav S, Aavula N, Shashikumara P, Singhal T, Sankar SM, Danakumara T, Hemanth S, Kapoor C, Singh N (2024) Deciphering the genetic variability for biochemical parameters influencing rancidity of pearl millet *(Pennisetum glaucum L. R. Br.)* flour in a set of highly diverse lines and their categorization using rancidity matrix. J Food Compos Anal 128. https://doi.org/10.1016/j.jfca.2024.106035

Bunkar DS (2021) Nutritional, functional role of kodo millet and its processing: a review. Int J Curr Microbiol App Sci 10(01):1972–1985. https://doi.org/10.20546/ijcmas.2021.1001.229

Chakraborty SK, Kotwaliwale N, Navale SA (2018) Rheological characterization of gluten free millet flour dough. J Food Meas Charact 12(2):1195–1202. https://doi.org/10.1007/s11694-018-9733-4

Chandrakar L, Shahi S (2023) Millets and their nutritional value: a review. J Adv Zool 44:1690–1697

Charu C, Vignesh S, Chidanand DV, Mahendran R, Baskaran N (2024) Impact of cold plasma on pearl and barnyard millets' microbial quality, antioxidant status, and nutritional composition. Food Humanity 2:100238

Chauhan M, Kalaivendan RGT, Eazhumalai G, Annapure US (2023) Atmospheric pressure pin-to-plate cold plasma effect on physicochemical, functional, pasting, thermal, and structural characteristics of proso-millet starch. Food Res Int 173. https://doi.org/10.1016/j.foodres.2023.113444

Chen Y, Chen Y, Jiang L, Huang Z, Zhang W (2024) Mechanism of improving the digestibility of coconut globulin by atmospheric cold plasma treatment: the perspective of protein structure. Food Hydrocoll 152. https://doi.org/10.1016/j.foodhyd.2024.109886

Cheng JH, Lv X, Pan Y, Sun DW (2020) Foodborne bacterial stress responses to exogenous reactive oxygen species (ROS) induced by cold plasma treatments. Trends Food Sci Technol 103:239–247. Elsevier Ltd. https://doi.org/10.1016/j.tifs.2020.07.022

Chizoba Ekezie FG, Sun DW, Cheng JH (2017) A review on recent advances in cold plasma technology for the food industry: current applications and future trends. Trends Food Sci Technol 69:46–58. Elsevier Ltd. https://doi.org/10.1016/j.tifs.2017.08.007

Conrads H, Schmidt M (2000) Plasma generation and plasma sources. Plasma Sources Sci Technol 9:441

Dasan BG, Mutlu M, Boyaci IH (2016) Decontamination of *Aspergillus flavus* and *Aspergillus parasiticus* spores on hazelnuts via atmospheric pressure fluidized bed plasma reactor. Int J Food Microbiol 216:50–59. https://doi.org/10.1016/j.ijfoodmicro.2015.09.006

Dey S, Saxena A, Kumar Y, Maity T, Tarafdar A (2022) Understanding the antinutritional factors and bioactive compounds of kodo millet *(Paspalum scrobiculatum)* and little millet *(Panicum sumatrense).* J Food Qual 2022. Hindawi Limited. https://doi.org/10.1155/2022/1578448

Du Z, Li X, Zhao X, Huang Q (2024) Multi-scale structural disruption induced by radio frequency air cold plasma accelerates enzymatic hydrolysis/hydroxypropylation of tapioca starch. Int J Biol Macromol 260:129572. https://doi.org/10.1016/j.ijbiomac.2024.129572

Gavahian M, Chu YH, Mousavi Khaneghah A, Barba FJ, Misra NN (2018) A critical analysis of the cold plasma induced lipid oxidation in foods. Trends Food Sci Technol 77:32–41. Elsevier Ltd. https://doi.org/10.1016/j.tifs.2018.04.009

Ge X, Shen H, Sun X, Liang W, Zhang X, Sun Z, Lu Y, Li W (2022) Insight into the improving effect on multi-scale structure, physicochemical and rheology properties of granular cold water-soluble rice starch by dielectric barrier discharge cold plasma processing. Food Hydrocoll 130. https://doi.org/10.1016/j.foodhyd.2022.107732

Ghimire B, Lamichhane P, Lim JS, Min B, Paneru R, Weltmann KD, Choi EH (2018) An atmospheric pressure plasma jet operated by injecting natural air. Appl Phys Lett 113(19). https://doi.org/10.1063/1.5055592

Gilmore BF, Flynn PB, O'Brien S, Hickok N, Freeman T, Bourke P (2018) Cold plasmas for biofilm control: opportunities and challenges. Trends Biotechnol 36(6):627–638. Elsevier Ltd. https://doi.org/10.1016/j.tibtech.2018.03.007

Gnananethri C, Jahan A, Kuna A, Kumar MN (2024) Cold plasma treatment on pearl millet grain and flour: impact on bio actives, physical and functional properties. Int J Agric Ext Social Dev 7(7S):118–124. https://doi.org/10.33545/26180723.2024.v7.i7sb.811

de Groot GJJB, Hundt A, Murphy AB, Bange MP, Mai-Prochnow A (2018) Cold plasma treatment for cotton seed germination improvement. Sci Rep 8(1). https://doi.org/10.1038/s41598-018-32692-9

Gupta RK, Guha P, Srivastav PP (2024) Physical action of nonthermal cold plasma technology for starch modification. Food Phys 1:100011. https://doi.org/10.1016/j.foodp.2024.100011

Harikrishna S, Anil PP, Shams R, Dash KK (2023) Cold plasma as an emerging nonthermal technology for food processing: a comprehensive review. J Agric Food Res 14. https://doi.org/10.1016/j.jafr.2023.100747

Ismail R, Tannous JH (2024) Free radical quantification in chemical systems: challenges and future perspectives. Chem Eng Res Des 210:97–111. https://doi.org/10.1016/j.cherd.2024.08.021

Jaddu S, Abdullah S, Dwivedi M, Pradhan RC (2022a) Multipin cold plasma electric discharge on hydration properties of kodo millet flour: modelling and optimization using response surface methodology and artificial neural network—genetic algorithm. Food Chem: Mol Sci 5. https://doi.org/10.1016/j.fochms.2022.100132

Jaddu S, Pradhan RC, Dwivedi M (2022b) Effect of multipin atmospheric cold plasma discharge on functional properties of little millet *(Panicum miliare)* flour. Innov Food Sci Emerg Technol 77. https://doi.org/10.1016/j.ifset.2022.102957

Jaddu S, Sahoo S, Sonkar S, Alzahrani K, Dwivedi M, Misra NN, Pradhan RC (2024a) Cold plasma treatment of little millet flour: impact on bioactives, antinutritional factors and functional properties. Plant Foods Hum Nutr 79(2):503–510. https://doi.org/10.1007/s11130-024-01171-0

Jaddu S, Sonkar S, Seth D, Dwivedi M, Pradhan RC, Goksen G, Kumar Sarangi P, ReŽekJambrak A (2024b) Cold plasma: unveiling its impact on hydration, rheology, nutritional, and anti-nutritional properties in food materials—an overview. Food Chem: X 22. Elsevier Ltd. https://doi.org/10.1016/j.fochx.2024.101266

Judée F, Simon S, Bailly C, Dufour T (2018) Plasma-activation of tap water using DBD for agronomy applications: identification and quantification of long lifetime chemical species and production/consumption mechanisms. Water Res 133:47–59. https://doi.org/10.1016/j.watres.2017.12.035

Katsigiannis AS, Bayliss DL, Walsh JL (2022) Cold plasma for the disinfection of industrial food-contact surfaces: an overview of current status and opportunities. Compr Rev Food Sci Food Safety 21(2):1086–1124. Wiley. https://doi.org/10.1111/1541-4337.12885

Kaur P, Annapure US (2023) Effects of pin-to-plate atmospheric cold plasma for modification of pearl millet *(Pennisetum glaucum)* starch. Food Res Int 169. https://doi.org/10.1016/j.foodres.2023.112930

Kaur P, Annapure US (2024a) Rheological and gelling properties of atmospheric pressure cold plasma treated finger millet *(Eleusine coracana)* starch. Food Res Int 187. https://doi.org/10.1016/j.foodres.2024.114418

Kaur P, Annapure US (2024b) Understanding the atmospheric cold plasma-induced modification of finger millet *(Eleusine coracana)* starch and its related mechanisms. Int J Biol Macromol 268. https://doi.org/10.1016/j.ijbiomac.2024.131615

Kaur C, Kaur J (2024) Impact of non-thermal techniques on millets. Nutr Food Sci 54(2):403–420. Emerald Publishing. https://doi.org/10.1108/NFS-04-2023-0082

Ke Z, Bai Y, Bai Y, Chu Y, Gu S, Xiang X, Ding Y, Zhou X (2022) Cold plasma treated air improves the characteristic flavor of dry-cured black carp through facilitating lipid oxidation. Food Chem 377. https://doi.org/10.1016/j.foodchem.2021.131932

Khlyustova A, Labay C, Machala Z, Ginebra MP, Canal C (2019) Important parameters in plasma jets for the production of RONS in liquids for plasma medicine: a brief review. Front Chem Sci Eng 13(2):238–252. https://doi.org/10.1007/s11705-019-1801-8

Kondeti VSSK, Phan CQ, Wende K, Jablonowski H, Gangal U, Granick JL, Hunter RC, Bruggeman PJ (2018) Long-lived and short-lived reactive species produced by a cold atmospheric pressure plasma jet for the inactivation of Pseudomonas aeruginosa and Staphylococcus aureus. Free Radic Biol Med 124(October 2017):275–287. https://doi.org/10.1016/j.freeradbiomed.2018.05.083

Kopuk B, Gunes R, Palabiyik I (2022) Cold plasma modification of food macromolecules and effects on related products. Food Chem 382. Elsevier Ltd. https://doi.org/10.1016/j.foodchem.2022.132356

Kříž P, Petr B, Zbyněk H, Jaromír K, Pavel O, Petr Š, Miroslav D (2015) Influence of plasma treatment in open air on mycotoxin content and grain nutriments. Plasma Med 5(4). https://doi.org/10.1615/PlasmaMed.2016015752

Lamichhane P, Paneru R, Nguyen LN, Lim JS, Bhartiya P, Adhikari BC, Mumtaz S, Choi EH (2020) Plasma-assisted nitrogen fixation in water with various metals. React Chem Eng 5(11):2053–2057. https://doi.org/10.1039/d0re00248h

Laroque DA, Seó ST, Valencia GA, Laurindo JB, Carciofi BAM (2022) Cold plasma in food processing: design, mechanisms, and application. J Food Eng 312. Elsevier Ltd. https://doi.org/10.1016/j.jfoodeng.2021.110748

Laroussi M (2005) Low temperature plasma-based sterilization: overview and state-of-the-art. Plasma Process Polym 2(5):391–400. https://doi.org/10.1002/ppap.200400078

Latha Ravi J, Rana SS (2024) Maximizing the nutritional benefits and prolonging the shelf life of millets through effective processing techniques: a review. ACS Omega. American Chemical Society. https://doi.org/10.1021/acsomega.4c03466

Leandro GC, Laroque DA, Monteiro AR, Carciofi BAM, Valencia GA (2024) Current status and perspectives of starch powders modified by cold plasma: a review. J Polym Environ 32(2):510–523. Springer. https://doi.org/10.1007/s10924-023-03027-1

Lebedev YA (2010) Microwave discharges: generation and diagnostics. J Phys Conf Ser 257(1). https://doi.org/10.1088/1742-6596/257/1/012016

Li L, Jiang J, Li J, Shen M, He X, Shao H, Dong Y (2014) Effects of cold plasma treatment on seed germination and seedling growth of soybean. Sci Rep 4. https://doi.org/10.1038/srep05859

Li L, Chen J, Wang H, Guo H, Li D, Li J, Liu J, Shao H, Zong J (2021) Cold plasma treatment improves seed germination and accelerates the establishment of centipedegrass. Crop Sci 61(4):2827–2836. https://doi.org/10.1002/csc2.20513

Liao X, Muhammad AI, Chen S, Hu Y, Ye X, Liu D, Ding T (2019) Bacterial spore inactivation induced by cold plasma. Crit Rev Food Sci Nutr 59(16):2562–2572. Taylor and Francis Inc. https://doi.org/10.1080/10408398.2018.1460797

Lim JS, Kim RH, Hong YJ, Lamichhane P, Adhikari BC, Choi J, Choi EH (2020) Interactions between atmospheric pressure plasma jet and deionized water surface. Results Phys 19:103569. https://doi.org/10.1016/j.rinp.2020.103569

Lin D, Ma Y, Qin W, Loy DA, Chen H, Zhang Q (2022) The structure, properties and potential probiotic properties of starch-pectin blend: a review. Food Hydrocoll 129. Elsevier B.V. https://doi.org/10.1016/j.foodhyd.2022.107644

Ling L, Jiangang L, Minchong S, Chunlei Z, Yuanhua D (2015) Cold plasma treatment enhances oilseed rape seed germination under drought stress. Sci Rep 5. https://doi.org/10.1038/srep13033

Liu H, Guo D, Feng X (2021) Plasma degradation of pesticides on the surface of corn and evaluation of its quality changes. Sustainability (Switzerland) 13(16). https://doi.org/10.3390/su13168830

Livingstone MC, Johnson NM, Roebuck BD, Kensler TW, Groopman JD (2018) Dysregulated microRNAs in aflatoxin-induced hepatocellular carcinoma: serum miR-182 as a potential predictive biomarker. Cancer Res 78(13_Supplement):5401

Lokeswari R, Sharanyakanth PS, Jaspin S, Mahendran R (2021a) Cold plasma effects on changes in physical, nutritional, hydration, and pasting properties of pearl millet *(Pennisetum Glaucum)*. IEEE Trans Plasma Sci 49(5):1745–1751. https://doi.org/10.1109/TPS.2021.3074441

Lokeswari R, Sharanyakanth PS, Mahendran R (2021b) Improvement in millet soaking by way of bubbled cold plasma processed air exposure; phytic acid reduction cum nutrient analysis concern. Front Adv Mater Res 3(2):1–16

Loureiro ADC, Souza FDCDA, Sanches EA, Bezerra JDA, Lamarão CV, Rodrigues S, Fernandes FA, Campelo PH (2021) Cold plasma technique as a pretreatment for drying fruits: evaluation of the excitation frequency on drying process and bioactive compounds. Food Res Int 147. https://doi.org/10.1016/j.foodres.2021.110462

Lu X, Keidar M, Laroussi M, Choi E, Szili EJ, Ostrikov K (2019) Transcutaneous plasma stress: from soft-matter models to living tissues. Mater Sci Eng R: Rep 138(April):36–59. https://doi.org/10.1016/j.mser.2019.04.002

Ma S, Jiang H (2024) The effect of cold plasma on starch: structure and performance. Carbohydr Polym 340. Elsevier Ltd. https://doi.org/10.1016/j.carbpol.2024.122254

Ma KK, Greis M, Lu J, Nolden AA, McClements DJ, Kinchla AJ (2022) Functional performance of plant proteins. Foods 11(4). MDPI. https://doi.org/10.3390/foods11040594

Maharajan T, Krishna TPA, Krishnakumar NM, Vetriventhan M, Kudapa H, Ceasar SA (2024) Role of genome sequences of major and minor millets in strengthening food and nutritional security for future generations. Agriculture 14(5):670. https://doi.org/10.3390/agriculture14050670

Majid A, Priyadarshini CG (2020) Millet derived bioactive peptides: a review on their functional properties and health benefits. Crit Rev Food Sci Nutr 60(19):3342–3351. Bellwether Publishing, Ltd. https://doi.org/10.1080/10408398.2019.1686342

Maurya R, Boini T, Misro L, Radhakrishnan T, Sreedharan AP, Gaidhani D (2023) Comprehensive review on millets: nutritional values, effect of food processing and dietary aspects. J Drug Res Ayurvedic Sci 8(Suppl 1):S82–S98. https://doi.org/10.4103/jdras.jdras_123_23

Mehta D, Yadav SK (2022) Recent advances in cold plasma technology for food processing. Food Eng Rev 14(4):555–578. Springer. https://doi.org/10.1007/s12393-022-09317-z

Mishra R, Jangra S, Mishra A, Pandey S, Prakash R, Chhabra M (2024) Enhancement of structural and functional characteristics of millet flours using non-equilibrium cold plasma. Rad Eff Defects Solids 179(7–8):881–887. https://doi.org/10.1080/10420150.2024.2378405

Mohandoss S, Mohan H, Balasubramaniyan N, Assadi AA, Loganathan S (2024) Pearl millet seed surface modification and improved germination by non-thermal plasma discharge: understanding the role of reactive species. Plasma Chem Plasma Process 44(2):1031–1051. https://doi.org/10.1007/s11090-024-10460-0

Moisan M, Wertheimer MR (1993) Comparison of microwave and rf plasmas: fundamentals and applications. Surf Coat Technol 59(1–3):1–13

Moisan M, Barbeau C, Claude R, Ferreira CM, Margot J, Paraszczak J, Sá AB, Sauvé G, Wertheimer MR (1991) Radio frequency or microwave plasma reactors? Factors determining the optimum frequency of operation. J Vacuum Sci Technol B 9(1):8–25. https://doi.org/10.1116/1.585795

Monica V, Anbarasan R, Mahendran R (2023) Influence of cold plasma in accelerating the germination and nutrient composition of foxtail millet *(Setaria italica L.)*. Plasma Chem Plasma Process 43(6):1843–1861. https://doi.org/10.1007/s11090-023-10368-1

Monica V, Anbarasan R, Mahendran R (2024) Cold plasma-induced changes in the structural and techno-functional properties of sprouted foxtail millet protein concentrate. Food Bioprocess Technol. https://doi.org/10.1007/s11947-024-03494-x

Niemira BA (2012) Cold plasma decontamination of foods. Annu Rev Food Sci Technol 3(1):125–142. https://doi.org/10.1146/annurev-food-022811-101132

Nigar S, Subrahmanyam K, Choudhury DB, Gul K, Sehrawat R (2024) Alginate-millet starch composite matrices as novel carriers for controlled delivery of grape seed polyphenols: preparation, characterization, and in vitro release behaviour. Food Struct 42. https://doi.org/10.1016/j.foostr.2024.100385

Niranjan T, Subrahmanyam K, Pradhan RC, Tiwari BK, Dwivedi M (2024) Modelling of millet kernel water absorption and dimensions during hydration at different temperatures using a non-destructive image analysis technique. Measurement: Food 13:100146. https://doi.org/10.1016/j.meafoo.2024.100146

Olatunde OO, Hewage A, Dissanayake T, Aluko RE, Karaca AC, Shang N, Bandara N (2023) Cold atmospheric plasma-induced protein modification: novel nonthermal processing technology to improve protein quality, functionality, and allergenicity reduction. Comprehensive Rev Food Sci Food Safety 22(3):2197–2234. Wiley. https://doi.org/10.1111/1541-4337.13144

Ollegott K, Wirth P, Oberste-Beulmann C, Awakowicz P, Muhler M (2020) Fundamental properties and applications of dielectric barrier discharges in plasma-catalytic processes at atmospheric pressure. Chemie-Ingenieur-Technik 92(10):1542–1558. https://doi.org/10.1002/cite.202000075

Pal M, Gizaw F, Abera F, Kumar Shukla P, Hazarika RA (2015) Mycotoxins: a growing concern to human and animal health. Beverage Food World 42(5): 42–50.

Pankaj SK, Bueno-Ferrer C, Misra NN, Milosavljević V, O'Donnell CP, Bourke P, Keener KM, Cullen PJ (2014) Applications of cold plasma technology in food packaging. Trends Food Sci Technol 35(1):5–17. https://doi.org/10.1016/j.tifs.2013.10.009

Pankaj SK, Wan Z, Keener KM (2018) Effects of cold plasma on food quality: a review. Foods 7(1). MDPI Multidisciplinary Digital Publishing Institute. https://doi.org/10.3390/foods7010004

Park BJ, Takatori K, Sugita-Konishi Y, Kim IH, Lee MH, Han DW, Chung KH, Hyun SO, Park JC (2007) Degradation of mycotoxins using microwave-induced argon plasma at atmospheric pressure. Surf Coat Technol 201(9–11 SPEC. ISS):5733–5737. https://doi.org/10.1016/j.surfcoat.2006.07.092

Perner J, Matoušek J, Auer Malinská H (2024) Cold plasma treatment influences the physiological parameters of millet. Photosynthetica 62(1):126–137. https://doi.org/10.32615/ps.2024.010

Ponraj SB, Sharp JA, Kanwar JR, Sinclair AJ, Kviz L, Nicholas KR, Dai XJ (2017) Argon gas plasma to decontaminate and extend shelf life of milk. Plasma Process Polym 14(11). https://doi.org/10.1002/ppap.201600242

Poore J, Nemecek T (2018) Reducing food's environmental impacts through producers and consumers. Science 360(6392):987–992

Rajan A, Boopathy B, Radhakrishnan M, Rao L, Schlüter OK, Tiwari BK (2023) Plasma processing: a sustainable technology in agri-food processing. Sustain Food Technol 1(1):9–49. Royal Society of Chemistry. https://doi.org/10.1039/d2fb00014h

Rao W, Li Y, Dhaliwal H, Feng M, Xiang Q, Roopesh MS, Pan D, Du L (2023) The application of cold plasma technology in low-moisture foods. Food Eng Rev 15(1):86–112. Springer. https://doi.org/10.1007/s12393-022-09329-9

Ratish Ramanan K, Sarumathi R, Mahendran R (2018) Influence of cold plasma on mortality rate of different life stages of *Tribolium castaneum* on refined wheat flour. J Stored Prod Res 77:126–134. https://doi.org/10.1016/j.jspr.2018.04.006

Rehman MU, Jawaid P, Uchiyama H, Kondo T (2016) Comparison of free radicals formation induced by cold atmospheric plasma, ultrasound, and ionizing radiation. Arch Biochem Biophys 605:19–25. https://doi.org/10.1016/j.abb.2016.04.005

Rotela S, Borkar S, Borah A (2021) Health benefits of millets and their significance as functional food: a review. Pharma Innov 10(5):158–162. https://doi.org/10.22271/tpi.2021.v10.i5c.6192

Sahu BB, Han JG (2016) Comparison of plasma excitation, ionization, and energy influx in single and dual frequency capacitive discharges. Phys Plasmas 23(12). https://doi.org/10.1063/1.4969088

Samtiya M, Aluko RE, Dhewa T (2020) Plant food anti-nutritional factors and their reduction strategies: an overview. Food Prod Process Nutr 2(1). BioMed Central Ltd. https://doi.org/10.1186/s43014-020-0020-5

Samtiya M, Soni K, Chawla S, Poonia A, Sehgal S, Dhewa T (2021) Key anti-nutrients of millet and their reduction strategies: an overview. Acta Scientific Nutr Health 5(12). (ISSN: 2582-1423)

Sarkar A, Niranjan T, Patel G, Kheto A, Tiwari BK, Dwivedi M (2023) Impact of cold plasma treatment on nutritional, antinutritional, functional, thermal, rheological, and structural properties of pearl millet flour. J Food Process Eng 46(5). https://doi.org/10.1111/jfpe.14317

Scholtz V, Pazlarova J, Souskova H, Khun J, Julak J (2015) Nonthermal plasma - A tool for decontamination and disinfection. Biotechnology advances 33(6):1108–1119. Elsevier Inc. https://doi.org/10.1016/j.biotechadv.2015.01.002

Sharath Kumar N, Dar AH, Dash KK, Kaur B, Pandey VK, Singh A, Fayaz U, Shams R, Mukarram SA, Kovács B (2024) Recent advances in cold plasma technology for modifications of proteins: a comprehensive review. J Agric Food Res 16). Elsevier B.V. https://doi.org/10.1016/j.jafr.2024.101177

Sharma S, Singh RK (2020) Cold plasma treatment of dairy proteins in relation to functionality enhancement. In: Trends in food science and technology, vol 102. Elsevier Ltd, pp 30–36. https://doi.org/10.1016/j.tifs.2020.05.013

Sharma S, Saxena DC, Riar CS (2018) Changes in the GABA and polyphenols contents of foxtail millet on germination and their relationship with in vitro antioxidant activity. Food Chem 245:863–870. https://doi.org/10.1016/j.foodchem.2017.11.093

Sharma V, Tomar MS, Sahoo S, Pradhan RC (2024) Development of barnyard millet flour rich in bioactive compounds with enhanced functional properties by application of multipin atmospheric cold plasma. J Food Process Eng 47(6). https://doi.org/10.1111/jfpe.14667

Shi H, Ileleji K, Stroshine RL, Keener K, Jensen JL (2017) Reduction of aflatoxin in corn by high voltage atmospheric cold plasma. Food Bioprocess Technol 10(6):1042–1052. https://doi.org/10.1007/s11947-017-1873-8

Singh A, Kumar M, Shamim M (2020) Importance of minor millets (Nutri cereals) for nutrition purpose in present scenario. Int J Chem Stud 8(1):3109–3113. https://doi.org/10.22271/chemi.2020.v8.i1au.9226

Sonkar S, Jaddu S, Pradhan RC, Dwivedi M, Seth D, Goksen G, Sarangi PK, Lorenzo JM (2023) Effect of atmospheric cold plasma (pin type) on hydration and structure properties of kodo-millet starch. LWT 182. https://doi.org/10.1016/j.lwt.2023.114889

Sruthi NU, Rao PS (2021) Effect of processing on storage stability of millet flour: a review. In: Trends in food science and technology, vol 112. Elsevier Ltd., pp 58–74. https://doi.org/10.1016/j.tifs.2021.03.043

Subrahmanyam K, Gul K, Sehrawat R, Allai FM (2023) Impact of in-package cold plasma treatment on the physicochemical properties and shelf life of button mushrooms *(Agaricus bisporus)*. Food Biosci 52. https://doi.org/10.1016/j.fbio.2023.102425

Subrahmanyam K, Gul K, Mir NA, Sehrawat R, Sarkar A (2024a) Biopolymeric formulations as delivery systems in food. In: Advances in biopolymers for food science and technology. Elsevier, pp 171–200. https://doi.org/10.1016/B978-0-443-19005-6.00008-6

Subrahmanyam K, Gul K, Sehrawat R, Tiwari BK, Sahoo S (2024b) Cold plasma-mediated inactivation of microorganisms for the shelf-life extension of animal-based foods: efficiency, mechanism of inactivation, and impact on quality attributes. In: Food control, vol 162. Elsevier Ltd. https://doi.org/10.1016/j.foodcont.2024.110464

Sun X, Saleh ASM, Lu Y, Sun Z, Zhang X, Ge X, Shen H, Yu X, Li W (2022a) Effects of ultra-high pressure combined with cold plasma on structural, physicochemical, and digestive properties of proso millet starch. Int J Biol Macromol 212:146–154. https://doi.org/10.1016/j.ijbiomac.2022.05.128

Sun X, Saleh ASM, Sun Z, Ge X, Shen H, Zhang Q, Yu X, Yuan L, Li W (2022b) Modification of multi-scale structure, physicochemical properties, and digestibility of rice starch via microwave and cold plasma treatments. LWT 153. https://doi.org/10.1016/j.lwt.2021.112483

Takhellambam RD, Chimmad BV, Prkasam JN (2016) Ready-to-cook millet flakes based on minor millets for modern consumer. J Food Sci Technol 53(2):1312–1318. https://doi.org/10.1007/s13197-015-2072-0

Taslikh M, Abbasi H, Mortazavian AM, Ghasemi JB, Naeimabadi A, Nayebzadeh K (2022) Effect of cold plasma treatment, cross-linking, and dual modification on corn starch. Starch/Staerke 74(5–6). https://doi.org/10.1002/star.202200008

Ten Bosch L, Pfohl K, Avramidis G, Wieneke S, Viöl W, Karlovsky P (2017) Plasma-based degradation of mycotoxins produced by Fusarium, Aspergillus and Alternaria species. Toxins 9(3):97

Thirumdas R, Kadam D, Annapure US (2017a) Cold plasma: an alternative technology for the starch modification. Food biophysics 12(1):129–139. Springer New York LLC. https://doi.org/10.1007/s11483-017-9468-5

Thirumdas R, Trimukhe A, Deshmukh RR, Annapure US (2017b) Functional and rheological properties of cold plasma treated rice starch. Carbohydr Polym 157:1723–1731. https://doi.org/10.1016/j.carbpol.2016.11.050

Tu Chau T, Chi Kao K, Blankt G, Madrid F (1996) Microwave plasmas for low-temperature dry sterilization. Biomateriak 17:1273–1277

Ucar Y, Ceylan Z, Durmus M, Tomar O, Cetinkaya T (2021) Application of cold plasma technology in the food industry and its combination with other emerging technologies. Trends Food Sci Technol 114:355–371. Elsevier Ltd. https://doi.org/10.1016/j.tifs.2021.06.004

Udeh HO, Duodu KG, Jideani AIO (2017) Finger millet bioactive compounds, bioaccessibility, and potential health effects - a review. Czech J Food Sci 35(1):7–17. Czech Academy of Agricultural Sciences. https://doi.org/10.17221/206/2016-CJFS

Vagin KY, Mamontova TV, Uryupin SA (2020) High-frequency skin effect in a photoionized inert gas plasma. J Phys Conf Ser 1692(1). https://doi.org/10.1088/1742-6596/1692/1/012005

Wang B, Mahoney NE, Khir R, Wu B, Zhou C, Pan Z, Ma H (2018) Degradation kinetics of aflatoxin B1 and B2 in solid medium by using pulsed light irradiation. J Sci Food Agric 98(14):5220–5224. https://doi.org/10.1002/jsfa.9058

Wang H, Fu Y, Zhao Q, Hou D, Yang X, Bai S, Diao X, Xue Y, Shen Q (2022a) Effect of different processing methods on the millet polyphenols and their anti-diabetic potential. Front Nutr 9. Frontiers Media S.A. https://doi.org/10.3389/fnut.2022.780499

Wang LH, Li Z, Qin J, Huang Y, Zeng XA, Aadil RM (2022b) Investigation on the impact of quality characteristics and storage stability of foxtail millet induced by air cold plasma. Front Nutr 9. https://doi.org/10.3389/fnut.2022.1064812

Winter J, Brandenburg R, Weltmann KD (2015) Atmospheric pressure plasma jets: an overview of devices and new directions. Plasma Sources Sci Technol 24(6). https://doi.org/10.1088/0963-0252/24/6/064001

Yan D, Sherman JH, Keidar M (2017) Cold atmospheric plasma, a novel promising anti-cancer treatment modality. Oncotarget 8(9):15977–15995. https://www.impactjournals.com/oncotarget/

Yodpitak S, Mahatheeranont S, Boonyawan D, Sookwong P, Roytrakul S, Norkaew O (2019) Cold plasma treatment to improve germination and enhance the bioactive phytochemical content of germinated brown rice. Food Chem 289:328–339. https://doi.org/10.1016/j.foodchem.2019.03.061

Yogendra K, Sanivarapu H, Avuthu T, Gupta SK, Durgalla P, Banerjee R, Raman A, Tyagi W (2024) Comparative metabolomics to unravel the biochemical mechanism associated with rancidity in pearl millet *(Pennisetum glaucum L.)*. Int J Mol Sci 25(21):11583. https://doi.org/10.3390/ijms252111583

Yousaf L, Hou D, Liaqat H, Shen Q (2021) Millet: a review of its nutritional and functional changes during processing. Food Res Int 142. https://doi.org/10.1016/j.foodres.2021.110197

Zhang K, Perussello CA, Milosavljević V, Cullen PJ, Sun DW, Tiwari BK (2019) Diagnostics of plasma reactive species and induced chemistry of plasma treated foods. Crit Rev Food Sci Nutr 59(5):812–825. https://doi.org/10.1080/10408398.2018.1564731

Zhu Q, Yao S, Wu Z, Li D, Ding T, Liu D, Xu E (2023) Hierarchical structural modification of starch via non-thermal plasma: a state-of-the-art review. Carbohydr Polym 311. Elsevier Ltd. https://doi.org/10.1016/j.carbpol.2023.120747

Zhu Y, Feng X, Zhang J, Shen R, Yao Y (2024) A comprehensive analysis of edible and processing quality changes in three whole grain flours induced by dielectric barrier discharge atmospheric cold plasma treatment. Int J Food Sci Technol 59(7):5169–5180. https://doi.org/10.1111/ijfs.17258

4 Exploring Tradition: Conventional Methods in Millet Processing—An In-Depth Comparison with Modern Approaches

Udipta Hazarika, Priyanka Barman, and Manisha Choudhury

4.1 Introduction

With the increasing issues of rising water scarcity, food insecurity, extreme climate changes, and population explosion, world governments, researchers, and scientists are attempting to scour long-term solutions for environmental sustainability, as well as agricultural and nutritional security. Ergo, millets are an integral part of the Sustainable Development Goals (SDGs) that can help attain the SDGs such as 'attaining zero hunger' (SDG 2) by promoting millet cultivation, 'climate action' (SDG 13) by being harsh climate-resistant and improving food production, 'good health and well-being' (SDG 3) as millets are rich in nutrients, and 'no poverty' (SDG 1) by providing livelihood stability and enhancing income generation activities (Antony Ceasar and Maharajan 2022). With increasing population and food insecurity, millets have been identified as one of the most important drought-resistant crops and can play a pivotal role to tackle these bottlenecks and, as such, have been dubbed as 'nutri-cereals' (Saleh et al. 2013). Millets are small-seeded, cereal grains belonging to the *Poaceae* family. They are excellent drought crops, which means that they can be grown in areas of minimal rainfall and in degraded land with deficit nutrients present in them (Kumar et al. 2021). Another great advantage is that these grains have a higher heat tolerance, as well as can easily grow and adapt in highly acidic, saline soils. Moreover, these crops have a higher resistance

U. Hazarika
Food, Agriculture and Bioresources, Asian Institute of Technology, Bangkok, Thailand

P. Barman
National Institute of Food Technology, Entrepreneurship and Management (NIFTEM-T, LO), Guwahati, Assam, India

M. Choudhury (✉)
Department of Food Science and Nutrition, College of Community Science, Assam Agricultural University, Jorhat, India
e-mail: manisha.choudhury@aau.ac.in

S. Jaddu, R. Chandra Pradhan (eds.), *Innovative Millet Processing*,
https://doi.org/10.1007/978-3-031-99864-5_4

towards pests and diseases and have a shorter growing period. Millets are classified as major and minor millets. Major millets include sorghum, pearl millet, and finger millet, whereas the minor millets include foxtail millet, kodo millet, barnyard millet, and proso millet (Anandharamakrishnan et al. 2022). The millet grains are the basis and backbone which are assisting humans in tackling food and nutritional insecurity, as well as improving the economic situation of farmers and presented in Table 4.1.

Millets are an important staple food crop in developing regions of the world which include Africa and Asia. With more than 93 countries that cultivate millets, these are one of the major crops that are grown for food and feed purposes (Kumar et al. 2021). Out of the total production of millets, 97% is produced in the continents of Africa and Asia. India is the largest producer of millets and contributes to about 26.6% of total millet cropping area around the world (Singh et al. 2018). In 2022, India produced around 18.1% of total millet, followed by China (6.6%) and Niger (6.5%). However, in recent years there has a continuous decline in the total area harvested for millets, with a total reduction of 25.7% from 1961 to 2018 (Yousaf et al. 2021). A magnitude of factors has played their part in this drastic fall, which includes lack of crop improvement efforts, a shift towards high value crops, low profitability, and low implementation of government policies. These factors, coupled with decrease in farmers' interest for cultivation, made millet an underutilized crop (Saleh et al. 2013; Anal et al. 2024). Sorghum is one of the most cultivated

Table 4.1 Millet characteristics and health benefits (Yousaf et al. 2021; Sharma et al. 2021; Anandharamakrishnan et al. 2022; Anal et al. 2024)

Name	Image	Scientific name	Size (cm)	Grain origin	Health benefits
Pearl millet		*Pennisetum glaucum*	0.3–0.4	West Africa	Antioxidant, anti-inflammatory
Finger millet		*Eleusine coracana*	0.1–0.2	East Central Africa	Effective against cardiovascular diseases
Foxtail millet		*Setaria italica*	0.2	China	Lowers cholesterol
Little millet		*Panicum sumatrense*	0.18–0.19	Southeast Asia	Antioxidant, anti-glycaemic effect
Kodo millet		*Paspalum scrobiculatum*	0.12–0.01	India, West Africa	Antioxidant, lowers cholesterol
Barnyard millet		*Echinochloa frumentacea*	0.2–0.3	Japan, India	Lowers blood glucose levels
Proso millet		*Panicum miliaceum*	0.2–0.3	Central and East Africa	Reduction in blood glucose levels

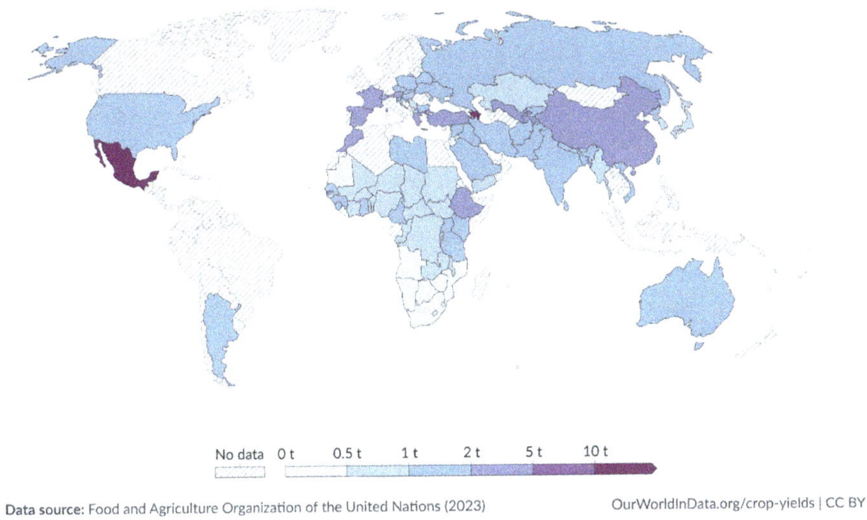

Fig. 4.1 Millet yield around the world (*Food and Agriculture Organization of the United Nations (2023)—with major processing by Our World in Data*) (FAO et al. 2023)

millet crops, with Sudan being its largest producer (7.1 M ha), followed by Nigeria (6.12 M ha) and India (4.9 M ha). The concentration of millet production in Asia is around India, China, and Nepal. In India, millets are mostly grown in arid, dry regions with low spells of rain, and the most widely cultivated millet is pearl millet (Kumar et al. 2021) (Fig. 4.1).

With their agronomic benefits, millets are excellent climate-resilient crops. Apart from this, millets are one of the most nutritionally significant and abundant crop in the world. They are an excellent source of macro- and micronutrients, and have an excellent antioxidant and amino acid profile (Sharma et al. 2021). With increasing food and nutritional insecurity, the quality of food that is consumed is vital for overall physical and mental well-being. Millets have a great potential of being a good source of major carbohydrates and proteins, including vitamins and minerals. Current research suggests the quality of protein in millets is comparable to other major cereal crops (Saleh et al. 2013). Apart from being rich in macro- and micronutrients, millets can also be developed, in conjecture with other food groups, into personalized food-based nutraceuticals and functional foods. Millets are rich in polyphenols, B vitamins, magnesium, calcium, iron, fibre, and a host of other vitamins and minerals, which shows that they have an untapped potential for use as nutraceuticals (Yousaf et al. 2021). Each type of millet has unique characteristics of its own which makes them a nutritious source for the nutritional needs of humans and has the potential to be used as a used in the form of various processed products such as flakes, noodles, extruded products, snacks, etc. With a complex composition of soluble fibre, insoluble fibre, and resistant starch, pearl millet is a great source of energy (Saleh et al. 2013). The protein characterization of foxtail millet showed that the quality of the protein has the potential to be an essential functional ingredient and can act as a supplementary protein source. Apart from this, finger millet has also been identified to have a host of health benefits, as it is rich in polyphenols, essential fatty acids, good-quality protein, and higher percentages of crude fibre and minerals. Barnyard millet is another minor millet which is rich in protein, crude fibre, and particularly iron. The products from barnyard millet can be suitable for diabetics, infants, and pregnant mothers as it is rich in iron (Renganathan et al. 2020).

As a group, millets offer a host of nutritional advantages to the consumer, being rich in almost all groups of macro- and micronutrients. However, one of the major reasons which has hindered the utilization of millets as a viable source for human consumption is the presence of antinutrients. The major antinutritional factors (ANFs) like phytates and oxalic acid interfere with the availability of minerals in the human body and can lead to disruptions in the uptake of the various minerals by the human body (Mundassery et al. 2024). ANFs are present in different concentrations in various millets, out of which phytic acid concentrations can be a metter of concern. Phytic acid (PA) is the organic form of phosphorus which is highly reactive and binds itself to mineral compounds like Ca, Fe, and Zn, thus interfering with their availability in the body. As humans lack the phytase enzyme, accumulation of PA can lead to serious consequences (Sheethal et al. 2022).

As the consumption of millets has increased over the last few years due to global interest and because of more scientific data availability, reduction of ANFs from millets has become paramount. Various conventional methods and modern approaches of processing of millets are available which aid in reduction of ANFs from millets and make them more palatable and provide a host of other advantages (Kheto et al. 2024). Traditional methods of millet processing include decortication, milling, germination, fermentation, malting, roasting, cooking, etc. which help reduce the concentration of ANFs considerably. Moreover modern apporaches such as thermal methods (microwave, ohmic heating) and non-thermal methods (cold plasma, pulsed electric field, high-pressure processing, ultrasonication) are also being investigated which can reduce the ANF concentrations (Joshi et al. 2023) (Fig. 4.2).

4.2 Nutritional and Antinutritional Profile of Millets

4.2.1 Carbohydrate

Millets generally contain high levels of carbohydrates, ranging from 60% to 75%, depending on the variety. Finger millet contains about 72–79.5% carbohydrates (Yousaf et al. 2021). Proso millet exhibits a high carbohydrate content of approximately 75.06% (Sharma et al. 2021). Little millet contains about 74.75% carbohydrates. Millets are rich in complex carbohydrates, which include dietary fibre and resistant starch (RS) (Table 4.2). These contribute to a low glycaemic index, making them suitable for diabetic and weight management diets (Sharma et al. 2021). The high fibre content in millets aids in digestion, reduces constipation, and promotes a healthy gut microbiome. Their low glycaemic index helps in managing blood sugar

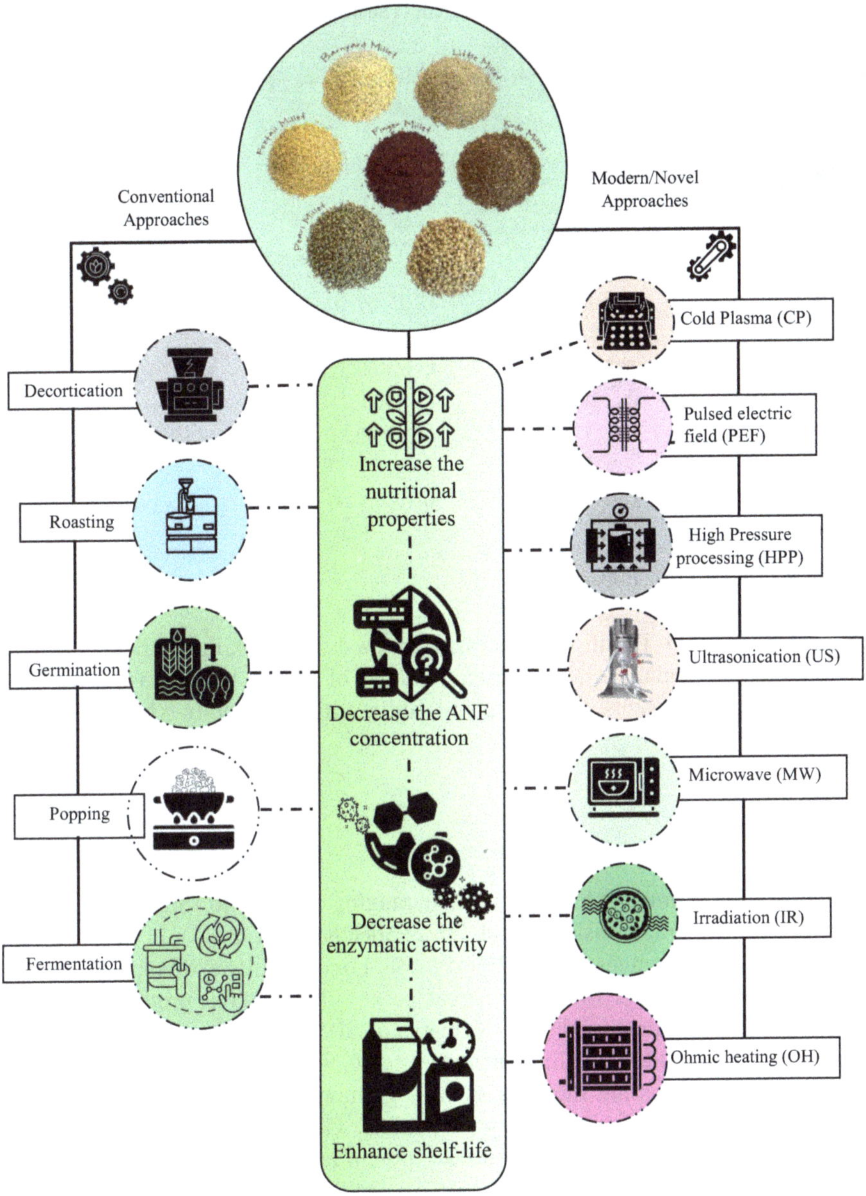

Fig. 4.2 Conventional and modern approaches towards millet processing

Table 4.2 Resistant starch values of millet varieties (Sharma and Gujral 2019; Kaimal et al. 2021; Sharma et al. 2021)

Millet	Image	Starch (g/100 g dw)	RS (g/100 g starch)
Pearl millet		53.00–68.02	9.01–16.52
Finger millet		55.01–65.00	1.90–2.07
Foxtail millet		56.21–73.11	35.22–51.00
Little millet		42.03–57.40	46.55–47.72
Kodo millet		47.62–60.32	37.50–51.21
Barnyard millet		48.21–60.21	41.40–45.02
Proso millet		58.05–77.90	8.00–18.96

dw dry weight, *RS* resistant starch

(Yousaf et al. 2021). Traditional processing methods like soaking, fermentation, and germination can enhance the nutritional quality of millets by increasing the bioavailability of nutrients and reducing antinutritional factors (Nanje Gowda et al. 2022).

4.2.2 Proteins

Millets contain significant levels of protein, ranging from 7% to 13%, depending on the type (Anandharamakrishnan et al. 2022; Sheethal et al. 2022). In finger millet, the protein content ranges from 6.7% to 8.0%. Foxtail millet exhibits protein content between 8.98% and 14.37%, whereas pearl millet contains protein levels that can vary, but typically around 10% (Kumar et al. 2021). Millets are rich in essential amino acids, particularly methionine and cysteine, which are often lacking in other staple cereals. Processing methods such as germination can enhance the protein digestibility and nutritional quality of millets. Germinated millet flours show improved protein efficiency ratios and net protein ratios (Mundassery et al. 2024). The balanced amino acid profile (Table 4.3) of millets contributes to improved human dietary nutrition. Millets combined with legumes provide a complete protein source, enhancing the nutritional value of diets, particularly in regions prone to malnutrition. Traditional and modern processing techniques, such as fermentation, can reduce ANFs like phytic acid and tannins, thereby increasing protein bioavailability and enhancing overall nutritional quality (Anal et al. 2024).

Table 4.3 Some essential amino acid profile (g/g) of some millet varieties (Kumar et al. 2021; Anandharamakrishnan et al. 2022; Anal et al. 2024)

Millet → Amino acids ↓	Finger millet	Proso millet	Foxtail millet	Little millet	Kodo millet	Pearl millet
Lysine	0.22	0.19	0.14	0.11	0.15	0.19
Arginine	0.30	0.29	0.22	0.25	0.27	0.30
Tryptophan	0.10	0.05	0.06	0.06	0.11	0.05
Histidine	0.13	0.11	0.13	0.12	0.14	0.12
Phenylalanine	0.31	0.31	0.33	0.33	0.29	0.43
Isoleucine	0.40	0.41	0.48	0.37	0.36	0.26
Valine	0.48	0.41	0.43	0.35	0.41	0.33
Leucine	0.696	0.754	1.03	0.76	0.64	0.74
Threonine	0.24	0.15	0.19	0.19	0.24	0.20
Tyrosine	0.22	–	–	–	–	0.20
Methionine	0.21	0.16	0.18	0.18	0.18	0.15
Cysteine	0.14	–	0.10	0.09	0.11	0.11

4.2.3 Lipids

Millets contain moderate amounts of lipids, generally ranging from 1.20% to 4.16% (Sharma et al. 2021) depending on the variety. Lipid content in finger millet ranges from 1.2% to 1.7%, and foxtail millet exhibits lipid content between 2.79% and 4.16% (Sharma et al. 2021). Millets are rich in essential fatty acids, particularly unsaturated fatty acids like linoleic and oleic acids, which are beneficial for cardiovascular health (Bora et al. 2019). The lipid profile of millets also includes a good balance of omega-3 and omega-6 fatty acids, contributing to their anti-inflammatory properties (Saleh et al. 2013). Millets have been shown to have hypolipidaemic effects, helping to manage hyperlipidaemia and obesity by reducing total cholesterol, triglycerides, and low-density lipoprotein (LDL) levels while increasing high-density lipoprotein (HDL) levels (Anitha et al. 2020; Mishra et al. 2024).

4.2.4 Vitamins and Minerals

Millets are rich in several vitamins, including water-soluble vitamins like B-complex vitamins (thiamine, riboflavin, niacin, and folic acid) and fat-soluble vitamins like vitamins A, D, E, and K (Mishra et al. 2024). Finger millet is known for its high content of calcium and vitamin D, which are crucial for bone health. Pearl millet contains significant amounts of B-complex vitamins, particularly niacin, which is essential for energy metabolism (Kumar et al. 2021). The rich vitamin content in millets contributes to various health benefits, including improved immune function, better skin health, enhanced energy metabolism, and reduced oxidative stress (Anal et al. 2024).

B-complex Vitamins: Thiamine (vitamin B1) is essential for energy metabolism and nervous system function. Millets provide a good source of thiamine, contributing to better energy levels and cognitive function (Mishra et al. 2024). Riboflavin (vitamin B2) is important for growth, red blood cell production, and skin health. Studies have shown that millets, particularly when processed correctly, retain significant amounts of riboflavin (Kumar et al. 2021). Niacin (Vitamin B3): It is vital for DNA repair and stress responses. Pearl millet is especially noted for its high niacin content. Folic Acid (Vitamin B9): Crucial for DNA synthesis and repair, and for preventing neural tube defects in pregnancy. Millets contribute significantly to dietary folic acid intake (Kumar et al. 2021).

Vitamin E: Millets are rich in vitamin E, particularly tocopherols and tocotrienols, which are potent antioxidants. These compounds help protect cells from oxidative damage, contributing to overall health and disease prevention (Ahirwar and Ravindra Singh 2023).

Vitamin C: Some millet varieties, particularly finger millet, contain appreciable levels of vitamin C, which is essential for immune function and skin health (Ahirwar and Ravindra Singh 2023)

Vitamin A: Proso millet is particularly rich in provitamin A carotenoids, which are converted into vitamin A in the body, essential for vision, immune function, and skin health (Ahirwar and Ravindra Singh 2023).

Millets are rich in several essential minerals, including calcium, phosphorus, magnesium, iron, zinc, copper, manganese, and potassium (Ahirwar and Ravindra Singh 2023; Mishra et al. 2024). Millets, especially finger millet, are known for their high calcium content, which is crucial for bone health. Finger millet contains approximately 344 mg of calcium per 100 grams. Millets are a good source of iron, which is essential to produce haemoglobin and the prevention of anaemia. Kodo millet and finger millet are particularly high in iron content, with values ranging from 5.0 to 9.3 mg per 100 grams (Anandharamakrishnan et al. 2022). Magnesium is vital for muscle and nerve function, blood glucose control, and bone health. Millets like pearl millet contain significant amounts of magnesium, enhancing their nutritional profile (Anitha et al. 2020; Kumar et al. 2021). Zinc plays a critical role in immune function, protein synthesis, and wound healing. Millets such as foxtail millet and pearl millet are rich in zinc, providing about 2.0–5.5 mg per 100 grams (Anitha et al. 2020). Phosphorus is essential for the formation of bones and teeth, and it plays a role in the body's energy production. Millets generally contain about 280–350 mg of phosphorus per 100 grams (Anandharamakrishnan et al. 2022). Potassium helps maintain fluid balance, nerve signals, and muscle contractions. Millets provide a good amount of potassium, which supports cardiovascular health. Copper and manganese which are trace minerals and are important for cardiovascular health and antioxidant defence are majorly found in millets like foxtail and little millets. Efforts to biofortify millets with higher mineral content are ongoing, with studies indicating successful enhancement of mineral concentrations through selective breeding and genetic improvement (Anitha et al. 2020).

4.2.5 Phytic Acid (PA)

Phytic acid (myo-inositol hexakisphosphate) is a major antinutritional factor in millets, found in significant amounts across different varieties such as finger millet, pearl millet, and sorghum. It binds minerals like iron, zinc, calcium, and magnesium, forming insoluble complexes that reduce their bioavailability (Reddy et al. 2022). Studies have quantified the phytic acid content in various millet species. For example, pearl millet contains approximately 987.19 mg/100 g, while finger millet can range from 571 mg/100 g to higher levels depending on the genotype (Reddy et al. 2022). Fermentation is highly effective in reducing phytic acid content. Traditional fermentation methods significantly reduced phytic acid content in sorghum, finger millet, and pearl millet by 62.9%, 34.1%, and 29.35%, respectively (Azeez et al. 2022). Germination enhances mineral bioavailability by reducing phytic acid levels. For example, germinating pearl millet for 96 hours resulted in a significant reduction of phytic acid, correlating with an increase in mineral extractability. Another study demonstrated that germination of finger millet decreased the molar ratio of phytic acid to iron, improving iron bioavailability. The combination of processing methods, such as soaking, autoclaving, and fermentation, can effectively reduce phytic acid (Azeez et al. 2022). Novel methods like plasma-processed air exposure have shown promising results. This technique reduced phytic acid content in pearl millet by up to 60.66% within an hour of treatment (Lokeswari et al. 2021). Kheto et al. (2024) reviewed the impact of pulsed electric field in reducing the concentration of ANFs which also contribute towards improving protein digestibility. The reduction can be possible due to a few factors which include phytate ring cleavage leading to formation of insoluble penta- or tetra-phosphate fractions, hydrolysis, and dissolving of H-bonds in the phytate ring. Reducing phytic acid enhances the bioavailability of essential minerals, addressing deficiencies such as iron and zinc, which are prevalent in populations relying on millet as a staple food (Reddy et al. 2022).

4.2.6 Tannins

These compounds are mainly located in the outer layers of the grain, such as the testa, which is the seed coat (Kumar et al. 2021). The tannin content varies among different millet species and even among varieties within a species. Foxtail millet showed the highest tannin value at 12.97 mg of tannic acid equivalent/g, compared to 6.67 mg/g in barnyard millet (Yousaf et al. 2021). Tannins in millets can be classified into hydrolysable tannins and condensed tannins. Condensed tannins, also known as proanthocyanidins, are more prevalent in millets and contribute to their astringent taste and antioxidant properties (Anandharamakrishnan et al. 2022). Tannins bind to dietary proteins and digestive enzymes, forming insoluble complexes that reduce the digestibility and bioavailability of proteins and minerals such as iron and zinc (Singh et al. 2024a). This binding can lead to decreased nutrient absorption and potential deficiencies, particularly in populations that rely heavily

Table 4.4 Some key antinutrients found in millets (Sharma et al. 2021; Dey et al. 2022)

Millets	Tannin	Phytic acid
Barnyard millet	1.30 mg TA/g	3.35–3.72 mg/g
Proso millet	–	7.1–7.22 mg/g
Finger millet	3.30–3.51 mg/ g	5.50–5.58 mg/g
Pearl millet	2.02–2.20 mg/g	9.10–9.21 mg/g
Kodo millet	1.01–1.20 mg/g	1.20–1.40 mg/g
Little millet	3.32–3.36 mg CE/g	–

TA tannic acid, *CE* catechol equivalent

on millets as a staple food (Yousaf et al. 2021). Fermentation significantly reduces tannin levels in millets. A study showed that fermentation could decrease tannin content by up to 90% in pearl millet, finger millet, and sorghum (Singh et al. 2017). Germination activates endogenous enzymes that degrade tannins, enhancing the nutritional quality of millets. During germination, tannin content in finger millet and pearl millet significantly decreased. Simple household techniques such as soaking and boiling can effectively reduce tannin content. Soaking followed by boiling was shown to lower tannin levels in millet varieties, making nutrients more bioavailable (Anandharamakrishnan et al. 2022). Research on minor millets indicated that phenolic compounds, including tannins, negatively correlated with protein digestibility. This relationship underscores the importance of processing methods to reduce tannins for improved nutritional outcomes (Sharma and Gujral 2019; Jaddu et al. 2024). Studies have explored the potential of biofortification to enhance the nutritional quality of millets while managing tannin levels. Biofortified millets with reduced tannin content have shown promise in improving mineral bioavailability without compromising their beneficial antioxidant properties (Sharma et al. 2021) (Table 4.4).

4.3 Digestibility of Millets

The digestibility of millets plays a crucial role in its consumption. Limited research regarding its starch and protein digestibility is available, which hinders towards its proper acceptance and how it can impact the quality of millet-based products. As millets are one of the most nutritionally prolific and climate-resistant grain, the nutritional profile of millets constitutes of 60–70% carbohydrates (mostly complex starches), 7–12% protein, 1.5–5.0% lipid profile, and 2.0–7.0% fibre, and is rich in flavonoids. Moreover, millets are gluten-free and are rich in various minerals such as Fe, Zn, Mn, K, etc. (Kumar et al. 2021). Apart from this, millet-based products are a viable alternative for people suffering from celiac disease and can be used for management of type II diabetes (Yousaf et al. 2021). Therefore, to increase its availability and make it nutritionally more enhanced, studies on starch and protein digestibility are required as there has been a gradual increase in the demand for millets as nutraceutical and functional food. Processing of millets plays a crucial role in improving its digestibility. As such, various studies have been conducted which have discussed and investigated the hypoglycaemic effect of millet. A study investigating the hypoglycaemic effect was done by preparing a multitude of food products using foxtail millet, barnyard millet, legume flour, and fenugreek powder. Dhokla, uppuma, and laddus were prepared using various combinations of this flour, and the hypoglycaemic effect was measured in terms of spike in glycaemic index (GI). GI was found to be 34.96, 23.52, and 17.60 for dhokla, laddu, and uppuma, respectively. Uppuma, which had the highest concentrations of millet flour, had the lowest GI spike (Jaybhaye et al. 2014). Another study by Sharma et al. (2019) investigated the correlation between the interaction between nutritional and antinutritional components with in vitro protein and starch digestibility. It was found that ANFs such as condensed tannins and PA were found in appreciable amounts and inhibited protein and starch digestibility. It was also found that phenolics and dietary fibre suppressed retrogradation. These studies show that both nutritional and antinutritional factors play a vital role in millet digestibility (Sharma and Gujral 2019).

Various processing techniques are also employed on millets, which largely influence the starch and protein digestibility of millets. These processes also influence the nutritional quality of millet-based foods and beverages. Thus, the research on these processes, i.e. germination, fermentation, decortication, and thermal processing, has become a primary focus of research. During decortication, the germ and pericarp are readily removed from the grain. Although this process reduces the amount of antinutrients, there is also loss of lipids, minerals, fibres, and some phenolic acids (Annor et al. 2017). Moreover, Annor et al. (2017) investigated the hypoglycaemic properties of decorticated kodo millet flour and how it impacted the GI when products were prepared using kodo millet flour. They compared the in vitro *starch digestibility* (IVSD) of the decorticated millet samples with rice and wheat flour. It was found that the millet samples had significant reduction in GI when compared to cooked rice flour. This reduction can be due to removal of proteins, lipids, and phenolic compounds after decortication. Another study investigated the combined effects of germination and fermentation on the IVSD and in in vitro *protein digestibility* (IVPD) of pearl millet. It was found that when samples were germinated for more than 24 h, there was three times increase in the IVSD. This was further enhanced when the germinated samples were fermented with bacterial population. This increase in IVSD can be due to breakdown of starch molecules during the fermentation process. Additionally, the combined processes also enhanced the IVPD, which was noticed during the treatment, which can be due to release of certain proteolytic enzymes during the fermentation process (Annor et al. 2017).

4.4 Conventional Methods of Millet Processing

4.4.1 Decortication

Decortication, also known as dehulling, is a traditional process used to remove the outer husk or bran layers of millet grains. This technique enhances the culinary quality of millets and reduces antinutritional factors. For an efficient retention of the nutritional qualities, dehulling of 12–30% of the kernel is considered suitable. Dehulling beyond this percentage often leads to loss of nutrients such as dietary fibre, phenolics, and bioactive compounds (Yousaf et al. 2021). Traditional decortication methods include:

- *Hand Pounding*: Hand pounding using mortar and pestle or grinding stones is one of the oldest methods of decortication. This labour-intensive process involves manually striking the grains to remove the outer husk. While effective, it can lead to significant grain breakage and variable decortication efficiency. Research on the decortication of pearl millet and sorghum revealed that traditional methods effectively reduced antinutritional factors such as phytates and tannins, thereby improving mineral bioavailability (Hama et al. 2011)
- *Stone Grinding*: Traditional stone mills, or querns, are also used for decorticating millets. The grains are ground between two stones, which helps to peel off the husk. This method, like hand pounding, is time-consuming and can cause loss of fine particles, reducing the overall yield (Babiker et al. 2018).
- *Mortar and Pestle*: The mortar-and-pestle method involves placing the millet grains in a wooden or stone mortar and using a pestle to pound them. This method is widely used in rural areas and helps in partially decorticating the grains, although it may not completely remove the bran (Babiker et al. 2018).

Traditional decortication can lead to significant nutrient losses, particularly in minerals and vitamins concentrated in the bran. Studies have shown that the content of calcium, phosphorus, and iron decreases after decortication due to the removal of the outer layers. Hama et al. (2011) investigated the impact of traditional decortication methods on the nutritional composition (iron, zinc, phytic acid, starch, and fibre) of pearl millet and white sorghum. It was found that no significant changes were found between the two methods (abrasive decortication vs. traditional decortication), as the type of grain used was more significant and had a greater impact on the nutrients. Abrasive decortication was done until the final moisture content of the grains was 10% each, with 20 g aliquots were taken for the final analysis, and stored at 4 °C. A study on decorticated finger millet examined the cooking characteristics and impact of decortication. It was found that soft texture of the decorticated millet was achieved within 6 minutes and the moisture content was 71 g/ 100 g, with a loss in solubility of 3.5 g/100 g. Moreover, the redness value was enhanced by 39% when exposed to air. Overall acceptability score was 8.0, thereby revealing decorticated finger millet as a convenient source for different food formulations (Yousaf et al. 2021). Decortication methods do not require expensive equipment, making them economically viable for small-scale farmers and rural communities. These methods provide a low-cost means of processing millets, which is essential for food security in resource-limited settings. Overall, decortication leads to an increase in the protein and starch digestibility, but has a decreasing impact on the protein, dietary fibre, fat, iron, and calcium content (Yousaf et al. 2021; Anandharamakrishnan et al. 2022).

4.4.2 Roasting

Roasting can have varied effects on the nutritional composition of millets. For instance, it has been observed that roasting enhances the fat content and amino acid profile while reducing the moisture content. The total phenolic content (TPC) and antioxidant activity of millets also increase significantly after roasting. Traditional roasting techniques effectively reduce antinutritional factors such as tannins, phytic acid, and polyphenols. This reduction enhances the bioavailability of essential nutrients like iron, zinc, and calcium (Yousaf et al. 2021). Mohd et al. (2023) investigated the impact of roasting on whole grain barnyard millet (WGBM) and its influence on the proximate, amino acid, TPC, and antioxidant activity. The results confirmed that roasting had significantly improved the total amino acid content, enhancing it from 116.71 ng to 123.51 ng. Moreover, enhancement in TPC and antioxidant activity was also observed in roasted WGBM. Roasting further enhances the functional properties of millets, such as water solubility index and oil absorption capacity. These improvements make roasted millet flour suitable for various food applications, including baked goods and snacks (Jaybhaye et al. 2014). Azad et al. (2019) compared the impact of four processing methods, i.e. roasting, steaming, puffing, and extrusion, on the antioxidant and phenolic profile of broomcorn millet flour. It was observed that roasting had the highest impact on the TPC, total flavonoids, as well as with DPPH and FRAP assays. It was concluded that roasting can be considered as a favourable method for enhancing the broomcorn millet flour functionality. Studies on the sensory qualities of traditionally roasted millets indicate that these methods produce grains with favourable sensory attributes, such as texture and flavour, making them more acceptable to consumers. For example, roasted barnyard millet-based products received high sensory ratings for their taste and texture (Sharma and Gujral 2019).

4.4.3 Germination/Malting

Malting significantly enhances the nutritional profile of millets. It increases the bioavailability of essential nutrients such as proteins, vitamins, and minerals. For instance, germination of finger millet increases protein availability by 17% and enhances the total energy (Yousaf et al. 2021). Malting effectively reduces antinutritional factors such as phytates, tannins, and polyphenols, which can inhibit the absorption of minerals. A study showed that malting reduced phytate content in pearl millet by 45% and tannin content by 46% (Singh et al. 2024b). The process of malting enhances the functional properties of millets, such as their solubility and digestibility. This makes malted millet flour suitable for a variety of food applications, including weaning foods and gluten-free products. For example, germinated barnyard millet showed improved dietary fibre content and increased levels of γ-aminobutyric acid (GABA), a beneficial bioactive compound (Sharma et al. 2021). The increase in GABA was studied in kodo millets, where an increase from 9.36 to 47.43 mg/100 g was observed. In addition to that, enhancement in dietary fibre, protein content, and mineral content was observed (Yousaf et al. 2021). Malting also enhances the flavour and texture of millet products. The enzymatic breakdown of complex carbohydrates during germination results in a sweeter taste and softer texture, which improves the palatability of millet-based foods (Kumar et al. 2021). Additionally, germination leads to biochemical changes and structural modifications due to enzymatic hydrolysis, which can assist in enhanced bioactivity as well as increased protein digestibility. Increase in protein digestibility can also be attributed to release of pepsin which occurs during sprouting. When combined with fermentation, germination can provide a distinctive nutritional approach to production of value-added production (Annor et al. 2017). Traditional malting include:

- *Soaking*: Soaking is the initial step in malting where millet grains are immersed in water to initiate the germination process. The duration of soaking can vary, typically ranging from 4 to 24 hours depending on the type of millet and desired outcome. Soaking softens the grain and initiates enzymatic activity, which is crucial for subsequent germination (Kate and Singh 2021).
- *Germination*: Germination involves placing the soaked grains in a humid environment to sprout. This process usually lasts between 24 and 72 hours. During germination, enzymes such as amylases and proteases are activated, breaking down starches and proteins into simpler, more digestible forms. The temperature and humidity are carefully controlled to optimize enzyme activity and nutrient release (Kate and Singh 2021).
- *Drying and Milling*: After germination, the sprouted grains are dried to halt enzymatic activity and prevent spoilage. This can be done using sun drying or mechanical dryers. Once dried, the malted grains are milled into flour, which can be used in various food products (Azeez et al. 2022; Mundassery et al. 2024).

4.4.4 Popping

Popping enhances the carbohydrate and energy content of millets. Studies have shown that popped millets have higher digestibility and bioavailability of nutrients compared to raw grains. For instance, popped foxtail millet has significantly higher carbohydrate and energy values. The popping process reduces antinutritional factors such as phytates and tannins, which inhibit mineral absorption. Popping decreases the levels of these compounds, enhancing the bioavailability of minerals like iron and zinc (Choudhury et al. 2011). Popping improves the sensory qualities of millets, making them more appealing for consumption. The process imparts a crisp texture and enhances the flavour, making popped millets a popular snack food (Jaybhaye et al. 2014).Traditional flaking involves manually beating and rolling millet grains to flatten them. This method is labour-intensive but effective in producing thin, crisp flakes that can be easily cooked or consumed as a ready-to-eat snack (Jaybhaye et al. 2014). Steam flaking involves steaming the grains to soften them, followed by rolling to create flakes. This method helps retain the nutritional value of millets while improving their texture and digestibility. Like mechanical popping, mechanical flaking uses specialized equipment to produce uniform flakes. This method is efficient for large-scale production and ensures consistency in the size and thickness of the flakes (Choudhury et al. 2011). Flaking extends the shelf-life of millets by reducing moisture content. This makes flaked millets more stable for storage and transport, ensuring a longer shelf-life without significant loss of nutritional quality (Mundassery et al. 2024). A study on popped foxtail millet showed significant increases in nutrient content and digestibility. The popped millet had higher protein and energy values, making it a nutritious snack option. Research on the bioavailability of minerals in flaked millets revealed that traditional flaking techniques improve the absorption of calcium, iron, and zinc (Anal et al. 2024).

4.4.5 Fermentation

Fermentation enhances the nutritional quality of millets by increasing the bioavailability of essential nutrients such as vitamins, amino acids, and minerals. For instance, the fermentation of pearl millet reduces phytic acid and tannin content, making minerals like iron and zinc more bioavailable (Sharma et al. 2021). Traditional fermentation significantly reduces antinutritional factors such as phytates, tannins, and enzyme inhibitors. This reduction improves nutrient absorption and enhances the nutritional value of millet-based foods (Azeez et al. 2022). For example, a study found that fermentation of pearl millet reduced phytate content by 75% and tannins by 83% (Anal et al. 2024). The fermentation process improves the functional properties of millets, such as solubility, digestibility, and texture. This makes fermented millet products more suitable for various culinary applications, including weaning foods and beverages. For example, fermented millet beverages exhibit better viscosity and flavour profiles, making them more acceptable to consumers. Fermentation enhances the sensory qualities of millet products by developing desirable flavours and aromas. The process of lactic acid fermentation, for instance, produces a sour taste and improved texture, which are appealing characteristics for traditional foods like *Koozh* and *ogi* (Jaybhaye et al. 2014). Different traditional methods of fermentation include:

- *Natural Fermentation*: Natural fermentation relies on the indigenous microflora present in millet grains. This method involves soaking the grains in water and allowing them to ferment spontaneously over a period. For example, in the preparation of *ben-saalga* a fermented gruel from Burkina Faso, pearl millet is soaked, ground, and left to ferment naturally before cooking (Anandharamakrishnan et al. 2022).
- *Lactic Acid Fermentation*: Lactic acid fermentation is a controlled process that uses lactic acid bacteria (LAB) to ferment millet grains. This method is common in the production of various traditional fermented foods. The LAB converts sugars into lactic acid, lowering the pH and preserving the food. This technique is used in making fermented beverages like *Boza* and *Koozh* (Yousaf et al. 2021).

- *Alcoholic Fermentation*: Alcoholic fermentation involves yeast fermentation, where sugars in millet grains are converted to alcohol and carbon dioxide. This method is used in the preparation of traditional alcoholic beverages like *Burukutu* and *Togwa*. Yeasts and sometimes moulds are used to initiate and sustain the fermentation process (Kumar et al. 2021).
- *Co-fermentation*: Co-fermentation involves fermenting millet grains along with other ingredients such as legumes or tubers to enhance nutritional value. For example, co-fermenting millet with groundnut improves the protein content and overall nutritional profile of the product. This method is used in making fortified traditional foods like *Lohoh* and improved versions of *ben-saalga* (Jaybhaye et al. 2014).

4.5 Modern Approaches Towards Millet Processing

Millets are one the most resilient crops that can adapt to harsh climatic conditions, ranging from shallow lands to soils having an uneven pH. Even nutritionally, millets are equivalent to other cereals and have numerous health benefits (Mishra et al. 2024). Despite of all these benefits, there are various challenges which are present regarding the adoption and processing techniques of millets and millet-based products, globally. The major challenges that are a bottleneck and significantly impact its nutritional attributes include limited shelf-life, high lipid content, and elevated enzymatic activity (Joshi et al. 2023). Once the millet grains have been processed (dehulled), factors such as enzymes, moisture, heat, microbes, etc. start deteriorating the quality of the milled grains. The enzymes responsible for lipid degradation are peroxidase, lipoxygenase, lipase, and polyphenol oxidase (PPO) (Kumar et al. 2021; Joshi et al. 2023). These enzymes affect the grains in a variety of ways:

- Due to its poor shelf-life, lipids present in millets, once decorticated, are acted decisively by lipase into fatty acids and glycerol (Joshi et al. 2023).
- Further oxidation by lipoxygenase and peroxidase takes place on the free fatty acids that are generated which lead to formation of unpleasant odour (Sarkar et al. 2023).

Thus, processing of millets is crucial for a longer shelf-life and to improve other characteristics such as palatability. However, processing techniques can also lead to loss and reductions in phenolic contents, dietary fibre, vitamins, and minerals (Niranjan and Dwivedi 2022). Therefore, technologies are required to minimize the loss of essential nutrients as well as improve the shelf-life. There are various thermal and non-thermal techniques which have been researched and investigated and have been able to enhance the shelf-life after processing.

4.5.1 Thermal Techniques of Millet Processing

4.5.1.1 Microwave Processing

Microwave processing (MW) involves the use of microwave heating and the ability of the food material to absorb this energy and convert it into heat. It is considered as one of the most promising technologies, as scalability is possible under both batch and continuous conditions and can also be combined with any other processes for treatment (Barba et al. 2019). MW processing employs two mechanisms for heating up the food materials—dipolar mechanism and ionic mechanism. Moreover, it also includes electromagnetic radiation in the wavelength range of 0.30–300 GHz. These combinations allow the MW to optimize resource utilization and allows for a low operational cost (Anal et al. 2024). MW processing is usually done to inactivate the antinutritional factors (ANFs) in millet and millet-based products, which has been successfully investigated by researchers. Covalent bond dissociation, peptide bond hydrolysis, and alterations in the disulphide bonds induce inactivation of ANFs in the grains (Joshi et al. 2023). Optimized processing of pearl millet, finger millet, and sorghum through mechanisms of MW processing, germination, soaking, and fermentation. There was a significant reduction in the concentration of ANFs among the selected millet grains. The MW processing parameters were kept at 900 W, 2450 MHz for 40–100 s (Singh et al. 2017). Another study investigated the action of MW treatment on pearl millet and its efficacy to reduce ANFs. It was found that when the microwave was kept at constant 900 W for 100 s, there was more than 90% reduction in the lipase activity of the pearl millet samples. The increase in temperature during the process also attributed to the reduction in lipase activity (Mundassery et al. 2024). The impact of MW roasting on sorghum was investigated by Sharanagat et al. (2019), and there were significant changes in the physical, thermal, functional, and pasting properties. The most prominent effect was seen in the total phenolic content, total flavonoid content, and the antioxidant content of sorghum as there was significant reduction among all these properties after treatment. Although there was a decrease in the bioactivity of sorghum, there was improvement in the digestibility as well as functional properties, which can be utilized for development of various functional products (Sharanagat et al. 2019).

Although MW processing can be used effectively on millet and millet-based products, various challenges still lie ahead in terms of its proper application. For instance, MW tends to cause thermal damage due to non-uniform heating. When bulk samples are used, there is low penetration of MW which leads to uneven heating (Prashanth et al. 2024). This can lead to formation of hotspots, where temperatures are higher than surrounding sample area. Therefore, further studies on the application of MW processing need to be done along with studies that investigate its impact on ANFs for proper utilization of this technique (Fig. 4.3).

4.5.1.2 Radiofrequency (RF) Processing

RF processing involves the use of non-ionizing electromagnetic waves, within the range of 1–300 MHz. Heat is generated volumetrically as the RF waves can pass through dipole rotation or ionic polarization. Because of this dipole rotation, the RF

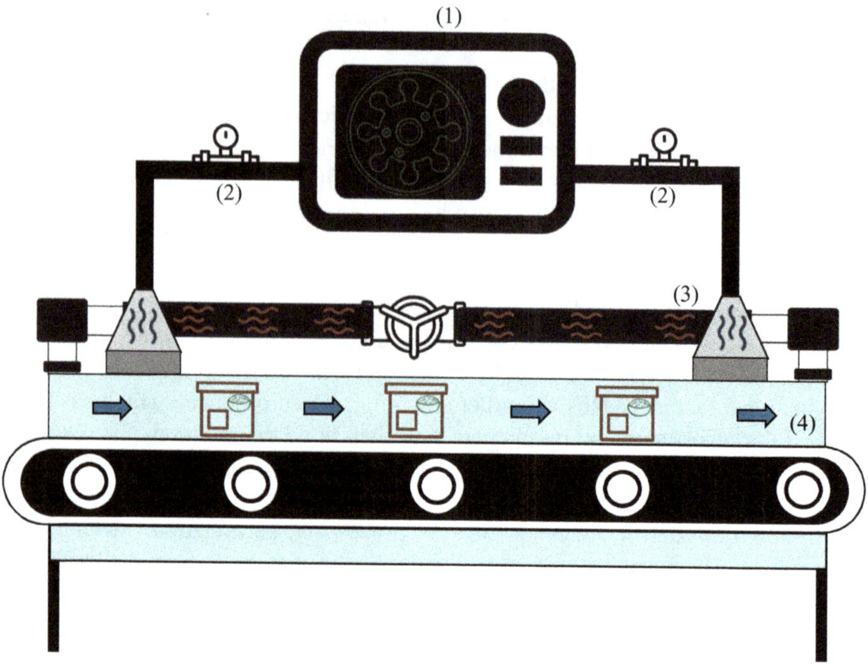

Fig. 4.3 Microwave processing mechanism: (1) = magnetron, (2) = waveguide, (3) = hot air blowing system, (4) = conveyor carrying the treated samples

waves can penetrate the food deeper and leads to uniform heating (Ling et al. 2019). RF has been used for quite some time in agriculture due to its effectiveness in enzyme inactivation, insect control, protein modification, drying, etc. in various grains like rice and wheat, and other food products such as walnuts (Joshi et al. 2023). As of now, only a handful of investigations regarding the effect and processing of millets through RF have been conducted. One of the hurdles of using RF for millet processing is the uneven heat distribution within the environment and the food product (Joshi et al. 2023). A study by Yarrakula et al. (2024) was done to study the functional and storage parameters of pearl millet treated with hot air-assisted radiofrequency (HARF). Pearl millet at different moisture levels (10.5%, 12.0%, 15.0%) were undertaken for this study. It was observed that there were significant reductions in the free fatty acid and peroxide values after HARF treatment among the untreated and treated samples. This significantly enhanced the storage stability of the pearl millet samples. Another study was conducted to study the effectiveness of RF in suppressing lipase activity and improve lipid stability of wheat germ. It was found that RF was able to enhance the water absorption capacity, higher protein solubility, higher antioxidant activity, and emulsifying properties (Ling et al. 2019). However, to study RF effectiveness among millet processing, further research is required that can study the impact of RF processing and provide more data on the pros and cons of using RF processing on millet-based products (Fig. 4.4).

4.5.1.3 Infrared (IR) Processing

The process of IR heating involves the use of IR radiation, where the wavelengths of the radiation are electromagnetic waves and fall between visible light and MW. In this process, the heat inside the grain rises rapidly when the wave is transmitted through it. The generation of heat inside the grain is caused by rapid internal friction between the grain molecules when the radiation falls on the grain (Singh et al. 2024b). Usually, there are three forms of IR heating, near IR, mid-IR, and far-IR with wavelengths in the region of 0.78–1.40 μm, 1.40–3.0 μm, 3–1000 μm, respectively. There are known advantages of using IR for processing purposes that include high heat transfer rate, lower energy consumption, uniform heating, and enhanced product quality (Joshi et al. 2023). Moreover, the IR treatment allows the attachment between the hull and endosperm to loosen up, which makes it easier to dehull the grain (Singh et al. 2024b). A study on the effect of IR treatment on sorghum flour for evaluation of action of free fatty acids (FFA) and lipase activity during storage conditions showed that at a temp. of 120 °C for 8.50 min led to lowering of FFA and lipase activity among the sample. The pasting properties were also improved during the IR treatment, thereby indicating that the sorghum flour can be used for caloric-dense food (Swaminathan et al. 2015). Another study investigated the effect of IR treatment on the nutritional and antinutritional factors of sorghum grain. It

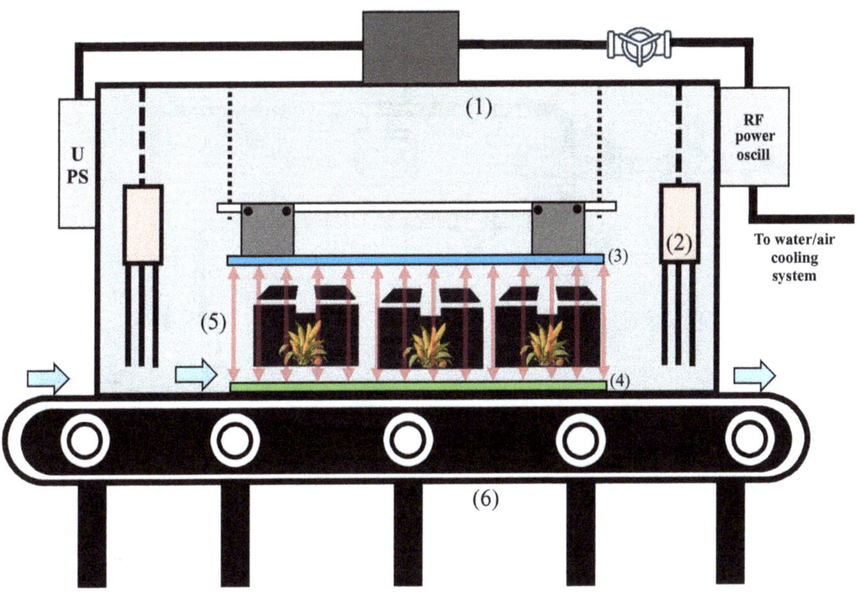

Fig. 4.4 Radiofrequency (RF) processing: (1) = radiofrequency cavity/holder, (2) = RF shield, (3) = top electrode (blue), (4) = bottom electrode (green), (5) = alternating electric field, (6) = conveyor

was found that the concentrations of PA and tannins were significantly reduced while total phenols increased compared to the control. However, protein digestibility was reduced which can be due to denaturation (Swaminathan et al. 2015). Therefore, these investigations show that IR processing can be used effectively to reduce the ANFs and enhance storage properties of millet grains. But more research is required in this aspect which can further verify its effectiveness and efficiency (Jindal and Nikhanj 2023) (Fig. 4.5).

4.5.1.4 Ohmic Heating

Ohmic heating (OH) is a novel thermal process which involves the use of alternating current to heat the product. When the material is exposed to OH process, the heat is generated volumetrically, which allows the heat to rise rapidly inside the material (Joshi et al. 2023). It is an advantageous technique, as it is highly energy efficient, allows for rapid cooking, and low energy is consumed. Depending on the composition and exposure time of the material, it can lead to changes in the moisture content, gelatinization of starch, or protein denaturation. OH has been used extensively for treatment of fruit and vegetable juices. Moreover, OH has also been applied for treatment of meat, milk, chicken, fish, etc.; most of these studies have successfully

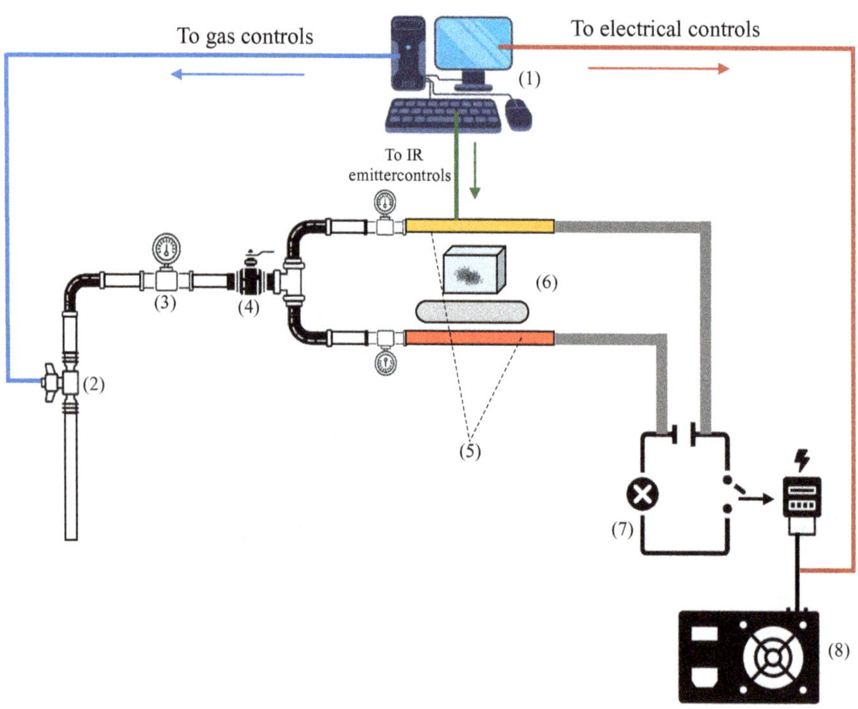

Fig. 4.5 IR processing mechanism: (1) = main system controls, (2) = natural gas supply, (3) = flow meter, (4) = flow divider, (5) = IR catalytic emitter, (6) = sample + turntable, (7) = electrical circuit, (8) = electric supply

investigated the reduction in pathogenic microbes from the samples and enhances cooking time (Sain et al. 2024). However, the use of OH for millet processing has been limited and has been less explored. Darvishi et al. (2020) investigated the effect of OH on the quality and quantitative attributes of black grape molasses. It was found that when samples were treated with OH, highest antioxidant activity was recorded when compared to conventional methods. Additionally, there was reduction in the use of energy (6.3-fold) as well as reduction in processing time when OH was used. The sensory characteristics of the molasses were intact, with low to minimal damage. This study therefore shows that OH is an effective technique which can enhance the quality of the product, as well as reduce energy consumption and enhance bioactive quality. This can be investigated in millet-based beverages, such as fermented drinks or millet milk, and the characteristics can be further studied. Another study was done which involved the use of two techniques, OH and conventional open-pan (CONV) on pearl millet grains, to study its effect on the cooking yield, colour characterization, functional properties, texture, and pasting properties. The results showed that the OH-treated samples had lower cooking yield than the CONV, which can be due to presence of the pericarp. Starch gelatinization also occurred and was comparable for both techniques employed (Dias-Martins et al. 2019). These studies show that if the volumetric heat is distributed evenly among the product, it can lead to higher processing rate than conventional methods. However, if the grains have not been dehulled and the pericarp layer is present, the effect on the grains will be less profound leading to lower processing of the grains (Singh et al. 2024b) (Fig. 4.6).

4.5.2 Non-thermal Techniques of Millet Processing

4.5.2.1 High-Pressure Processing (HPP)

HPP is one of the most promising methods of non-thermal processing for millets. HPP is one of the few such methods which allow for precise maintenance of the quality and safety of the food product, without causing any deterioration and side effects which can be induced by high temperatures. HPP allows for maintenance of the fresh-like quality of the food after processing, without causing any changes in its nutritional quality. Additionally, it extends the shelf-life of the food product. Although the main objective of using HPP is maintenance of the quality of food and prolonging the shelf-life of foods, it adds another layer of protection to the processed products in the form of safety against microbial spores. Various vegetative pathogenic bacteria can be inactivated at normal HPP parameters. The process of HP is usually done at pressures in the range of 300–700 MPa, either at an ambient temperature or in the range of 10.0–30.0 °C (Tsevdou et al. 2019). However, the application of HPP has been limited in case of millet processing. Sharma et al. (2018) investigated the effect of HP soaking on various functional and biochemical properties of germinated and non-germinated foxtail millet grains. It was found that this process significantly enhanced the starch gelatinization degree, total phenolic content, and antioxidant activity of the germinated flours. Moreover, antinutrient

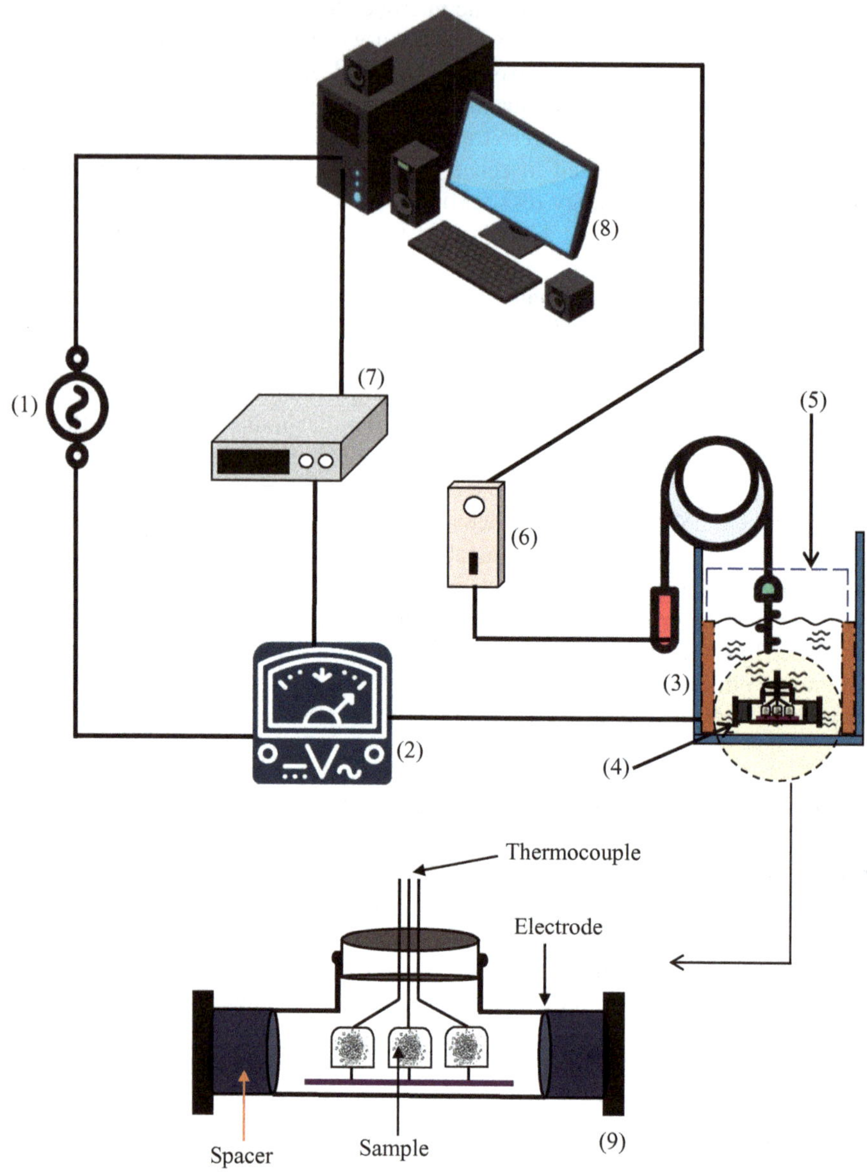

Fig. 4.6 Working of an ohmic heating system: (1) = AC supply, (2) = voltage corrector, (3) = ohmic heating system, (4) = sample, (5) = electrode, (6) = thermocouple system, (7) = data logger, (8) = computer, (9) = enlarged sample

concentration of PA and tannin decreased in the flour. Another study investigated the effect of HPP on protein digestibility of cooked proso millet, and it was found that HPP did not degrade the quality of proso millet proteins that can hamper its digestibility (Gulati et al. 2018). These studies show that HPP can be used effectively to prolong the shelf-life as well as maintain the quality of treated millet products. Further studies focussing on the safety aspect of HPP and how it can prevent microbial contamination can be investigated (Sruthi and Rao 2021) (Fig. 4.7).

4.5.2.2 Ultrasound (US) Processing

Another emerging non-thermal processing technique is ultrasound or ultrasonication (US). It is one of the few technologies that ensures that quality and safety is not hampered, and the product is minimally processed without any major characteristic changes. In the process of US, high-energy sound waves are generated which are allowed to pass through the food, thereby enhancing the extraction efficiency, and functional properties of the product (Sruthi and Rao 2021). Modification in the food matrix occurs due to formation of rarefaction and compression cycles which enter the food matrix and create characteristic changes. This process is called acoustic cavitation. Furthermore, this process leads to changes in pressure and temperature. US treatment falls under two categories: low-frequency, high-intensity ultrasound with range between 20 kHz and 100 kHz and high-frequency, low-intensity ultrasound in the range of 100 kHz–1 MHz. This is a promising novel technology which allows for physical modifications of the food, including quicker extraction rate and increase in final yield. However, properties such as flavour, colour, etc. can be

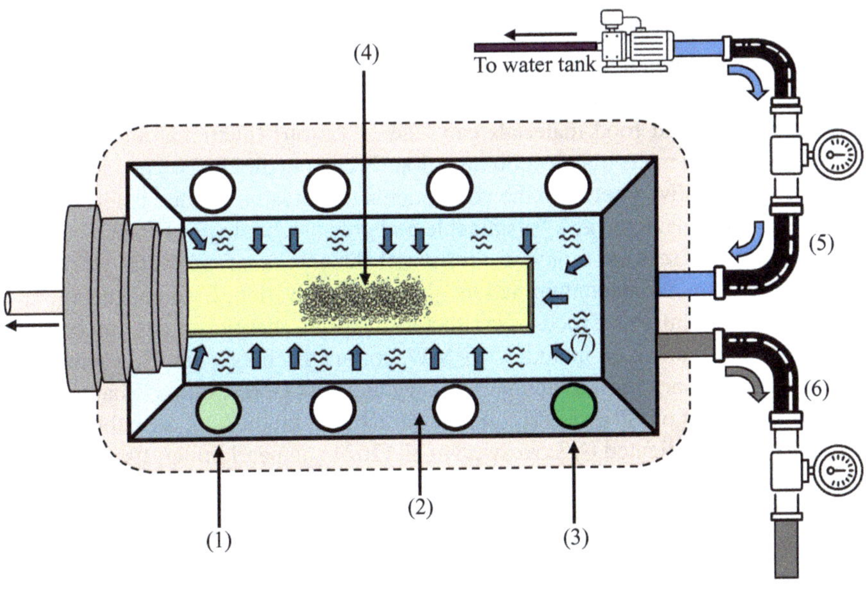

Fig. 4.7 HPP mechanism: (1) = cooling inlet, (2) = pressurized cylinder, (3) = cooling outlet, (4) = sample, (5) = water inlet, (6) = water outlet, (7) = pressurized water cabin

impacted negatively. Studies on treatment of food samples with US have shown that it enhances the emulsifying properties of proteins (Joshi et al. 2023). A study by Sullivan et al. (2018) investigated the effect of US on the structural, physico-chemical, and changes in protein structure of sorghum, particularly kafirin, the major protein in sorghum. When the sample was treated for 10 mins at 40% amplitude, it was found that protein solubility was enhanced in sorghum. Moreover, there was improved hydrolysis and antioxidant activity in the sample. The effect of US on finger millet starch isolates and its antinutrients were also investigated. The samples underwent US-assisted hydration, with an amplitude of 66% for 26 mins. The samples were found to have reduction in phytate and tannin content, with the starch isolates having increased crystallinity. Moreover, this technique allowed for homogenous distribution of starch particles (Yadav et al. 2021). US treatment on millet protein concentrate (MPC) was also investigated, and there was overall enhancement in the functional properties, increase in protein solubility, and increase in structural properties of the US-treated MPC sample (Nazari et al. 2018). Millet bran, which is a good source of dietary fibre, was also investigated for changes in its physico-chemical properties after US treatment. Functional properties were improved like swelling capacity, and there were enhancements in the water-holding and oil-holding capacity. This can be due to loosening of the fibre structure which makes it porous. With improvements in the thermal stability also observed in the fibres, US treatment can be utilized to improve availability of soluble dietary fibre of millet bran (Yadav et al. 2021) (Fig. 4.8).

4.5.2.3 Cold Plasma (CP)

Cold plasma (CP) is a novel non-thermal technique, which involves the use of plasma, a highly energized energy state, for various food-based applications, including processing. When a highly energized gas such as plasma is allowed to interact with gases in the atmospheric pressure and low temperatures, the interaction between the gas and food materials can lead to various functional and structural changes, and even microbial decontamination (Sonkar et al. 2023). The interaction, however, is short-lived between the components and thus no residue is left behind. Thus, it can be considered as a 'sustainable and green technology'. CP has found its way into various sectors, which includes soil decontamination, water treatment, food preservation and decontamination, enzyme inactivation, food packaging, and other such agriculture and food-based sectors. Various energy sources can be used to generate plasma, which includes RF, MW, corona discharge, etc. However, optimization is necessary for maximum efficiency and cost-effectiveness (Sharma et al. 2024). The effects of CP processing via pre-treatment methods of air bubbling and soaking were investigated by Lokeswari et al. (2021) on pearl millet. Its impacts on the physico-chemical, functional, and ANFs (PA) were also studied. The samples were air bubbled at 180 V with airflow rate of 10 litres/h. It was found that upon air bubbling for 1 h and 2 h, PA concentrations reduced significantly. Other functional properties were also enhanced. There were enhancements in the properties as well as reduction in ANFs. Another study used CP treatment to investigate its effect on the functional properties of little millet flour. The flour was treated for 10, 20, and

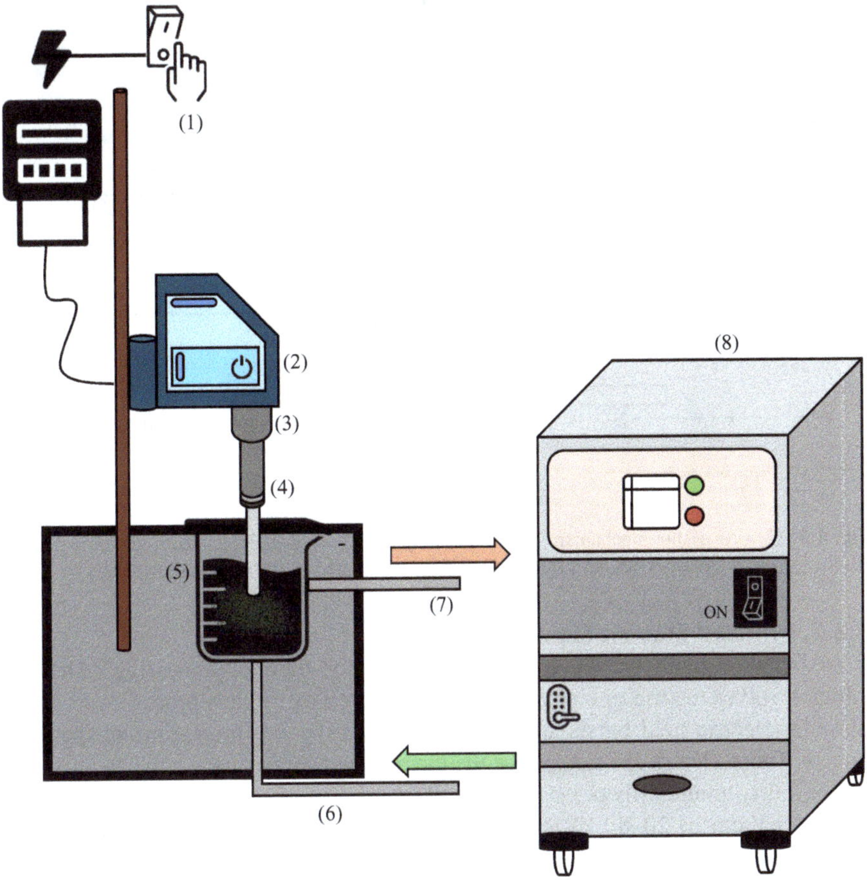

Fig. 4.8 US processing mechanism: (1) = power source and generator, (2) = transducer, (3) = fixed horn, (4) = detachable probe, (5) = sample, (6) = fluid inlet, (7) = fluid outlet, (8) = temperature control system

30 mins at 13 W and 24 W, respectively. Microstructural changes in the form of broken starch granules and decreased crystallinity were observed. Enhancement in the functional properties of the flour, which included swelling capacity, solubility index, oil absorption capacity, and water-holding capacity, were also observed. The microstructural changes that occurred due to the CP treatment may have led to enhancement in its characteristics (Jaddu et al. 2022). As research suggests, there are various ways in which CP can be effectively used to improve the characteristics of millet and millet-based products. However, hindrances in the form of high cost of investments, lack of control over process variables, regulatory issues, and toxicity-related issues persist which makes its adoption difficult. Despite all the issues, CP has the potential to be used extensively with clear environmental and sustainability benefits (Jaddu et al. 2023) (Fig. 4.9).

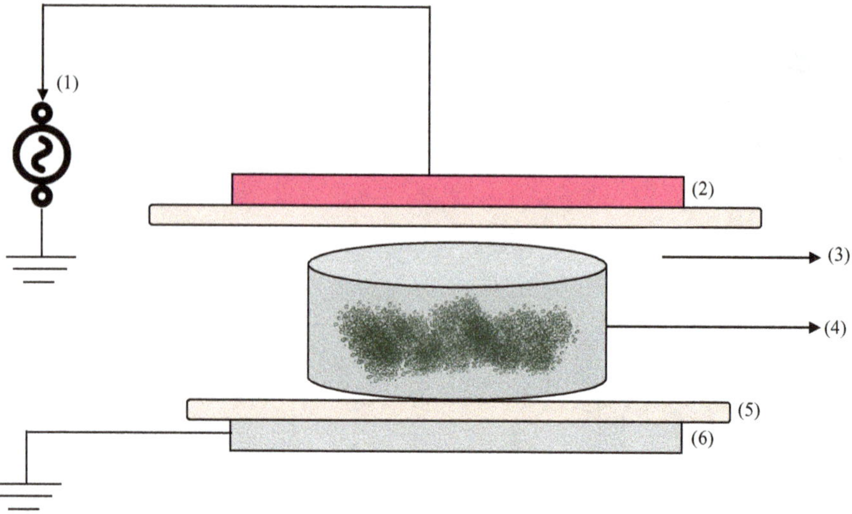

Fig. 4.9 CP processing mechanism: (1) = A/C supply, (2) = high-voltage electrode, (3) = plasma discharge, (4) = sample with packaging, (5) = dielectric material, (6) = ground electrode

4.5.2.4 Pulsed Electric Field (PEF)

Pulsed electric field (PEF) is a relatively novel, non-thermal processing technique which involves the use of electrical pulses in short burst. In this process, very short duration electric field bursts are directed towards the cell membrane of the food matrix. These electrical fields pierce the membrane which causes it to lose its semi-permeability temporarily or permanently. The pulse amplitude lies in the range from 0.1–0.3 kV/cm to 20–80 kV/cm (Vorobiev and Lebovka 2019). This method of electroporation has significant potential in food and agricultural applications. Processes such as cold extraction, selective extraction, food preservation, and inactivation of enzymes and pathogens can be fulfilled by this technology (Vorobiev and Lebovka 2019; Joshi et al. 2023). Moreover, PEF also has the potential to be used for secondary structural changes in protein modification. But the intensity of electric field must be higher in order to cause permanent denaturation (Joshi et al. 2023). However, studies related to use of PEF for millet processing and preservation are quite limited. A study by Lohani et al. (2016) assessed the effects of PEF at different treatment times on sorghum flour and how it impacted the release of bioactive compounds from the sample. The flour was treated at 45% flour-to-water ratio, with PEF (electric field intensity) at 2 kV/cm, for 875 μs. It was observed that the TPC and antioxidant activity were 24.80% and 33.90%, respectively, higher than the control sample. Microstructural analysis showed that PEF led to cell collapse, resulting in the release of phenolics from within the structure (Lohani and Muthukumarappan 2016). Another study attempted to extract bioactive compounds from brewers' spent grains (BSG), which are a by-product of beer production. It was observed that when PEF extraction was used as pre-treatment (14.5 s, 50 Hz, 2.5 kV/cm), the total free and bound phenolic compound concentrations were enhanced by 2.70 and 1.70

times, respectively (Martín-García et al. 2020). Moreover, PEF-treated BSG has the potential to show more antimicrobial activity against microbial pathogens such as *S. typhimurium* and *L. monocytogenes* (Kumari et al. 2019). These studies show that PEF can be used effectively to process millet-based products such as millet milk. The products can be treated along with other processing techniques such as germination, malting, fermentation, etc. and can be coupled with PEF to enhance the availability of bioactive compounds (Nair et al. 2020). Hence, more research on the impact of PEF is required to understand its advantages and challenges (Fig. 4.10).

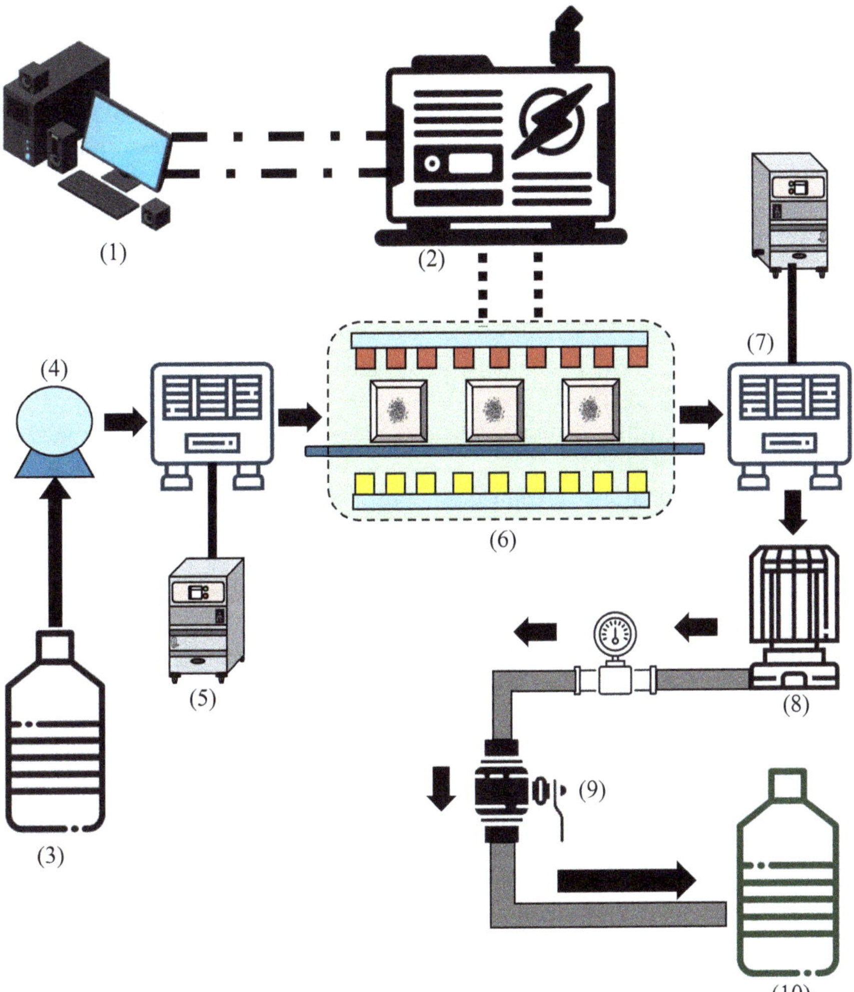

Fig. 4.10 PEF treatment mechanism: (1) = monitoring system, (2) = pulse generator, (3) = raw sample, (4) = pump, (5) = temperature chamber, (6) = treatment chamber, (7) = temperature chamber (after treatment), (8) = cooling area, (9) = control valve, (10) = treated product

4.6 Conventional Approaches Versus Modern/Novel Approaches of Millet Processing

Processing of millets is an integral and crucial step that can help in easier access of these nutritious grains, and can assist in making these grains an affordable, complete, and healthier meal option among the human population. As discussed, millets are one of the few cereal crops that are rich in complex carbohydrates, dietary fibre, protein, essential lipids, vitamins, minerals, bioactive polyphenols, and antioxidants (Saleh et al. 2013; Sharma et al. 2021). Processing allows for bioavailability of nutrients, palatability, and a reduction in ANFs which can hamper nutrient absorption. All the processing methods that can be applied on millets have their own unique characteristics which influence and enhance the bioavailability of nutrients. However, millets should be subjected to multiple processing methods, as the processes collectively enhance the nutritional properties (Yousaf et al. 2021). Be it conventional or novel approaches towards millet processing, all these aim for the fulfilment of the following criteria (Nanje Gowda et al. 2022):

- Enhancement in the overall palatability and organoleptic properties.
- Enhancement in the total nutrient content availability after processing, including bioactivity and antioxidant characteristics (Renganathan et al. 2020; Nanje Gowda et al. 2022).
- Reduction in the ANFs of millets, which are a characteristic of these cereals (Sharma et al. 2021).
- Extend the shelf-life, texture, and taste and reduce microbial contamination (Nanje Gowda et al. 2022).
- Improve the starch and protein digestibility of millet-based products (Annor et al. 2017; Kumar et al. 2023).

Conventional approaches towards millet processing include decortication, milling, soaking, germination, malting, fermentation, roasting, cooking, and puffing, and include processes in combination with each other to enhance the nutritional characteristics. However, some processes can also lead to loss of some nutritional properties. Decortication, which involves removal of about 12–30% of husk, bran, and germ portion, also leads to loss of proteins and amino acids such as lysine, arginine, etc. (Yousaf et al. 2021; Nanje Gowda et al. 2022). Despite that, germination and fermentation processes have shown to improve the protein and starch digestibility (Gulati et al. 2017). Starch, which is a principal nutrient in millets, is also impacted by these processes (Kumar et al. 2023). It was found that germination of pearl millets led to a non-significant increase in the first 24 and 48 h but decreased significantly after 72 h (Saleh et al. 2013). Germination also had a significant impact on the fibre content, as increasing germination time enhanced the fibre content in foxtail millet (Nanje Gowda et al. 2022). Methods like cooking effect the protein quality and digestibility as well. Different cooking methods like microwave cooking and pan cooking interact with the protein concentration differently, which can lead to various changes in its structure and digestibility (Yousaf et al. 2021). As a result, it can be said that the conventional approaches of millet processing assist in providing improved quality millet-based products, which can be utilized for preparation of various food applications such as weaning foods, millet mix, millet-based noodles, millet flour, etc.

The novel/modern approaches to millet processing have been introduced quite recently. These methods mostly target for reducing the lipid activity of millets, reduction of ANF activity, as well as providing a stable shelf-life to the millet products. These methods which include CP, PEF, HPP, etc. can also be considered as green, sustainable forms of processing technology as they involve minimal use of energy and minimal residue formation (Vorobiev and Lebovka 2019; Pallarés et al. 2021). Moreover, these techniques are also targeted towards enhancing the quality of nutritional and functional properties of millets with a focus on minimal nutrient loss. However, their application in the sector of millet processing has been quite limited, and further studies are required to study the impact of these methods on the quality and the changes that are caused by these techniques.

Therefore, comparing both the conventional and modern approaches towards millet processing, it can be said that both approaches are required and necessary for optimum maintenance of nutritional quality as well as to provide a good-quality product, be it in the form of formulations, flours, dried products, or mixes. A combination of both conventional and modern approach is the need of the hour, so that millets can be consumed by every individual and take advantage of its nutritional benefits.

4.7 Future Perspective and Conclusion

Millets are one of the hardiest, climate-resilient crops which are quite abundant and rich in macronutrients and micronutrients. Their nutritional quality is comparable to that of rice or wheat. With the increasing scenario of food insecurity, growing population, and reducing food sources, millet can serve as a viable alternative towards providing a more secure and sustainable future. Millets have been in use and regular consumption in countries around the Asian and African continent predominantly. However, its interest among the Western countries has somehow decreased owing to lack of innovative processing technologies. Moreover, millets are a rich source of lipids which leads to its poor shelf-life and lack of products. Therefore, these conventional and modern processing technologies need to work parallelly so that there is enhancement in the nutritional and functional properties. These approaches should focus specifically on reduction of lipid activity and ANFs, enhancement of shelf-life, improving digestibility, and checking nutrient loss. Moreover, owing to lack of public awareness about millets, the lack of millet products has led to limited consumption of this nutritious grain. Although the conventional methods have been in practice and have been competent in providing useful data regarding enhancement of nutritional characteristics, the use of novel processing approaches needs to be emphasized as research has been limited and thereby resulting in low data generation. However, the available studies have shown that these novel approaches are reliable and can improve the nutritional qualities of millet. Another focus should be on focusing research on exploring the advantages of minor millets—proso, barnyard, kodo, brown top, and little millet—as each millet is structurally a bit different and specific operations and techniques may be required for processing.

References

Ahirwar NK, Singh R (2023) Assessment of nutritional components and nutraceutical benefits of millets: an integrative review. Southeast Asian J Case Rep Rev 10:50–56. https://doi.org/10.18231/j.sajcrr.2023.011

Anal AK, Singh R, Rice D et al (2024) Millets as supergrains: a holistic approach for sustainable and healthy food product development. Sustain Food Technol 2:908. https://doi.org/10.1039/d4fb00047a

Anandharamakrishnan C, Rawson A, Sunill CK (2022) Handbook of millets - processing, quality, and nutrition status. Springer, New York

Anitha S, Govindaraj M, Kane-Potaka J (2020) Balanced amino acid and higher micronutrients in millets complements legumes for improved human dietary nutrition. Cereal Chem 97:74–84. https://doi.org/10.1002/cche.10227

Annor GA, Tyl C, Marcone M et al (2017) Why do millets have slower starch and protein digestibility than other cereals? Trends Food Sci Technol 66:73–83. https://doi.org/10.1016/j.tifs.2017.05.012

Antony Ceasar S, Maharajan T (2022) The role of millets in attaining united nation's sustainable developmental goals. Plants People Planet 4:345–349. https://doi.org/10.1002/ppp3.10254

Azad MOK, Jeong DI, Adnan M et al (2019) Effect of different processing methods on the accumulation of the phenolic compounds and antioxidant profile of broomcorn millet (Panicum miliaceum L.) flour. Foods 8. https://doi.org/10.3390/foods8070230

Azeez SO, Chinma CE, Bassey SO et al (2022) Impact of germination alone or in combination with solid-state fermentation on the physicochemical, antioxidant, in vitro digestibility, functional and thermal properties of brown finger millet flours. LWT 154:112734. https://doi.org/10.1016/j.lwt.2021.112734

Babiker E, Abdelseed B, Hassan H, Adiamo O (2018) Effect of decortication methods on the chemical composition, antinutrients, Ca, P and Fe contents of two pearl millet cultivars during storage. World J Sci Technol Sustain Dev 15:278–286. https://doi.org/10.1108/wjstsd-01-2018-0005

Barba FJ, Roselló-Soto E, Marszałek K et al (2019) Green food processing: concepts, strategies, and tools. In: Green food processing techniques: preservation, transformation and extraction. Elsevier, pp 1–21

Bora P, Ragaee S, Marcone M (2019) Characterisation of several types of millets as functional food ingredients. Int J Food Sci Nutr 70:714–724. https://doi.org/10.1080/09637486.2019.1570086

Choudhury M, Das P, Baroova B (2011) Nutritional evaluation of popped and malted indigenous millet of Assam. J Food Sci Technol 48:706–711. https://doi.org/10.1007/s13197-010-0157-3

Darvishi H, Koushesh Saba M, Behroozi-Khazaei N, Nourbakhsh H (2020) Improving quality and quantity attributes of grape juice concentrate (molasses) using ohmic heating. J Food Sci Technol 57:1362–1370. https://doi.org/10.1007/s13197-019-04170-1

Dey S, Saxena A, Kumar Y, et al (2022) Understanding the Antinutritional Factors and Bioactive Compounds of Kodo Millet (Paspalum scrobiculatum) and Little Millet (Panicum sumatrense). J Food Qual 2022:1–19. https://doi.org/10.1155/2022/1578448

Dias-Martins AM, Cappato LP, da Costa Mattos M et al (2019) Impacts of ohmic heating on decorticated and whole pearl millet grains compared to open-pan cooking. J Cereal Sci 85:120–129. https://doi.org/10.1016/j.jcs.2018.11.007

FAO, IFAD, UNICEF, WFP and WHO (2023) The State of Food Security and Nutrition in the World 2023. Urbanization, agrifood systems transformation and healthy diets across the rural–urban continuum. FAO, Rome. https://doi.org/10.4060/cc3017en

Gulati P, Li A, Holding D et al (2017) Heating reduces proso millet protein digestibility via formation of hydrophobic aggregates. J Agric Food Chem 65:1952–1959. https://doi.org/10.1021/acs.jafc.6b05574

Gulati P, Sabillón L, Rose DJ (2018) Effects of processing method and solute interactions on pepsin digestibility of cooked proso millet flour. Food Res Int 109:583–588. https://doi.org/10.1016/j.foodres.2018.05.005

Hama F, Icard-Vernière C, Guyot JP et al (2011) Changes in micro- and macronutrient composition of pearl millet and white sorghum during in field versus laboratory decortication. J Cereal Sci 54:425–433. https://doi.org/10.1016/j.jcs.2011.08.007

Jaddu S, Pradhan RC, Dwivedi M (2022) Effect of multipin atmospheric cold plasma discharge on functional properties of little millet (Panicum miliare) flour. Innov Food Sci Emerg Technol 77:102957. https://doi.org/10.1016/j.ifset.2022.102957

Jaddu S, Abdullah S, Dwivedi M, Pradhan RC (2023) Optimization of functional properties of plasma treated kodo millet (open air multipin) using response surface methodology (RSM) and artificial neural network with genetic algorithm (ANN-GA). J Food Process Eng 46:e14207. https://doi.org/10.1111/jfpe.14207

Jaddu S, Sahoo S, Sonkar S et al (2024) Cold plasma treatment of little millet flour: impact on bioactives, antinutritional factors and functional properties. Plant Foods Hum Nutr 79:503–510. https://doi.org/10.1007/s11130-024-01171-0

Jaybhaye RV, Pardeshi IL, Vengaiah PC, Srivastav PP (2014) Processing and technology for millet based food products: a review. J Ready Eat Food 1:32–48

Jindal P, Nikhanj P (2023) A review on processing technologies for value added millet products. J Food Process Eng 46. https://doi.org/10.1111/jfpe.14419

Joshi TJ, Singh SM, Rao PS (2023) Novel thermal and non-thermal millet processing technologies: advances and research trends. Eur Food Res Technol 249:1149–1160. https://doi.org/10.1007/s00217-023-04227-8

Kaimal AM, Mujumdar AS, Thorat BN (2021) Resistant starch from millets: recent developments and applications in food industries: resistant starch from millets. Trends Food Sci Technol 111:563–580. https://doi.org/10.1016/j.tifs.2021.02.074

Kate A, Singh A (2021) Processing technology for value addition in millets. In: Millets and millet technology. Springer Singapore, pp 239–254

Kheto A, Manikpuri S, Sarkar A et al (2024) How pulsed electric field treatment affects anti-nutritional factors and plant protein digestibility: a concise review. Food Biosci 61:104849. https://doi.org/10.1016/j.fbio.2024.104849

Kumar A, Tripathi MK, Joshi D, Kumar V (2021) Millets and millet technology. Springer Singapore

Kumar SR, Tangsrianugul N, Sriprablom J et al (2023) Effects of single and dual modifications with debranching and heat-moisture treatments on physicochemical, rheological, and digestibility properties of proso millet starch. Carbohydr Polym Technol Appl 6:100399. https://doi.org/10.1016/j.carpta.2023.100399

Kumari B, Tiwari BK, Walsh D et al (2019) Impact of pulsed electric field pre-treatment on nutritional and polyphenolic contents and bioactivities of light and dark brewer's spent grains. Innov Food Sci Emerg Technol 54:200–210. https://doi.org/10.1016/j.ifset.2019.04.012

Ling B, Ouyang S, Wang S (2019) Radio-frequency treatment for stabilization of wheat germ: storage stability and physicochemical properties. Innov Food Sci Emerg Technol 52:158–165. https://doi.org/10.1016/j.ifset.2018.12.002

Lohani UC, Muthukumarappan K (2016) Application of the pulsed electric field to release bound phenolics in sorghum flour and apple pomace. Innov Food Sci Emerg Technol 35:29–35. https://doi.org/10.1016/j.ifset.2016.03.012

Lokeswari R, Sharanyakanth PS, Mahendran R (2021) Improvement in millet soaking by way of bubbled cold plasma processed air exposure; phytic acid reduction cum nutrient analysis concern. Front Adv Mater Res:1–16. https://doi.org/10.34256/famr2121

Martín-García B, Tylewicz U, Verardo V et al (2020) Pulsed electric field (PEF) as pre-treatment to improve the phenolic compounds recovery from brewers' spent grains. Innov Food Sci Emerg Technol 64:102402. https://doi.org/10.1016/j.ifset.2020.102402

Mishra S, Kumar S, Srivastava CR (2024) Genetic improvement of small millets. Springer Nature Singapore

Mohd MS, Shafie NH, Ahmad F, Shafie SR (2023) Effect of roasting on whole grain barnyard millet to the proximate composition, amino acid profile, Total phenolic content and antioxidant activity. Malays J Med Health Sci 19:69–76. https://doi.org/10.47836/mjmhs.19.6.10

Mundassery A, Ramaswamy J, Natarajan T et al (2024) Modern and conventional processing technologies and their impact on the quality of different millets. Food Sci Biotechnol 33:2441–2460. https://doi.org/10.1007/s10068-024-01579-z

Nair UKA, Hema V, Sinija VR, Hariharan S (2020) Millet milk: a comparative study on the changes in nutritional quality of dairy and non-dairy milks during processing and malting. J Food Process Eng 43. https://doi.org/10.1111/jfpe.13324

Nanje Gowda NA, Siliveru K, Vara Prasad PV et al (2022) Modern processing of Indian millets: a perspective on changes in nutritional properties. Foods 11. https://doi.org/10.3390/foods11040499

Nazari B, Mohammadifar MA, Shojaee-Aliabadi S et al (2018) Effect of ultrasound treatments on functional properties and structure of millet protein concentrate. Ultrason Sonochem 41:382–388. https://doi.org/10.1016/j.ultsonch.2017.10.002

Niranjan T, Dwivedi M (2022) Moisture-dependent engineering properties of underutilized browntop millet (unhulled and hulled) and characterization of milled flour. J Food Process Preserv 46:e17088. https://doi.org/10.1111/jfpp.17088

Pallarés N, Berrada H, Ferrer E et al (2021) Ultrasound processing: a sustainable alternative. In: Sustainable production technology in food. Elsevier, pp 155–164

Prashanth P, Joshi TJ, Singh SM, Rao PS (2024) Impact of hydrothermal treatment on functional properties of pearl millet flour: process modelling and optimisation. J Food Meas Charact 18:7627–7640. https://doi.org/10.1007/s11694-024-02754-w

Reddy BHR, Thankachan P, Hatakayama M et al (2022) A natural low phytic acid finger millet accession significantly improves iron bioavailability in Indian women. Front Nutr 8. https://doi.org/10.3389/fnut.2021.791392

Renganathan VG, Vanniarajan C, Karthikeyan A, Ramalingam J (2020) Barnyard millet for food and nutritional security: current status and future research direction. Front Genet 11. https://doi.org/10.3389/fgene.2020.00500

Sain M, Minz PS, John H, Singh A (2024) Effect of Ohmic heating on food products: an in-depth review approach associated with quality attributes. J Food Process Preserv 2025937:1–12. https://doi.org/10.1155/2024/2025937

Saleh ASM, Zhang Q, Chen J, Shen Q (2013) Millet grains: nutritional quality, processing, and potential health benefits. Compr Rev Food Sci Food Saf 12:281–295. https://doi.org/10.1111/1541-4337.12012

Sarkar A, Niranjan T, Patel G et al (2023) Impact of cold plasma treatment on nutritional, antinutritional, functional, thermal, rheological, and structural properties of pearl millet flour. J Food Process Eng 46:e14317. https://doi.org/10.1111/jfpe.14317

Sharanagat VS, Suhag R, Anand P et al (2019) Physico-functional, thermo-pasting and antioxidant properties of microwave roasted sorghum [Sorghum bicolor (L.) Moench]. J Cereal Sci 85:111–119. https://doi.org/10.1016/j.jcs.2018.11.013

Sharma B, Gujral HS (2019) Influence of nutritional and antinutritional components on dough rheology and in vitro protein & starch digestibility of minor millets. Food Chem 299:125115. https://doi.org/10.1016/j.foodchem.2019.125115

Sharma N, Goyal SK, Alam T et al (2018) Effect of high pressure soaking on water absorption, gelatinization, and biochemical properties of germinated and non-germinated foxtail millet grains. J Cereal Sci 83:162–170. https://doi.org/10.1016/j.jcs.2018.08.013

Sharma R, Sharma S, Dar BN, Singh B (2021) Millets as potential nutri-cereals: a review of nutrient composition, phytochemical profile and techno-functionality. Int J Food Sci Technol 56:3703–3718. https://doi.org/10.1111/ijfs.15044

Sharma V, Tomar MS, Sahoo S, Pradhan RC (2024) Development of barnyard millet flour rich in bioactive compounds with enhanced functional properties by application of multipin atmospheric cold plasma. J Food Process Eng 47:e14667. https://doi.org/10.1111/jfpe.14667

Sheethal HV, Baruah C, Subhash K et al (2022) Insights of nutritional and anti-nutritional retention in traditionally processed millets. Front Sustain Food Syst 5. https://doi.org/10.3389/fsufs.2021.735356

Singh A, Gupta S, Kaur R, Gupta HR (2017) Process optimization for anti-nutrient minimization of millets. Asian J Dairy Food Res. https://doi.org/10.18805/ajdfr.dr-1215

Singh RB, Khan S, Chauhan AK et al (2018) Millets as functional food, a gift from Asia to Western world. In: The role of functional food security in global health. Elsevier, pp 457–468

Singh DK, Kurichh R, Bist Y et al (2024a) Effect of microwave roasting on the physicochemical, functional, rheological, and antioxidant properties of Kodo millet flour. J Food Process Preserv 2024. https://doi.org/10.1155/2024/6861190

Singh SM, Jayasree Joshi T, Rao PS (2024b) Technological advancements in millet dehusking and polishing process; an insight into pretreatment methods, machineries and impact on nutritional quality. Grain Oil Sci Technol 7:186

Sonkar S, Jaddu S, Dwivedi M, Pradhan RC (2023) Impact of multi-pin atmospheric cold plasma on dynamic rheological characteristics of kodo millet starch. J Food Process Eng 46:e14485. https://doi.org/10.1111/jfpe.14485

Sruthi NU, Rao PS (2021) Effect of processing on storage stability of millet flour: a review. Trends Food Sci Technol 112:58–74. https://doi.org/10.1016/j.tifs.2021.03.043

Sullivan AC, Pangloli P, Dia VP (2018) Impact of ultrasonication on the physicochemical properties of sorghum kafirin and in vitro pepsin-pancreatin digestibility of sorghum gluten-like flour. Food Chem 240:1121–1130. https://doi.org/10.1016/j.foodchem.2017.08.046

Swaminathan I, Guha M, Hunglur UH, Rao DB (2015) Optimization of infrared heating conditions of sorghum flour using central composite design. Food Sci Biotechnol 24:1667–1671. https://doi.org/10.1007/s10068-015-0216-7

Tsevdou M, Gogou E, Taoukis P (2019) High hydrostatic pressure processing of foods. In: Green food processing techniques: preservation, transformation and extraction. Elsevier, pp 87–137

Vorobiev E, Lebovka N (2019) Pulsed electric field in green processing and preservation of food products. In: Green food processing techniques: preservation, transformation and extraction. Elsevier, pp 403–430

Yadav S, Mishra S, Pradhan RC (2021) Ultrasound-assisted hydration of finger millet (Eleusine Coracana) and its effects on starch isolates and antinutrients. Ultrason Sonochem 73:105542. https://doi.org/10.1016/j.ultsonch.2021.105542

Yarrakula S, Mummaleti G, Toleti KS, Saravanan S (2024) Functional characteristics and storage stability of hot air assisted radio frequency treated pearl millet. Food Prod Process Nutr 6. https://doi.org/10.1186/s43014-024-00228-6

Yousaf L, Hou D, Liaqat H, Shen Q (2021) Millet: a review of its nutritional and functional changes during processing. Food Res Int 142:110197. https://doi.org/10.1016/j.foodres.2021.110197

5 Pressure Points: High-Pressure Processing in Millet Transformation: Efficiency, Quality Enhancement, and Sustainability

Shagolshem Mukta Singh, N. Meena, Yaghuvendra Kumar, Abhishek Kumar, Aparna S. Krishna, Heena Sudhakar Kamdi, Prashant Kumar, T. Jayasree Joshi, and P. Srinivasa Rao

5.1 Introduction

Millets are climate-resistant crops commonly cultivated in the semiarid regions of Africa and Asia. They are also termed as nutri-cereals and serves as a vital source of energy and protein for approximately 500 million people (Yadav et al. 2012). In 2010, the global production of millet grains was 762,712 metric tons, while India led with the production of 334,500 tons (Jaybhaye et al. 2014). Although millets are emerging as an alternative food source, there are significant challenges associated with their storage. Millets contain more fat than other cereals, making them more susceptible to oxidative rancidity during prolonged storage periods. This can impact the quality and reduce the shelf life of millets, posing a challenge to their widespread adoption and use (Dekka et al. 2023). Rancidity could be explained as when the grains are milled into flour; it exposes the lipids present in the pericarp and germ to environmental factors. This exposure triggers lipolysis and the oxidative breakdown of the resulting de-esterified unsaturated fatty acids. As a result, these chemical changes can lead to the formation of off-flavors, particularly during storage, especially in environments with high moisture and oxygen levels. In millet products, this process causes the flour to become increasingly rancid and less acceptable over time (Sruthi and Rao 2021).

Moreover, millets are processed to prolong the shelf life of their products while preserving their nutritional benefits (Yadav et al. 2022). Moisture content, fat, and water activity are the major factors that affect the storage stability of millet-based processed food products. Water activity significantly influences storage stability, as

S. M. Singh · N. Meena · Y. Kumar · A. Kumar · A. S. Krishna
H. S. Kamdi · P. Kumar · T. Jayasree Joshi · P. S. Rao (✉)
Agricultural and Food Engineering Department, Indian Institute of Technology Kharagpur, Kharagpur, India

S. Jaddu, R. Chandra Pradhan (eds.), *Innovative Millet Processing*,
https://doi.org/10.1007/978-3-031-99864-5_5

lower levels can inhibit microbial growth. However, chemical and enzymatic reactions may still occur, leading to product spoilage (Singh et al. 2020). Various processing techniques, including thermal and nonthermal methods, have been utilized to extend the shelf life of millet products while preserving their nutritional content. Thermal treatments include conventional techniques like dry heat treatments, moist heat treatments, superheated steam treatments, and novel techniques like infrared heating, ohmic heating, microwave processing, radiofrequency heating, etc. Applying thermal treatments may induce alterations, such as denaturation and other chemical reactions, in the molecular composition of the foodstuff. These procedures could potentially have unfavorable implications on the nutritional integrity of the product, as noted by Yadav et al. (2022). Furthermore, they can promote racemization reactions, generating undesirable compounds, as highlighted by Sá et al. (2020). HPP is a promising alternative for traditional thermal processing approaches for inactivating food-borne microorganisms and enzymes. HPP is a technique where a sample is exposed to uniform pressure, typically ranging from 100 to 800 MPa. The process can increase the shelf life of food products without the need for chemical preservatives or additives, and it operates at notably lower temperatures compared to standard thermal methodologies, as reported by Kaushik et al. (2016). High-pressure application leads to the death of microorganisms in the food substances (Joshi et al. 2023). Hydrostatic pressures between 300 and 600 MPa can inactivate yeasts, molds, and most vegetative bacteria. Bacterial spores require very high pressures (>1000 MPa) for inactivation (Smelt 1998). Moreover, high pressure can change the structure of large molecules while keeping small molecules, such as vitamins, intact with minimal heat generation in the product during processing. Furthermore, soaking under high pressure can eliminate the external bran layers of grains (hydrophobic lipid layer), consequently improving the longevity of flour (Zhu et al. 2016). The following sections explore how HPP impacts the treatment and processing of millets and products derived from them.

5.2 Mechanism of Action

HPP is a food preservation technique that employs high levels of hydrostatic pressure to eliminate microorganisms and enzymes (Mirzababaee et al. 2022; Sharma et al. 2018). The working mechanism of HPP is based on three fundamental principles: the isostatic principle, Le Chatelier's principle, and the principle of microscopic ordering. The isostatic principle states that when pressure is applied to a material, uniform pressure is exerted throughout the sample, and the product is compressed uniformly, irrespective of size and geometry. Le Chatelier's principle states that a system at equilibrium will adjust to counteract any disturbance and restore equilibrium. In HPP, applying high pressure causes a decrease in volume, but the sample regains its original shape once the pressure is released unless the sample contains entrapped air. Reactions favored by volume decrease are accelerated, while reactions favored by volume increase are suppressed (Tahir-Kheli 2020). The principle of microscopic ordering asserts that as pressure rises, the molecular order at a consistent temperature also rises, impacting the molecular arrangement and chemical responses (Naik et al. 2013).

The HPP system comprises a high-pressure vessel, heating/cooling jacket, pressure generator, water reservoir, and control panel (Fig. 5.1). The product must be sealed in flexible water-impermeable packaging because the product needs to be immersed in water during processing. The product is then placed in a high-pressure

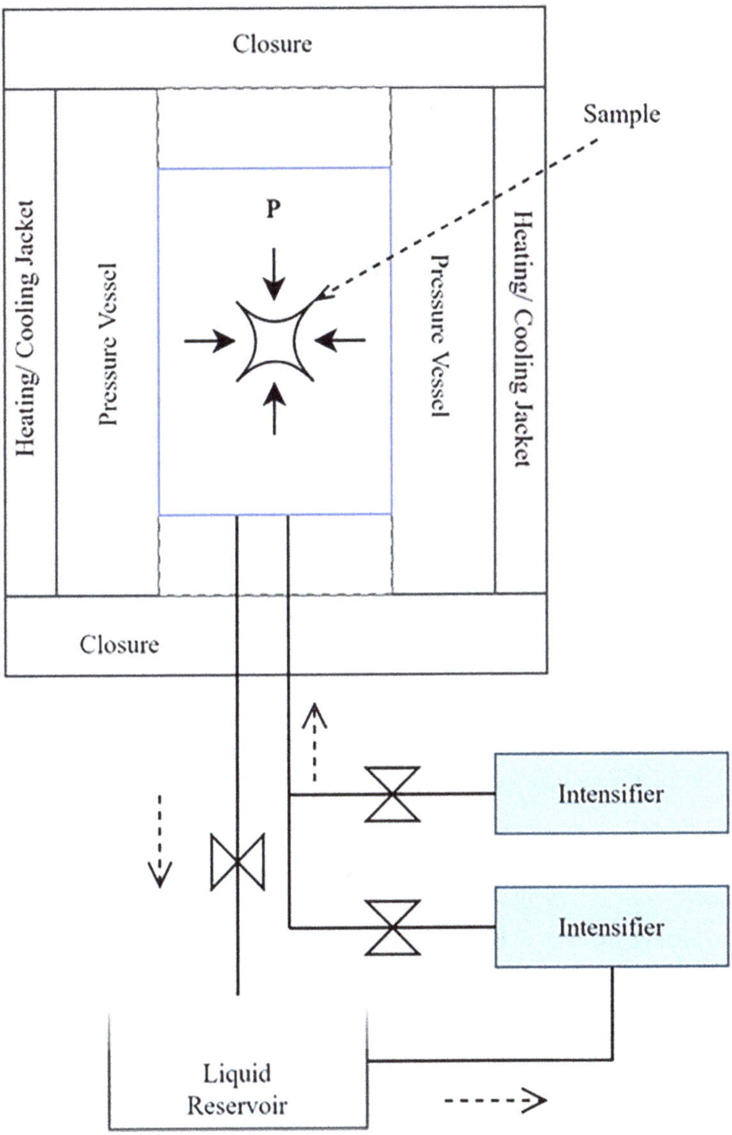

Fig. 5.1 Schematic diagram for the working principle of HPP

vessel and exposed to hydrostatic pressure between 100 and 600 MPa (up to 87,000 psi), transmitted by water for a few seconds to several minutes. After the pressure is released, the product is taken out. This pressure is evenly applied throughout the product, ensuring uniform treatment. High pressure ruptures the cell wall and expels cellular material, impairing cellular functions and ultimately killing microorganisms (Bansal et al. 2019). The effects of HPP on cell membrane permeability, cell morphology, biochemical reactions, and genetic mechanisms collectively lead to the destruction of microorganisms (Sehrawat et al. 2021). HPP can denature the protein of enzymes, leading to its inactivation, thus preventing spoilage of food products. Since the process is mainly mechanical and not thermal, it better preserves the sensory and nutritional qualities of the food than conventional thermal processing. The treated product can be further processed as per the requirements. HPP primarily alters the non-covalent bonds in biological macromolecules, changing how substances bind and reorganizing their structure (Zhang et al. 2022a).

5.2.1 Effects on Non-covalent Bonds

The primary structure of proteins remains unaffected during HPP treatment as the covalent bonds, like peptide and disulfide bonds, are significantly stronger than non-covalent bonds and preserve the amino acid sequence. Conversely, HPP impacts non-covalent bonds, such as hydrogen and ionic bonds, which are essential for maintaining the higher-order structure and function of macromolecules. Hydrogen bonds form weak interactions that stabilize the three-dimensional structures of proteins. HPP can disrupt these bonds, causing a rearrangement of the molecular structure, leading to changes in the folding and stability of proteins. Ionic bonds between charged amino acids contribute to the overall stability of macromolecules. HPP can alter these bonds by affecting the electrostatic interactions, leading to modifications in the tertiary and quaternary structures of the protein. HPP has been observed to enhance the binding between two molecules, with a 32.8% increase in binding rate when treated at 500 MPa for 15 min. This improvement is largely attributed to the effect of high pressure on non-covalent bonds. HPP demonstrates significant potential for boosting the functionality and stability of food ingredients (Hou et al. 2022).

5.2.2 Destruction and Reorganization of Bonds

HPP can change the way molecules and ions bind to each other. The high pressure can cause the breakage and reformation of bonds, leading to the reorganization of molecular structures and affecting the overall functionality and stability of the macromolecules (Zhang et al. 2022a). HPP can reduce the content of α-helices and random coils in proteins, which form the secondary structure. This reduction is due to the pressure-induced unfolding and reorganization of the protein structure, leading to a more compact and ordered configuration (Zhang et al. 2022a).

5.2.3 Modification of Surface Charge Distribution and Reduction in Hydrophobicity

HPP can influence the distribution of charges on the surface of proteins, altering their solubility and how proteins interact with each other and other molecules. This increases protein aggregation by promoting interactions between partially unfolded protein molecules, forming larger protein particles. HPP treatment significantly decreases the hydrophobicity of proteins, making them more hydrophilic. This improves the solubility and interaction of proteins with water-soluble and hydrophilic substances, enhancing their functionality in aqueous environments. The decreased hydrophobicity of HPP-treated proteins enables the direct use of millet gliadin in food production, improving its applicability and versatility as a food ingredient. It also allows packaging and delivering water-soluble and hydrophilic bioactive substances (Zhang et al. 2022a). This capability is valuable for developing functional foods where the stability and bioavailability of active ingredients are crucial.

5.2.4 Formation of Disulfide Bonds

HPP can promote the formation of disulfide bonds between cysteine residues, which can further stabilize the protein structure. These new covalent bonds help to maintain the more orderly structure formed under high pressure, increasing the thermal stability and denaturation temperature of proteins (Zhang et al. 2022a). This makes food products more robust for various food processing applications.

5.3 Impact of HPP on the Nutritional Quality of Millets

HPP is gaining attention for its capacity to maintain the nutritional integrity of foods, eliminate hazardous microorganisms, and extend shelf life (Amsasekar et al. 2022). HPP can affect the nutritional content of millets, affecting factors such as vitamins, minerals, phytochemicals, and overall sensory characteristics. Millets are nutritionally notable for their abundant dietary fiber, essential amino acids, minerals (calcium, iron, and magnesium), vitamins, and diverse bioactive substances such as phenolic and flavonoid compounds. An overview of the steps involved in HPP processing and its key features is shown in Fig. 5.2. The following section discusses the effect of HPP treatment on the nutritional quality of millet and millet products.

5.3.1 Retention of Vitamins

Vitamins are vital organic compounds crucial in numerous physiological functions within the human body. Millets, a rich source of vitamins, have multiple health benefits. The vitamin B complex, particularly pyridoxine, niacin, and folic acid, is

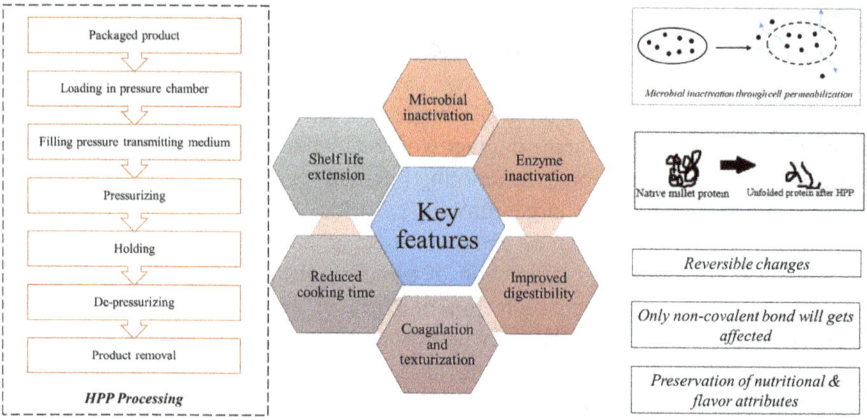

Fig. 5.2 Overview of HPP processing and its key features

abundant in millets (Mohite and Sharma 2024). Consuming these grains lowers the levels of triglycerides (pearl millet), reduces the risks of gastrointestinal tract disorders and diabetes mellitus (finger millet), reduces blood glucose and lipid levels (barnyard millet), protects against D-galactosamine-induced liver injury (proso millet), lowers cholesterol levels (little millet), and acts as anti-lipidemic and antihyperglycemic agents in diabetic conditions (foxtail millet), etc. (Nithiyanantham et al. 2019). A recent study by Kundgol et al. (2013) found that the alpha-tocopherol (vitamin E) content in whole little millet ranged from 1.75 to 5.44 mg per 100 g. However, the levels dropped to 0.83–1.60 mg per 100 g after decortication. Milling or sieving during millet processing also reduces the vitamin level as it removes the outer bran layer and the germ portion of the grain (Vashishth and Maurya 2024). Moreover, thermal processing reduces the heat-liable vitamins present in millet, such as vitamins B3, B9, B12, etc. (Choudhary et al. 2023).

Studies have shown that HPP retains higher levels of heat-sensitive vitamins, such as vitamin C and some vitamin B complex, compared to traditional thermal processing methods like boiling and steaming. HPP does not involve high temperatures that can degrade these vitamins. Balakrishna and Farid (2020) reported that an increase in pressure from 450 to 600 MPa led to a rise in the thiamine content of rice. However, increased processing time reduced the thiamine content, possibly due to the destruction of the thiamine structure under high pressure. HPP also inactivates the microorganisms present in the grain, preserving its structural integrity and nutritional profile. Estrada-Girón et al. (2005) mentioned that the HPP preserved vitamin B1 and vitamin B6 levels, but ultra-HPP exhibited no significant effects on the loss of thiamin and pyridoxal.

5.3.2 Preservation of Minerals

Minerals are essential inorganic elements crucial in maintaining body processes, such as promoting bone health, supporting nerve function, and facilitating enzyme activity. Millets are notably rich in minerals like calcium, iron, phosphorus, and magnesium. Unlike thermal processing, which can leach minerals into cooking water, HPP minimizes mineral loss by maintaining the integrity of the millet grains (Cilla et al. 2019). Thus, millet products processed using HPP will likely retain higher levels of essential minerals, contributing to their nutritional quality. Wu et al. (2023) reported that the iron, calcium, and zinc content of the millet grains increased with the application of HPP. The increase is mainly caused by the liberation of metal-chelated macromolecules or the strengthening of binding to active sites. Reducing anti-nutrients like phytic acid, tannin, trypsin inhibitors, etc. contributes to increased mineral availability. Xia et al. (2017a) reported that HPP treatment made it easier for cells to access calcium and copper in vitro than untreated brown rice samples. This finding suggests that HPP might be able to improve mineral bioaccessibility. A different study by Xia et al. (2017b) has shown that the levels of magnesium, manganese, and phosphorus in brown rice were highest when subjected to high hydrostatic pressure treatment at 300 MPa pressure but reduced when treated at 500 MPa. The rise in mineral levels results from the cell's breakdown, which releases minerals.

5.3.3 Impact on Anti-Nutritional Factors

Like any other grains and seeds, millets contain anti-nutritional factors, namely, tannins, phytates, and enzyme inhibitors. These compounds can interfere with mineral absorption and reduce nutrient bioavailability. Researchers have reported that HPP partially degrades these factors, potentially enhancing the bioavailability of minerals like zinc and iron in millets. This reduction in anti-nutrients can improve the overall nutrient absorption of millet-based foods, thereby enhancing their nutritional quality. HPP exerts extreme pressure uniformly throughout the food matrix, disrupting cellular structures. This disruption affects the integrity of cell membranes and walls, leading to the leakage of cellular contents. When it comes to the inactivation of protein-based anti-nutrients, such as lectins and protease inhibitors, it occurs through pressure-induced denaturation of these proteins. This denaturation disrupts the ability of anti-nutrients to bind with the nutrients and thus increases the bioavailability of these nutrients (Liu et al. 2023). Kim et al. (2015) reported that the phytic acid content decreases as the processing pressure levels increase, while Linsberger-Martin et al. (2013) mentioned that the phytic acid content decreases as the processing duration and temperature increase. They also mentioned that the treatment can significantly decrease trypsin inhibitor activity by applying high pressure at room temperature. Furthermore, differences in cultivars, growth, and storage periods impact trypsin inhibitor activity. HPP can make trypsin inhibitors less effective by breaking disulfide and non-covalent bonds. Furthermore, it can help turn on endogenous phytase, an enzyme that turns phytic acid into phosphate and inositol derivatives (Guerrero-Beltrán et al. 2009). Lee et al. (2018) noted that the hydrolytic enzyme activity of phytase, the formation of insoluble complexes between phytic acid and other components like phytate-protein and phytate-protein-mineral complexes, or the hydrolysis of inositol hexaphosphate with penta- and tetra-phosphates causes the reduction of phytic acid. They also mentioned that high-pressure treatment significantly decreases the tannin and phytic acid content. Sharma et al. (2018) experimented by soaking foxtail millet under high pressure and germinating it to analyze its effects on anti-nutritional factors. The study revealed a reduction in the phytic acid and tannin content from 8.89 to 5.42 mg/g and 3.04 to 1.67 mg/100 g, respectively. The study found that the breakdown of phytate phosphorus into inositol monophosphate caused the phytic acid level to drop. On the other hand, the tannin level decreased due to the breakdown of tannin compounds in water that occurred throughout the soaking and germination processes.

5.3.4 Retention of Dietary Fiber

Dietary fiber is crucial for digestive health, promoting satiety and potentially reducing the risk of developing chronic diseases like diabetes and cardiovascular disorders. Millets are excellent sources of dietary fiber, both soluble and insoluble (Nithiyanantham et al. 2019; Zhu et al. 2018). Refining whole grains into flour resulted in a loss of 80% of the fiber content. Refining also increases the loss of minerals, vitamins, and phytonutrients (Jonnalagadda et al. 2011). The fiber accumulates primarily in the outer layer of the grain. The grain's bran contained more fiber than the inner portion. The exposure of coarse and fine bran of wheat to a pressure of 600 MPa for 5 or 15 min at 36 °C increased its water-holding capacity. This finding confirms that HPP has a more pronounced impact on the hydration capabilities of bran (Marti et al. 2014). Martín-Diana et al. (2021) also found that the wheat bran treated at 608 MPa pressure for 5 min increases the extractability of the antioxidant and anti-inflammatory compounds. The application of HPP can modify the structure of the insoluble fiber content. This modification enhances the ability to bind bile salts (Lin et al. 2020). HPP does not significantly affect the fiber content of millets, as fiber is inherently stable under pressure (Zhang et al. 2023). Thus, millet products processed using HPP retain their dietary fiber content, providing the same digestive benefits as their raw counterparts.

5.4 Impact on Bioactive Components

Millets, often regarded as ancient grains, are increasingly recognized for their nutritional value and high level of bioactive components, particularly antioxidants (Prashanth et al. 2024). It contains a significant amount of bioactive components, including polyphenols (sinapic, vanillic, and coumaric acids), flavonoids (quercetin, catechin, myricetin, and luteolin), tannins, and phytosterols. These components offer a multitude of health benefits for maintaining a healthy lifestyle. Millet consumption provides various health benefits due to the presence of phytochemicals. These are natural compounds that possess multiple kinds of health-promoting characteristics. It can have the potential to reduce the risk of cardiovascular disease, cancer, diabetes, hypertension, a high cholesterol level, Parkinson's disease, inflammation, and a condition called metabolic syndrome. Millet also helps delay gastric emptying and fat absorption (Akanbi et al. 2019). The variety of bioactive components in millet contributes significantly to its nutritional and health benefits, making it a desirable inclusion in a balanced diet.

HPP generates interest due to its capacity to preserve nutrients and bioactive compounds. Studies suggest that subjecting foods to HPP at moderate temperatures can effectively maintain the levels of vitamins and polyphenols. The method offers an appealing alternative to standard thermal processing procedures by preserving a significant amount of their bioactivity (Rastogi et al. 2007). Altogether, moderate heating and HPP increase the phenolic content of cereals by releasing the insoluble-bound section, thereby improving their bioaccessibility (Shahidi et al. 2021). The bioactive components exhibit greater reliability at high pressure than thermal processing, which may be attributed to the fact that high pressure predominantly affects non-covalent bonds, such as ionic, hydrogen, and hydrophobic bonds. When subjected to high pressure, the influence on non-covalent bonds can result in the unfolding of macromolecules, such as protein chains. However, it minimally impacts the chemical elements related to desirable food quality attributes, notably flavor, color, and nutritional composition (Hogan et al. 2005). The impact of pressure-induced cell wall destruction can be utilized to improve the bioactive characteristics of pressurized food products. Exposing germinated brown rice to pressures of 30 MPa has effectively induced structural damage in the cells. This damage increases the efficiency of enzymatic hydrolysis in the modified substrate, leading to higher generation of antioxidants (γ-oryzanol), arabinoxylans, tricin 40-O-(threo-β-guaiacylglyceryl) ether (TTGE), γ-aminobutyric acid (GABA), and vitamins B and E (Gutiérrez et al. 2023). Overall, HPP is an effective technique for preserving food's bioactive components while maintaining nutritional and sensory attributes.

HPP effectively retains bioactive components in various foods; its application to millets has become an area of significant interest. Research has demonstrated the impact of HPP on the bioactive constituents of millets. Sharma et al. (2018) applied HPP treatment to germinated and non-germinated foxtail millet grains. They used pressure levels 200, 400, and 600 MPa; temperature levels 20, 40, 60, and 80 °C; and treatment time: 30, 60, 90, and 120 min. Subjecting germinated foxtail millet grains to HPP led to a 17.09% increase in the phenolic content of the resultant flour, with the pressure and treatment time elevated at 20 °C. The possible explanation for this rise is that elevated pressure accelerates the mass transfer rate, enhancing the capacity of solvent extraction of the grains by disrupting hydrophobic bonds and cellular walls. But in the case of bread developed from high-pressure treated oats, sorghum, and finger millet flours at 350 MPa at 20 °C for 10 min, Angioloni and

Collar (2012) found slightly lesser phenols and flavonoids than the gluten-added bread.

Meanwhile, the phenolic acids, including hydroxybenzoic and hydroxycinnamic acids, were higher in high-pressure treated sorghum bread than in conventional sorghum bread. The HPP treatment had adversely affected the bio-accessibility of phenol, irrespective of the flour utilized. The treatment may form a compact structure by causing interactions between existing biopolymers and other components. This structure efficiently traps phytochemicals, reducing their solvent extractability and limiting the enzymatic accessibility. The food matrix releases several phenolic acids and flavonoids that have antiradical activity. This is crucial because substantial research indicates that several of these molecules are bioactive and can potentially treat health issues associated with free radicals (Pérez-Jiménez et al. 2008). The antiradical activity of phenolic compounds, including cinnamic acid, syringic acid, gallic acid, caffeic acid, chlorogenic acid, and catechin, is enhanced by the use of HPP, which aligns with the increased antiradical power observed in high-pressure treated sorghum flour samples (Angioloni and Collar 2012).

Research reveals that while HPP does not significantly degrade phytochemicals, its impact can vary based on the particular component and processing conditions. Overall, HPP is considered mild compared to thermal processing methods, thereby preserving the phytochemical profile of millets to a large extent. HPP at a pressure of 250 MPa for 30 min did not impact the total phenolic content (TPC) and total flavonoid content (TFC). However, applying a higher pressure up to 550 MPa and extending the treatment time reduced the phytochemical contents. Covalent bonds, primarily characterized by intramolecular bonding, can withstand high pressures, but non-covalent bonds cannot (Hou et al. 2022). Oxidoreductase action and/or leaching with the exudate may be responsible for decreased phytochemicals (Saikaew et al. 2018). Multiple studies have indicated that HPP results in an increase of phytochemical levels. Plant cells rupture due to HPP's accelerated diffusion, which boosts solvent accessibility, extractability, and leaching of soluble compounds. The damage to the cells resulted in the liberation of the phenolic compounds from the cellular membrane (Cao et al. 2011; García-Parra et al. 2016). The millet gliadin was subjected to high hydrostatic pressure at a temperature of 25 °C, with a ompression rate of 200 MPa/min, for 10 min, under 100, 300, and 500 MPa pressure. The antioxidant capacity of millet gliadin treated with high hydrostatic pressure was higher than that of the control group, with maximum antioxidant activity observed at 50 min of treatment with 300 MPa. High hydrostatic pressure modifies the characteristics and structure of gliadin, permitting pepsin to identify specific amino acid sequences during shearing. This produces small peptides with effective antioxidant capabilities (Zhang et al. 2022b). Tannins and phytic acid reduced with increased temperature, pressure, and soaking time during HPP of germinated foxtail millets (Sharma et al. 2018). Overall, HPP can be a beneficial approach that may preserve and possibly improve the bioactive components in millets.

5.5 Effect on Functional Properties

The functional properties of grains are vital in defining the quality and appeal of processed food products. Key attributes like viscosity, solubility, emulsification, water-holding capacity, swelling power, elasticity, foaming, and gelation influence the final product's characteristics and consumer acceptance. These physicochemical properties are fundamental in functionality and performance of shaping grain-based foods after various processing treatments. The functionality of flour is primarily determined by the complex interactions between its various components, including proteins, starches, lipids, and phenols. For example, the quality of pasta and noodles significantly depends on water absorption and swelling capacity. At the same time, foaming and emulsion stability are vital for optimal batter performance in cakes and pastries (Singh and Sharma 2017). Starch gelatinization, involving hydration, granule swelling, and amylose release, is characterized by the transition from a structured to an unstructured state of starch granules. High hydrostatic pressure treatment affects starch gelatinization differently than moisture and heat treatments. Under high pressure, starch granules exhibit reduced swelling compared to heat treatment, preserving their granular shape while impacting both amorphous and crystalline structures. Consequently, a higher proportion of amylose remains within the granules (Buckow et al. 2007; Mirzababaee et al. 2022). Overall, HPP induces amylose release, starch degradation, rearrangement, aggregation, and the generation of new semicrystalline regions (Liu et al. 2016).

Liu et al. (2016) examined the influence of high hydrostatic pressure on the physicochemical characteristics of sorghum starch (SS). Their findings reported that HPP treatment enhanced the functional properties of SS, including increased water absorption capacity (WAC), solubility, and oil absorption capacity (OAC). Pressure treatments ranging from 120 to 600 MPa at ambient conditions for 20 min showed that at 600 MPa, starch granules were deformed and amalgamated, indicating complete gelatinization. This treatment significantly increased the WAC of SS from 0.70 g/g at 120 MPa to 4.30 g/g at 600 MPa, suggesting enhanced hydrophilicity due to stronger interactions between hydroxyl groups and water in HPP-treated samples caused by the severing of hydrogen bonds between crystalline and amorphous regions of starch. To prepare proso millet starch for culinary applications, Li et al. (2018) subjected it to ultrahigh pressures (150–600 MPa for 15 min). This process increased starch gelatinization, with complete gelatinization achieved at 600 MPa.

Additionally, the researchers tested the solubility and swelling power of the modified proso millet starch across a temperature range of 50–90 °C. At lower temperatures, swelling power increases with higher pressure. The solubility of starch (HPP-600) was highest, while the swelling power was lowest at 70 °C. This was attributed to the disruption of hydrogen bonds and the generation of new bonds between water and starch hydroxyl groups (Wani et al. 2012). The effects of soaking foxtail millet grains under high pressure were explored by Sharma et al. (2018). It was found that soaking the grains at pressures of 200, 400, and 600 MPa and temperatures ranging from 20 to 80 °C significantly increased the WAC, regardless of whether the grains were non-germinated or germinated. Both germinated and non-germinated grains showed improved water uptake as pressure levels increased. Additionally, the gelatinization degree of starch in foxtail millet flour increased with high-pressure soaking, reaching up to 64.93%. A maximum value of 6.77×10^{-9} m^2 s^{-1} was observed at 200 MPa and 60 °C for the effective diffusion coefficient of water. These changes were attributed to membrane disruption, starch crystalline structure changes, and enhanced grain permeability due to cracks, creating pathways for improved diffusion (Tan et al. 2009).

Mirzababaee et al. (2022) examined the effect of high hydrostatic pressure on the physicochemical, structural, thermal, and rheological properties of pearl millet starch. When subjected to 200, 400, and 600 MPa pressures for 10, 20, and 30 min at ambient temperature, the pearl millet starch exhibited decreased swelling power and solubility with increased pressure. The production of amylose-lipid complexes during high-pressure treatment hinders the release of amylose, reducing the solubility of starch and swelling power (Oh et al. 2008). WAC of millet starch significantly increased with HPP treatment, especially at higher pressure levels, except at 200 MPa. It rose from 0.66% for native starch to 2.19% for starch treated at 600 MPa for 30 min. This increase is due to the generation of pores and gaps on the granule surface, facilitating water penetration, hydrogen bonding with amylose and amylopectin hydroxyl groups, and enhancing water absorption (Vasanthan and Hoover 1992). In contrast, the OAC of millet starch decreased with increasing intensity of pressure and duration of HPP treatment due to the destruction of the granule structure at high pressures, reducing starch's ability to retain oil as the granules became fragmented and less capable of oil absorption (Heydari et al. 2021).

White finger millet starch (WFMS) was modified using high pressure at 200, 400, and 600 MPa for 10 min. Subsequently, the impact of HPP on the physicochemical, structural, and functional properties of WFMS was examined (Sherin et al. 2024). The WFMS became more soluble after HPP treatment, especially at higher temperatures, indicating that water-soluble components leached out during pressurization. The swelling power of HPP-treated WFMS also increased with temperature, with the 600 MPa sample showing the highest swelling at 50 °C compared to other samples. However, at 90 °C, the native (untreated) WFMS exhibited greater swelling power than the HPP-modified samples, due to the generation of amylose-lipid complexes at lower temperatures that inhibited granule swelling (Table 5.1).

5.6 Effect on Starch and Protein Digestibility

Millets are highly nutritious and have several health benefits. They contain many carbohydrates, proteins, dietary fibers, minerals, calcium, iron, magnesium, and phosphorus, along with several trace elements. It is a sustainable protein source that can be used easily to meet the needs of the ever-growing population, as millets are easy to grow and are less expensive. Moreover, millet crops have gluten-free, hypoallergenic properties, and a desirable amino acid profile (Renganathan et al. 2020; Sachdev et al. 2023). Like other staple foods like wheat and rice, the total

Table 5.1 Effect of HPP on functional properties of millet

Millet	Processing parameter	Major findings	Reference
Sorghum	Pressure: 120–480 MPa Time: 20 min	WAC increased up to 4.30 g/g at 600 MPa OAC was reduced to 0.55 g/g at 600 MPa Swelling power peaked at 50–60 °C and declined at 70–90 °C	Liu et al. (2016)
Proso millet	Pressure: 150–600 MPa Time: 15 min	At 70 °C, starch samples (HPP-600) showed the highest solubility and lowest SP	Li et al. (2018)
Foxtail millet	Pressure: 200–600 MPa Temperature: 20–80 °C Time: 30–120 min	Increased water absorption and degree of starch gelatinization up to 64.93% An increase of 6.77×10^{-9} m^2 s^{-1} was seen in the effective diffusion coefficient at 200 MPa and 60 °C	Sharma et al. (2018)
Pearl millet starch	Pressure: 200–600 MPa Time: 10–30 min	SP and solubility reduced as the pressure rose from 200 to 600 MPa After starch was subjected to 600 MPa for 30 min, WAC rose to 2.19% OAC declined as the time and pressure progressed	Mirzababaee et al. (2022)
White finger millet starch	Pressure: 200–600 MPa Time: 10 min	WAC increased to 3.105 g/g at 600 MPa OAC increased to 4.39 g/g at 600 MPa	Sherin et al. (2024)

protein content of different millets ranges from 7.52% to 12.1%. Adding to its value, essential and nonessential amino acids like leucine, proline, phenylalanine, serine, tyrosine, aspartic acid, and glutamic acid are seen in higher concentrations in millets. Also, because of their hypoglycemic nature, millets can potentially be used to treat type II diabetes. Despite being nutritious grains, they have become less popular because of their coarse texture and significant quantities of anti-nutritional components, making them not easily digestible. Anti-nutritional components, like phytic acid and condensed tannins, present in them prevent protein and starch from being hydrolyzed by digestive enzymes (Sharma and Gujral 2019; Sá et al. 2020; Anju and Sarita 2010). The following session discusses in detail the effect of HPP treatment on protein and starch digestibility.

5.6.1 Effect of HPP on Protein Digestibility

Plant proteins, especially those found in grains, generally have lower digestibility than animal tissue. Their internal arrangement in distinct protein bodies, their relative insolubility, the poor permeability of plant cell walls, and their protective seed coating could all contribute to this (Becker and Yu 2013). This problem could be overcome by giving sufficient physical or chemical treatments to bring about physical and/or chemical changes in them, like HPP. Several factors, including the treatment temperature, the amount and duration of pressure, and the physiochemical parameters of the material (such as pH and ionic strength), influence the functionality of the protein modified by HPP. This method uses the idea that pressure rises when volume decreases and vice versa. Proteins rupture, aggregate, or denature due to the volume change (Yang and Powers 2016).

Studies show that HPP can denature proteins by changing their three-dimensional structure, which may increase the proteins' solubility and accessibility to digestive enzymes, increasing digestibility. The proteins may precipitate or dissolve when high pressure (200–700 MPa) is applied during denaturation. Generally, these alterations are permanent above 300 MPa and reversible below 300 MPa pressure range. Denaturation could result from molecules unfolding and breaking hydrophobic and ion pair bonds. Oligomeric proteins tend to split into subunits under increased pressure, making them more susceptible to proteolysis (Knorr 2003; Rastogi et al. 2007; Farkas 2016). Sharma et al. (2023) found that altering the bonds and breaking down the protein inhibitors can change how plant proteins, such as those found in sorghum, finger millet, and amaranth, are digested. This correlates with improved digestibility as proteins are more readily available for enzymatic breakdown. High-pressure processing can improve digestibility, reduce allergic reactions, increase hydrophobicity, decrease solubility, and improve the emulsifying properties of millet proteins.

5.6.2 Effect of HPP on Starch Digestibility

Starch comprises tiny granules whose properties vary depending on their plant origin, including size, morphology, shape, and size distribution. It has been categorized into slowly digestible starch (SDS), rapidly digestible starch (RDS), and resistant starch (RS) based on the hydrolysis of α-amylase. One of the most serious conditions affecting health is diabetes. Resistant starch can prevent the onset of diabetes. Furthermore, resistant starch makes the person feel fuller because it passes intestinal digestion and enters the colon, where it gets fermented and serves as a prebiotic for good bacteria that aid digestion. The millets are highly favored as they contain substantial resistant starch. HPP can promote the formation of resistant starch in millet, which is beneficial for health. It can be used to modify starch as it is an environmentally friendly and green technique that can change non-covalent chemical bonds with much less impact on covalent bonds to have the desired properties (Annor et al. 2017 & Castro et al. 2020).

HPP can affect the gelatinization and retrogradation of starch in millet, which influences its digestibility. Gelatinization and physical alteration of food biopolymers, mainly starches, can be achieved nonthermally by using high hydrostatic pressure. It was found that variations in the kind of starch and the HPP processing parameters, like the combination of temperature and pressure, holding time, starch concentration, and suspending media, affect the degree of gelatinization, pasting qualities, and crystallinity (Kim et al. 2012). Ultrahigh-pressure treatment significantly altered the functional attributes of millet starch, where primarily the resistant starch increased along with a higher swelling capacity and solubility, denoting increased digestibility (Li et al. 2018).

5.7 Storage Stability of HPP-Processed Millet Products

Beyond its nutritional implications, HPP effectively eliminates pathogenic microorganisms such as bacteria and viruses without compromising the nutritional quality of millets. This ensures food safety and extends shelf life of millet products, making them safer for longer periods (Kaushik et al. 2016). The various methods to prolong the shelf life of grain flour include separating the lipid-rich bran and germ, deactivating lipolytic enzymes, and regulating bacterial and fungal development. HPP may effectively reduce flour spoilage by decreasing the microbial load while preserving nutritional attributes of the flour and functional capabilities (Padmaja et al. 2024). High pressure causes macromolecules to denature while keeping small molecules, like vitamins, unaffected. In addition, the product experiences less heat during the treatment process. High-pressure soaking eliminates the outer bran layers of the grain by breaking down the cell membranes, which consist of a hydrophobic lipid layer. This increases the stability of flour (Zhu et al. 2016). HPP is an appealing technology for both the food industry and consumers due to its ability to maintain nutritional value while improving safety. High-pressure immersion has a significant impact on the exterior bran layer of grains, particularly affecting the hydrophobic lipid layer, which enhances the shelf stability of millet flour. Sharma et al. (2018) observed that subjecting non-germinated foxtail millet flour to high-pressure soaking at 400 MPa and 60 °C for 60 min decreased the bound water percentage. This method reduced the sorption capacity while enhancing the functional properties associated with the shelf life of millet flour. However, HPP has not been used to inactivate lipase and other enzymes that trigger rancidity in millet flour. The findings suggest this technique can prevent flour deterioration by controlling key factors (Sruthi and Rao 2021).

HPP treatment helps in the shelf-life extension of beverages by inactivating enzymes/microorganisms and retaining volatile compounds. In a study by Zhou et al. (2022) examining cloudy apple juice, various sterilization methods were applied, including conventional pasteurization, microwave sterilization, ultrasonic sterilization, and high-pressure sterilization. Among these methods, high-pressure sterilization proved to be the most effective in preserving volatile compounds. Similar results were found in another experiment where the mango juice was treated with HPP (Pan et al. 2021). Researchers used those results as a reference and subjected millet composite beverages to 300 MPa and 600 MPa HPP treatments for 3 min. They observed that applying HPP treatment notably increased the shelf life of the beverages (Zhou et al. 2022). In a study Ilgaz (2014) investigated the impact of HPP on boza, a traditional millet-based fermented beverage. The study applied pressures of 150 MPa, 250 MPa, and 350 MPa to boza at 5 °C for 5 min, examining the effects on extending shelf life. It was found that applying HPP at 350 MPa and 5 °C for 5 min substantially decreased the bacterial load by approximately 3.0 log cycles. Additionally, all evaluated HPP treatments effectively eradicated yeast and mold. The shelf life of boza was extended to 19 days at 350 MPa, 15 days at 250 MPa, and 12 days at 150 MPa, compared to only 6–7 days at room temperature. Furthermore, the study explored the effects of HPP on foxtail millet takju. Treatment at 400 MPa for 10 min at room temperature successfully deactivated yeast and lactic acid bacteria in the takju. Complete sterilization was achieved with either 400 MPa for 60 min at 66 °C or 600 MPa for 10 min at 66 °C. The α-amylase and glucoamylase activities decreased by 59.7% and 20.5%, respectively, following the 400 MPa treatment for 10 min at 66 °C (Jwa et al. 2001).

Microbial inactivation occurs through cell membrane permeabilization, disruption of non-covalent bonds, and potential effects on the cell's DNA and RNA. A hydrostatic pressure of approximately 300 MPa at room temperature is needed to inactivate vegetative cells. However, to inactivate spores, a higher pressure of 600 MPa and elevated temperature are essential (Ilgaz 2014). Applying high pressure combined with mild heat treatment can break the non-covalent bonds within microbial cells, affecting proteins, nucleic acids, and lipids. This process inactivates microorganisms while preserving covalent bonds, which helps maintain color, aroma, flavor, and nutritional value of food (Meng et al. 2018). Moreover, secondary lipid oxidation consistently forms volatile low molecular weight compounds such as aldehydes and ketones. According to Le Chatelier's principle, HPP can potentially inhibit these reactions (Wang et al. 2018). However, despite its benefits, several obstacles currently hinder the widespread adoption of HPP in the food industry. The technology remains more expensive than conventional methods, and transitioning from traditional processing to HPP largely depends on the cost-benefit ratio and feasibility (Norton and Sun 2008). Although various inactivation models are being developed and studied, the absence of a comprehensive model for predicting enzyme behavior under HPP poses challenges for understanding the inactivation of key food enzymes (Cheung et al. 2023). As a result, further research is necessary to minimize the existing issues in its adoption.

5.8 Conclusion

HPP holds significant prominence in enhancing the shelf life and maintaining the nutritional integrity of millet-based products. HPP also mitigates the risks of rancidity and spoilage by effectively deactivating microorganisms and enzymes. This is particularly important for millets due to its higher fat content and vulnerability to oxidative rancidity. Research indicates that HPP can improve the stability and quality of millet flour and beverages by inhibiting lipase activity and preserving essential volatile compounds. However, the widespread adoption of HPP in the food industry is hindered by its high costs, the necessity for further investigation into enzyme inactivation mechanisms, and the absence of comprehensive predictive models for enzyme behavior under HPP conditions. Future studies should aim to refine HPP parameters to maximize its efficacy while minimizing costs and preserving the nutritional and sensory properties of millets. The ability of HPP to enhance the shelf life and safety of millet products highlights its potential as a valuable technology, especially in regions where millet is a dietary staple, thereby contributing to food security and improved nutritional outcomes.

References

Akanbi TO, Timilsena Y, Dhital S (2019) Bioactives from millet: properties and effects of processing on bioavailability. In: Bioactive factors and processing technology for cereal foods, pp 171–183. https://doi.org/10.1007/978-981-13-6167-8_10

Amsasekar A, Mor RS, Kishore A, Singh A, Sid S (2022) Impact of high-pressure processing on microbiological, nutritional and sensory properties of food: a review. Nutr Food Sci 52(6):996–1017. https://doi.org/10.1108/NFS-08-2021-0249

Angioloni A, Collar C (2012) Effects of pressure treatment of hydrated oat, finger millet, and sorghum flours on the quality and nutritional properties of composite wheat breads. J Cereal Sci 56(3):713–719. https://doi.org/10.1016/j.jcs.2012.08.001

Anju T, Sarita S (2010) Suitability of foxtail millet (Setaria italica) and barnyard millet (Echinochloa frumentacea) for development of low glycemic index biscuits. Malays J Nutr 16(3):361–368

Annor GA, Tyl C, Marcone M, Ragaee S, Marti A (2017) Why do millets have slower starch and protein digestibility than other cereals? Trends Food Sci Technol 66:73–83. https://doi.org/10.1016/j.tifs.2017.05.012

Balakrishna AK, Farid M (2020) Enrichment of rice with natural thiamine using high-pressure processing (HPP). J Food Eng 283:110040. https://doi.org/10.1016/j.jfoodeng.2020.110040

Bansal V, Jabeen K, Rao PS, Prasad P, Yadav SK (2019) Effect of high pressure processing (HPP) on microbial safety, physicochemical properties, and bioactive compounds of whey-based sweet lime (whey-lime) beverage. J Food Meas Charact 13:454–465. https://doi.org/10.1007/s11694-018-9959-1

Becker PM, Yu P (2013) What makes protein indigestible, from tissue-related, cellular, and molecular aspects? Mol Nutr Food Res 57(10):1695e1707. https://doi.org/10.1002/mnfr.201200592

Buckow R, Heinz V, Knorr D (2007) High-pressure phase transition kinetics of maize starch. J Food Eng 81(2):469–475. https://doi.org/10.1016/j.jfoodeng.2006.11.027

Cao X, Zhang Y, Zhang F, Wang Y, Yi J, Liao X (2011) Effects of high hydrostatic pressure on enzymes, phenolic compounds, anthocyanins, polymeric color and color of strawberry pulps. J Sci Food Agric 91(5):877–885. https://doi.org/10.1002/jsfa.4260

Castro MG, Alexandre EMC, Saraiva JA, Pintado M (2020) Impact of high pressure on starch properties: a review. Food Hydrocoll 106:105877. https://doi.org/10.1016/j.foodhyd.2020.105877

Cheung LKY, Sanders AD, Mandal R, Dee DR, Pratap-Singh A, Yada RY (2023) Effect of high-pressure processing on enzyme activity in roots, cereals, nuts, and their products. In: Effect of high-pressure technologies on enzymes. Academic Press, pp 299–329. https://doi.org/10.1016/B978-0-323-98386-0.00015-4

Choudhary C, Vignesh S, Chidanand DV, Baskaran N (2023) Effect of different processing methods on nutritional composition and microbial quality of pearl millet. Research Square. 10.21203/rs.3.rs-3614114/v1

Cilla A, Barberá R, López-García G, Blanco-Morales V, Alegría A, Garcia-Llatas G (2019) Impact of processing on mineral bioaccessibility/bioavailability. In: Innovative thermal and non-thermal processing, bioaccessibility and bioavailability of nutrients and bioactive compounds. Woodhead Publishing, pp 209–239. https://doi.org/10.1016/B978-0-12-814174-8.00007-X

Dekka S, Paul A, Vidyalakshmi R, Mahendran R (2023) Potential processing technologies for utilization of millets: an updated comprehensive review. J Food Process Eng 46(10):e14279. https://doi.org/10.1111/jfpe.14279

Estrada-Girón Y, Swanson BG, Barbosa-Cánovas GV (2005) Advances in the use of high hydrostatic pressure for processing cereal grains and legumes. Trends Food Sci Technol 16(5):194–203. https://doi.org/10.1016/j.tifs.2004.10.005

Farkas DF (2016) A short history of research and development efforts leading to the commercialization of high-pressure processing of food. In: High pressure processing of food. Springer, pp 19–36. https://doi.org/10.1007/978-1-4939-3234-4_2

García-Parra J, González-Cebrino F, Delgado J, Cava R, Ramírez R (2016) High pressure assisted thermal processing of pumpkin purée: effect on microbial counts, color, bioactive compounds and polyphenoloxidase enzyme. Food Bioprod Process 98:124–132. https://doi.org/10.1016/j.fbp.2016.01.006

Guerrero-Beltrán JA, Estrada-Girón Y, Swanson BG, Barbosa-Cánovas GV (2009) Pressure and temperature combination for inactivation of soymilk trypsin inhibitors. Food Chem 116(3):676–679. https://doi.org/10.1016/j.foodchem.2009.03.001

Gutiérrez ÁL, Rico D, Ronda F, Caballero PA, Martín-Diana AB (2023) The application of high-hydrostatic-pressure processing to improve the quality of baked products: a review. Foods 13(1):130

Heydari A, Razavi SMA, Hesarinejad MA, Farahnaky A (2021) New insights into physical, morphological, thermal, and pasting properties of HHP-treated starches: effect of starch type and industry-scale concentration. Starch 73(7–8):2000179. https://doi.org/10.1002/star.202000179

Hogan E, Kelly AL, Sun D (2005) High pressure processing of foods: an overview. In: Emerging technologies for food processing. Elsevier Academic Press, pp 3–27

Hou Z, Luan L, Hu X, Chen S, Ye X (2022) Beneficial effects of high pressure processing on the interaction between RG-I pectin and cyanidin-3-glucoside. Food Chem 383:132373. https://doi.org/10.1016/j.foodchem.2022.132373

Ilgaz NN (2014) The effects of high hydrostatic pressure (HHP) treatment on shelf life and quality parameters of conventionally produced boza. Middle East Technical University (Master thesis)

Jaybhaye RV, Pardeshi IL, Vengaiah PC, Srivastav PP (2014) Processing and technology for millet based food products: a review. J Ready Eat Food 1(2):32–48

Jonnalagadda SS, Harnack L, Liu RH, McKeown N, Seal C, Liu S, Fahey GC (2011) Putting the whole grain puzzle together: health benefits associated with whole grains-summary of American society for nutrition 2010 satellite symposium 1–3. J Nutr 141(5):1011S–1022S. https://doi.org/10.3945/jn.110.132944

Joshi TJ, Singh SM, Rao PS (2023) Novel thermal and non-thermal millet processing technologies: advances and research trends. Eur Food Res Technol 249(5):1149–1160. https://doi.org/10.1007/s00217-023-04227-8

Jwa M-K, Lim S-B, Mok C-K, Park Y-S (2001) Inactivation of microorganisms and enzymes in foxtail millet taken by high hydrostatic pressure treatment. Korean J Food Sci Technol 33(2):226–230

Kaushik N, Rao PS, Mishra HN (2016) Process optimization for thermal-assisted high pressure processing of mango (Mangifera indica L.) pulp using response surface methodology. LWT Food Sci Technol 69:372–381. https://doi.org/10.1016/j.lwt.2016.01.062

Kim HS, Kim BY, Baik MY (2012) Application of ultra-high pressure (UHP) in starch chemistry. Crit Rev Food Sci Nutr 52:123–141. https://doi.org/10.1080/10408398.2010.498065

Kim MY, Lee SH, Jang GY, Li M, Lee YR, Lee J, Jeong HS (2015) Influence of applied pressure on bioactive compounds of germinated rough rice (Oryza sativa L.). Food Bioprocess Technol 8:2176–2181. https://doi.org/10.1007/s11947-015-1565-1

Knorr D (2003) Impact of high-pressure processing on food safety and quality. Food Technol 57(2):72–78

Kundgol NG, Kasturiba B, Math KK, Kamatar MY, Usha M (2013) Impact of decortication on chemical composition, antioxidant content and antioxidant activity of little millet landraces. Int J Eng Res Technol 2(10):1705–1720

Lee H, Ha MJ, Shahbaz HM, Kim JU, Jang H, Park J (2018) High hydrostatic pressure treatment for manufacturing of red bean powder: a comparison with the thermal treatment. J Food Eng 238:141–147. https://doi.org/10.1016/j.jfoodeng.2018.06.016

Li W, Gao J, Saleh ASM, Tian X, Wang P, Jiang H (2018) The modifications in physicochemical and functional properties of proso millet starch after ultra-high pressure (UHP) process. Starch 70(5–6):1700235. https://doi.org/10.1002/star.201700235

Lin T, O'Keefe S, Duncan S, Fernández-Fraguas C (2020) Manipulation of the dry bean (Phaseolus vulgaris L.) matrix by hydrothermal and high-pressure treatments: impact on in vitro bile salt-binding ability. Food Chem 310:125699. https://doi.org/10.1016/j.foodchem.2019.125699

Linsberger-Martin G, Weiglhofer K, Phuong TPT, Berghofer E (2013) High hydrostatic pressure influences anti-nutritional factors and in vitro protein digestibility of split peas and whole white beans. LWT Food Sci Technol 51(1):331–336. https://doi.org/10.1016/j.lwt.2012.11.008

Liu H, Fan H, Cao R, Blanchard C, Wang M (2016) Physicochemical properties and in vitro digestibility of sorghum starch altered by high hydrostatic pressure. Int J Biol Macromol 92:753–760. https://doi.org/10.1016/j.ijbiomac.2016.07.088

Liu H, Lin Y, Sun M, Zhang X, Zhao X, Zhao X (2023) The effects of high-pressure processing on the nutritional quality of sprouts: a review. Food Biosci:103384. https://doi.org/10.1016/j.fbio.2023.103384

Marti A, Barbiroli A, Bonomi F, Brutti A, Iametti S, Marengo M, Pagani MA (2014) Effect of high-pressure processing on the features of wheat milling by-products. Cereal Chem 91(4):318–320. https://doi.org/10.1094/CCHEM-07-13-0133-CESI

Martín-Diana AB, Tomé-Sánchez I, García-Casas MJ, Martínez-Villaluenga C, Frías J, Rico D (2021) A novel strategy to produce a soluble and bioactive wheat bran ingredient rich in ferulic acid. Antioxidants 10(6):969. https://doi.org/10.3390/antiox10060969

Meng L, Zhang W, Wu Z, Hui A, Gao H, Chen P, He Y (2018) Effect of pressure-soaking treatments on texture and retrogradation properties of black rice. LWT Food Sci Technol 93:485–490. https://doi.org/10.1016/j.lwt.2018.03.079

Mirzababaee SM, Ozmen D, Hesarinejad MA, Toker OS, Yeganehzad S (2022) A study on the structural, physicochemical, rheological and thermal properties of high hydrostatic pressurized pearl millet starch. Int J Biol Macromol 223:511–523. https://doi.org/10.1016/j.ijbiomac.2022.11.044

Mohite AM, Sharma N (2024) Influence of moisture content on nutritional and processing properties of millets. In: Nutriomics of millet crops. CRC Press, pp 125–139

Naik L, Sharma R, Rajput Y, Manju G (2013) Application of high-pressure processing technology for dairy food preservation-future perspective. J Anim Prod Adv 3(8):232–241

Nithiyanantham S, Kalaiselvi P, Mahomoodally MF, Zengin G, Abirami A, Srinivasan G (2019) Nutritional and functional roles of millets—a review. J Food Biochem 43(7):e12859. https://doi.org/10.1111/jfbc.12859

Norton T, Sun D-W (2008) Recent advances in the use of high pressure as an effective processing technique in the food industry. Food Bioprocess Technol 1:2–34. https://doi.org/10.1007/s11947-007-0007-0

Oh HE, Pinder DN, Hemar Y, Anema SG, Wong M (2008) Effect of high-pressure treatment on various starch-in-water suspensions. Food Hydrocoll 22(1):150–155. https://doi.org/10.1016/j.foodhyd.2007.01.028

Padmaja PG, Venkateswarlu R, Singh SP, Tonapi VA (2024) Enhancing shelf life of pearl millet flour. In: Pearl millet in the 21st century: food-nutrition-climate resilience-tmproved livelihoods. Springer, pp 549–566. https://doi.org/10.1007/978-981-99-5890-0_2

Pan X, Wu J, Zhang W, Liu J, Yang X, Liao X, Hu X, Lao F (2021) Effects of sugar matrices on the release of key aroma compounds in fresh and high hydrostatic pressure processed Tainong mango juices. Food Chem 338:128117. https://doi.org/10.1016/j.foodchem.2020.128117

Pérez-Jiménez J, Arranz S, Tabernero M, Díaz-Rubio ME, Serrano J, Goñi I, Saura-Calixto F (2008) Updated methodology to determine antioxidant capacity in plant foods, oils and beverages: extraction, measurement and expression of results. Food Res Int 41(3):274–285. https://doi.org/10.1016/j.foodres.2007.12.004

Prashanth P, Joshi TJ, Singh SM, Rao PS (2024) Hydrothermal treatment of pearl millet grains: effects on nutritional composition, anti-nutrients and flour properties. Grain Oil Sci Technol 7(2):87–95. https://doi.org/10.1016/j.gaost.2024.04.002

Rastogi NK, Raghavarao KSMS, Balasubramanium VM, Niranjan K, Knorr D (2007) Opportunities and challenges in high pressure processing of foods. Crit Rev Food Sci Nutr 47:69–112. https://doi.org/10.1080/10408390600626420

Renganathan VG, Vanniarajan C, Karthikeyan A, Ramalingam J (2020) Barnyard millet for food and nutritional security: current status and future research direction. Front Genet 11:500. https://doi.org/10.3389/fgene.2020.00500

Sá AGA, Moreno YMF, Carciofi BAM (2020) Plant proteins as high-quality nutritional source for human diet. Trends Food Sci Technol 97:170–184. https://doi.org/10.1016/j.tifs.2020.01.011

Sachdev N, Goomer S, Singh LR, Pathak VM, Aggarwal D, Chowhan RK (2023) Current status of millet seed proteins and its applications: a comprehensive review. Appl Food Res 3(1):100288. https://doi.org/10.1016/j.afres.2023.100288

Saikaew K, Lertrat K, Meenune M, Tangwongchai R (2018) Effect of high-pressure processing on colour, phytochemical contents and antioxidant activities of purple waxy corn (Zea mays L. var. ceratina) kernels. Food Chem 243:328–337. https://doi.org/10.1016/j.foodchem.2017.09.136

Sehrawat R, Kaur BP, Nema PK, Tewari S, Kumar L (2021) Microbial inactivation by high pressure processing: principle, mechanism and factors responsible. Food Sci Biotechnol 30:19–35. https://doi.org/10.1007/s10068-020-00831-6

Shahidi F, Danielski R, Ikeda C (2021) Phenolic compounds in cereal grains and effects of processing on their composition and bioactivities: a review. J Food Bioactiv 15:10.31665/JFB.2021.15281

Sharma B, Gujral HS (2019) Influence of nutritional and anti-nutritional components on dough rheology and in vitro protein & starch digestibility of minor millets. Food Chem 299:125115. https://doi.org/10.1016/j.foodchem.2019.125115

Sharma N, Goyal SK, Alam T, Fatma S, Chaoruangrit A, Niranjan K (2018) Effect of high pressure soaking on water absorption, gelatinization, and biochemical properties of germinated and non-germinated foxtail millet grains. J Cereal Sci 83:162–170. https://doi.org/10.1016/j.jcs.2018.08.013

Sharma N, Sahu JK, Bansal V, Esua OJ, Rana S, Bhardwaj A, PuniaBangar S, Adedeji AA (2023) Trends in millet and pseudomillet proteins - characterization, processing and food applications. Food Res Int 164:112310. https://doi.org/10.1016/j.foodres.2022.112310

Sherin AJ, Sunil CK, Chidanand DV, Venkatachalapathy N (2024) Structural, physicochemical and functional properties of high-pressure modified white finger millet starch. Int J Biol Macromol 261:129919. https://doi.org/10.1016/j.ijbiomac.2024.129919

Singh A, Sharma S (2017) Bioactive components and functional properties of biologically activated cereal grains: a bibliographic review. Crit Rev Food Sci Nutr 57(14):3051–3071. https://doi.org/10.1080/10408398.2015.1085828

Singh R, Singh K, Singh Nain M, Singh R (2020) Storage stability of popped pearl millet based ready to eat breakfast cereal. Indian J Agric Sci 90(10):1915–1920. https://doi.org/10.56093/ijas.v90i10.107902

Smelt JPPM (1998) Recent advances in the microbiology of high pressure processing. Trends Food Sci Technol 9(4):152–158

Sruthi NU, Rao PS (2021) Effect of processing on storage stability of millet flour: a review. Trends Food Sci Technol 112:58–74. https://doi.org/10.1016/j.tifs.2021.03.043

Tahir-Kheli R (2020) Le Châtelier Principle. In: General and statistical thermodynamics. Springer, pp 309–327

Tan FJ, Dai WT, Hsu KC (2009) Changes in gelatinization and rheological characteristics of japonica rice starch induced by pressure/heat combinations. J Cereal Sci 49(2):285–289. https://doi.org/10.1016/j.jcs.2008.11.006

Vasanthan T, Hoover R (1992) A comparative study of the composition of lipids associated with starch granules from various botanical sources. Food Chem 43(1):19–27. https://doi.org/10.1016/0308-8146(92)90236-U

Vashishth N, Maurya NK (2024) Millet bran: unveiling its potential health benefits. Int J Nutr 1(2):11–19

Wang H, Zhu S, Ramaswamy HS, Hu F, Yu Y (2018) Effect of high pressure processing on rancidity of brown rice during storage. LWT Food Sci Technol 93:405–411. https://doi.org/10.1016/j.lwt.2018.03.042

Wani AA, Singh P, Shah MA, Schweiggert-Weisz U, Gul K, Wani IA (2012) Rice starch diversity: effects on structural, morphological, thermal, and physicochemical properties—a review. Compr Rev Food Sci Food Saf 11(5):417–436. https://doi.org/10.1111/j.1541-4337.2012.00193.x

Wu X, Tan M, Zhu Y, Duan H, Ramaswamy HS, Bai W, Wang C (2023) The influence of high pressure processing and germination on anti-nutrients contents, in vitro amino acid release and mineral digestibility of soybeans. J Food Compos Anal 115:104953. https://doi.org/10.1016/j.jfca.2022.104953

Xia Q, Tao H, Huang P, Wang L, Mei J, Li Y (2017a) Minerals in vitro bioaccessibility and changes in textural and structural characteristics of uncooked pre-germinated brown rice influenced by ultra-high pressure. Food Control 71:336–345. https://doi.org/10.1016/j.foodcont.2016.07.018

Xia Q, Wang L, Xu C, Mei J, Li Y (2017b) Effects of germination and high hydrostatic pressure processing on mineral elements, amino acids and antioxidants in vitro bioaccessibility, as well as starch digestibility in brown rice (Oryza sativa L.). Food Chem 214:533–542. https://doi.org/10.1016/j.foodchem.2016.07.114

Yadav DN, Anand T, Kaur J, Singh AK (2012) Improved storage stability of pearl millet flour through microwave treatment. Agric Res 1:399–404. https://doi.org/10.1007/s40003-012-0040-8

Yadav GP, Dalbhagat CG, Mishra HN (2022) Effects of extrusion process parameters on cooking characteristics and physicochemical, textural, thermal, pasting, microstructure, and nutritional properties of millet-based extruded products: a review. J Food Process Eng 45(9):e14106. https://doi.org/10.1111/jfpe.14106

Yang J, Powers JR (2016) Effects of high pressure on food proteins. In: High pressure processing of food. Food Engineering Series. Springer, pp 353–389. https://doi.org/10.1007/978-1-4939-3234-4_18

Zhang Y, Jing X, Chen Z, Wang X (2022a) Purification and identification of antioxidant peptides from millet gliadin treated with high hydrostatic pressure. LWT Food Sci Technol 164:113654. https://doi.org/10.1016/j.lwt.2022.113654

Zhang Y, Zhang X, Zhang Z, Chen Z, Jing X, Wang X (2022b) Effect of high hydrostatic pressure treatment on the structure and physicochemical properties of millet gliadin. LWT Food Sci Technol 154:112755. https://doi.org/10.1016/j.lwt.2021.112755

Zhang H, Feng X, Liu S, Ren F, Wang J (2023) Effects of high hydrostatic pressure on nutritional composition and cooking quality of whole grains and legumes. Innovative Food Sci Emerg Technol 83:103239. https://doi.org/10.1016/j.ifset.2022.103239

Zhou Z, Shang X, Wang Z, Sun C, Zhang L (2022) Effect of high-pressure treatment on the quality of Hericium erinaceus: millet composite beverage. J Food Qual 2022(1):2456921. https://doi.org/10.1155/2022/2456921

Zhu SM, Hu FF, Ramaswamy HS, Yu Y, Yu L, Zhang QT (2016) Effect of high pressure treatment and degree of milling on gelatinization and structural properties of brown rice. Food Bioprocess Technol 9:1844–1853. https://doi.org/10.1007/s11947-016-1770-6

Zhu Y, Chu J, Lu Z, Lv F, Bie X, Zhang C, Zhao H (2018) Physicochemical and functional properties of dietary fiber from foxtail millet (Setaria italica) bran. J Cereal Sci 79:456–461. https://doi.org/10.1016/j.jcs.2017.12.011

6 Radiant Transformations: Microwave Heating in Millet Processing (Rapid and Controlled Millet Treatment for Optimal Results)

Addanki Mounika, Perumal Thivya, Tamminana Jeeviteswara Rao, and Nalla Bhanu Prakash Reddy

6.1 Introduction

In recent years, there has been a growing public interest in alternative food crops or other sources of staple foods with a higher nutritional profile, functional functions, and illness-preventive rather than curative therapies (Nithiyanantham et al. 2019). According to the United Nations (UN), the world population is currently 8 billion and is anticipated to reach 9.7 billion by 2050 (United Nations Population Division (UNPD) 2019). To fulfill the need for a growing population, additional staple crops, rice, wheat, and maize are required to combat world hunger and provide food security. Millets are a tiny seeded crop utilized as a staple diet for humans for thousands of years, making it an excellent choice (Mounika et al. 2024). As time passes, millet spread worldwide in various varieties, including Asia, Africa, Europe, and America, and they have been adapted to local growing without any pest controllers or fertilizers (Tripathi and Vyas 2023). According to India Science, Technology and Innovation Portal (2023), millets are divided into two types based on their sizes, i.e., major or large and minor or small millets: under major millets sorghum, pearl, finger, foxtail, and proso millets, whereas in minor millets kodo, barnyard, browntop, and little millets. Even though all comes under the category of millets, each has its own nutritional and health benefits that are explained individually in Table 6.1.

A. Mounika · T. J. Rao · N. B. P. Reddy (✉)
National Institute of Food Technology, Entrepreneurship and Management, (NIFTEM-T), Thanjavur, Tamil Nadu, India

P. Thivya
Department of Food Process Engineering, National Institute of Technology, Rourkela, Odisha, India

S. Jaddu, R. Chandra Pradhan (eds.), *Innovative Millet Processing*,
https://doi.org/10.1007/978-3-031-99864-5_6

Table 6.1 Important nutritional properties and key health benefits of various millets (per 100 g)

S. no.	Millets	Protein (%)	Carbohydrates (%)	Energy (kcal)	Total dietary fiber (%)	Health benefits	References
1	Sorghum (cereal)	11	70.8	339	2.7	Improve gut health Reduces chronic diseases	Dahlberg et al. (2004); Xiong et al. (2019)
2	Pearl millet	11	73	756	8.5	Curing stomach ulcers Decrease gall stones	Malik (2015)
3	Finger millet	7.7	72.6	328	19.1	Decrease cardiac diseases, colon problems, constipation Lowers blood sugar levels	Devi et al. (2014); Shirsat (2020)
4	Foxtail millet	12.3	60.9	331	8	Increase metabolism of cholesterol, sugar Helps in preparing low GI foods	Hariprasanna (2016)
5	Proso millet	11.02	72.8	378	8.50	Low glycemic index Ideal food for diabetes and cardiovascular diseases	Das et al. (2019)
6	Kodo millet	10.6	59.2	353	10.2	Fight multiple diseases Low glycemic index food product preparation	Bunkar (2021)
7	Barnyard millet	10.1	68.8	398	12.5	Reduce blood glucose and lipid levels Prevent celiac diseases	Kaur and Sharma (2020)
8	Browntop millet	8.9	71.3	338	12.5	Maintenance of normal blood sugar and cholesterol level Improved bowel movements	Singh et al. (2022)

(continued)

Table 6.1 (continued)

S. no.	Millets	Protein (%)	Carbohydrates (%)	Energy (kcal)	Total dietary fiber (%)	Health benefits	References
9	Little millet	7.7	67	341	7.6	Improving bone health and hemoglobin Regulates body temperature, wound healing	Banerjee and Maitra (2020); Paschapur et al. (2021)

6.2 Fundamentals of Microwave Processing

Microwave (MW) processing is a nonconventional thermal technology designed to overcome limitations of traditional heating methods. Its use in food processing is growing globally due to benefits like faster heating, improved energy efficiency, and shorter processing times. In 1887–1888, Heinrich Hertz first generated electromagnetic frequency radiation deliberately. Later, in 1949, Percy L. Spencer demonstrated that this radiation could heat dielectric materials, and in 1954, Von Hippel explained how microwaves interact with matter (Orsat et al. 2017). Today, large-scale MW facilities are employed for drying, precooking meat, pasteurizing prepared foods, and tempering fish and meat. Rising consumer demand for high-quality, minimally processed foods has driven innovations in MW techniques, including pasteurization, blanching, thawing, baking, and extraction of bioactive compounds (Adedeji et al. 2008; Hong and Wang 2015; Shaheen et al. 2012).

Microwaves heat through dielectric mechanisms involving conductive, dielectric, and magnetic losses, depending on the electromagnetic field and material properties. A material's dielectric properties and the penetration depth of microwaves determine how effectively it is heated (Abdel-Hay 2022; Sun et al. 2016b). Understanding these properties is key to grasping MW-food interactions. Electromagnetic waves, unlike mechanical waves, propagate without a medium and are distinguished by their frequency, typically ranging from 10 to 10^3 meters in wavelength within the dielectric range. These waves cause dipolar rotation in materials, where an electric field distorts the charge cloud around atoms, inducing vibrations as the atoms return to equilibrium (Abdel-Hay 2022).

Microwaves, a subset of electromagnetic waves, range from 300 MHz to 300 GHz (wavelengths of 1 mm to 1 m), with 915 MHz and 2.45 GHz commonly used in industrial, scientific, and medical settings. When microwaves pass through a medium, they raise its temperature, enabling widespread applications in food, agriculture, and daily life. Compared to conventional heating, which relies on conduction and convection to transfer heat from the surface inward—creating temperature gradients and requiring more time and energy—MW heating is volumetric, occurring throughout the food quickly. This results in rapid heating, reduced processing times, and minimal temperature differences between a food's interior and exterior GHz (Guo et al. 2017).

MW heating generates thermal energy uniformly across a food's volume via interactions between polar water molecules and charged ions. This selective penetration increases atomic motion, converting kinetic energy into heat, enhancing moisture absorption, and maintaining consistent temperatures. Food, seen as a porous substance with proteins, carbohydrates, and water, contains liquid water, vapor, and air in its pores. Energy transfers through the surface or is absorbed volumetrically, with heat driven by pressure from water and vapor. Liquid water moves due to pressure differences, and as it transitions to vapor, internal pressure builds. In low-permeability materials, this can push liquid water outward, while vapor disperses through pores filled with gas. Unlike traditional heating, MW heating's internal evaporation creates high pressure and moisture flux, moving water to the surface. This, influenced by the food's porous structure and airflow, can lead to surface wetness or shrinkage from excessive water loss (Varith et al. 2007; Wei et al. 2021).

MW heating's advantages—speed, efficiency, and uniformity—make it a valuable tool in modern food processing, meeting demands for quality and convenience while offering a stark contrast to the slower, less efficient conventional methods. Integrating MW with traditional or alternative heating techniques can further enhance product quality (Fig. 6.1).

6.2.1 Overview of Microwaves

Microwaves are a type of electromagnetic radiation, similar to radio waves, visible light, and X-rays. It is a nonionizing radiation that induces molecular motion through the movement of ions and the rotation of dipoles, yet it does not alter the

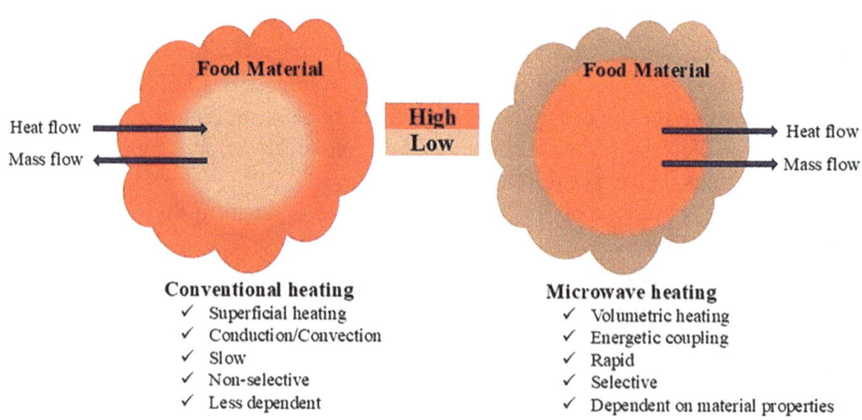

Fig. 6.1 Heat flow in conventional and microwave heating systems

molecular structure. MW energy is a segment of the electromagnetic spectrum that falls in the electromagnetic spectrum between infrared radiation and radio waves and is positioned within the wavelength range of 1 mm to 1 m and the frequency range of 300 MHz to 300 GHz, as illustrated in Fig. 6.2. The full spectrum of MW frequencies is categorized into several distinct bands. Microwaves are employed for the transmission of long-distance telephone signals, television broadcasts, and computer data across terrestrial and satellite networks. Additionally, microwaves are utilized in the detection of speeding vehicles. However, microwaves are perhaps most recognized as the energy source used for cooking and heating food. Wave energy undergoes polarity shifts from positive to negative with each wave cycle. In the case of microwaves, these polarity reversals occur millions of times per second. Food molecules, particularly water molecules, possess both positive and negative ends, like a bar magnet that has distinct north and south poles. When microwaves at an appropriate frequency interact with food, they induce movement and rotation in the polar molecules at the same frequency, also occurring millions of times per second. This molecular agitation generates friction, which subsequently heats the food. Notably, microwaves do not affect the molecules of paper, plastic, or glass; therefore, only the food is subjected to heating (Nguyen et al. 2021).

6.2.2 Concept of Heating with Microwave Frequency

When microwaves (type of electromagnetic waves) move in a void, the electric and magnetic fields oscillate perpendicular to one another; the transmission direction is at a right angle to both vibrating fields as depicted in Fig. 6.2. During microwave heating of materials, electromagnetic energy is transmitted directly to the substances via molecular interactions. This mechanism facilitates rapid volumetric heating, resulting in decreased processing times and energy savings (Abdel-Hay 2022). The applications of these waves can be categorized into two distinct classes, depending on whether they are utilized for the transmission of energy or information. The first category encompasses radio astronomy, satellite and terrestrial communication links, radar systems, MW thermography, and the measurement of material absolute

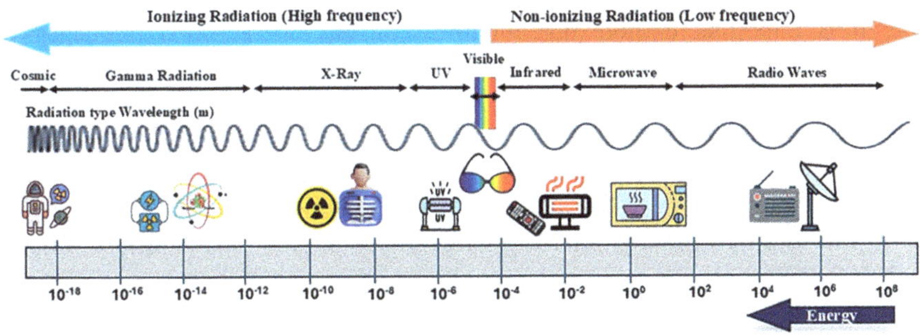

Fig. 6.2 Electromagnetic spectrum

permittivity. Conversely, the second category pertains to applications associated with wireless energy transmission and MW heating (Muthukumarappan and Swamy 2019). In MW heating, there is usually no apparent change, and the electromagnetic wave quickly interacts with both liquid and solid substances. When electromagnetic waves encounter a medium, they can be reflected, absorbed, transferred, or a combination of these three interactions. Multiple factors contribute to the interactions between microwaves and matter, including dielectric, conductive, and magnetic losses. These losses come from the characteristics of the electromagnetic fields and the materials that regulate and oversee common forms of heating like conductive and dielectric methods. The electromagnetic field can be divided into electric and magnetic fields, and each interacts with materials in distinct ways (Abdel-Hay 2022; Crane et al. 2013).

Microwaves are only utilized in a small range of frequency bands for thermal heating, as they are also used in convection industries. The frequently utilized frequency range is between 922 and 2450 MHz for food applications and industrial MW heating. Most microwave ovens operate at a frequency of 2450 MHz, whereas larger industrial systems operate at a frequency of 922 MHz. The frequency of these types depends on the size of the oven cavities that supply MW energy for heating. The short wavelengths of high frequencies allow them to be used in smaller oven cavities. The frequencies utilized in MW heating are contingent upon the dimensions of the oven cavities designed to deliver MW energy for heating purposes. Higher frequencies possess shorter wavelengths, enabling their use in oven cavities with smaller dimensions. Consequently, domestic MW ovens employ these higher frequencies, rendering their sizes suitable for typical heating applications. One notable characteristic of the frequency range used in MW heating, as compared to others, is its minimal absorption by most glass, plastics, or ceramics.

Metals reflect microwaves, which accounts for the sparks observed when metal objects are placed in a microwave oven. This reflection occurs because the electronic waves do not penetrate the interior of the conductor, given that the conductivity of metals is effectively infinite. All materials consist of molecules that exhibit dipole characteristics, possessing a positive charge on one side and a negative charge on the other. Within an electromagnetic field, these molecules undergo reorientation: the positive charge aligns with the negative pole, while the negative charge aligns with the positive pole. This molecular movement generates heat through friction. It is important to note that microwave ovens operate at a frequency of 2500 MHz; thus, microwaves at this frequency alter electromagnetic fields 2.5 billion times per second. Consequently, microwave ovens demonstrate significantly high thermal efficiency. Generally, microwaves possess coherent properties and polarizability; they can be absorbed, transmitted, or reflected depending on the structural characteristics of the material involved (Song et al. 2016). The equation below illustrates the relationship between the frequency and wavelength of electromagnetic waves:

$$C = \lambda v$$

where λ is wavelength (m), ν is frequency (Hz), and C is wave velocity (m/s).

6.2.3 Basic Principles of Microwave Food Processing

The well-known characteristics of electromagnetic waves like radio and light are (i) absence of mass, (ii) unrestricted movement at light speed, (iii) no resistance to motion, (iv) produced intense-high energy, and (v) bounced off the metal surface. Conversely, MW irradiation of dielectric materials results in various interactions based on the nature of the electromagnetic wave used. The impact of microwaves on dielectric materials is outlined as follows:

When interacting with electromagnetic waves in the MW region, a dielectric material having an electric dipole experiences orientation polarization. Moreover, space charge polarization causes Joule heating with the state of ions in solution. For example, dielectric heating, Joule heating, and MW heating of an electrolyte/water mix all happen at the same time as compared to pure water, resulting in extremely high exothermic efficiency. In electrolyte/water media, orientation polarization of water improves dielectric heating, and the Joule process improves resistance heating (Muthukumarappan and Swamy 2019).

6.2.4 Dielectric Properties of Food Materials

Dielectric heating is when high-frequency electromagnetic radiation heats materials with electric dipoles through the electric field component. One instance of this occurrence is when water is heated in the MW, demonstrating dielectric heating by dipole polarization. In addition to the electric field, microwaves possess a magnetic field component that interacts with various materials to generate heat (Sun et al. 2016a). Additionally, there are other types of MW heating, including magnetic loss heating and conductive loss heating. Additionally, understanding MW systems, a unique form of heating materials, involves a complex process when metals and metal compounds are subjected to a MW field. Notably, massive/bulk metals do not immediately react with MW energy, but they easily reflect the waves. In recent years, metal molecules have been able to be warmed up in a MW region if they are of a small enough size, like the process of sintering powdered metals and alloys (Sun et al. 2016a).

In addition to the effects of reflection and heating, electric discharges generated by microwaves can manifest in various forms, such as coronas, sparks, arcs, and/or weak discharges (Tiwari et al. 2020). This phenomenon occurs when metals with sharp edges, tips, and submicroscopic irregularities are subjected to a MW field. The charges accumulating on the metal surface at both macroscopic and microscopic levels can escape the material once they acquire sufficient kinetic energy, leading to ionization and disruption of the surrounding medium. As an immediate consequence, significant heat is generated during the discharge process, leading to the formation of localized high-temperature hot spots due to the transient and concentrated nature of the discharge. At a microscopic level, these hot spots are classified as plasmas (Tiwari et al. 2020; Zamri et al. 2021). Intensive generation of these discharges/plasmas could have significant implications for a variety of processes. Therefore, the specific effect of MW-metal interaction must not be overlooked and could play a crucial role in various applications. It is crucial to discover the unique bonding processes and potential effects between MW and metal substances. The dielectric material properties play a crucial role in showing how materials respond to the MW field (Tao et al. 2020).

Various factors like frequency of the applied electric field, moisture content, the temperature of the food, density and state, and chemical composition (specifically with the permanent dipole moments associated with water) and structure can impact the dielectric properties of food substances. In liquid foods, several factors influence the dielectric properties, whereas in hygroscopic substances, the water content is typically a primary determinant of their dielectric properties (Tao et al. 2020). Microwaves can directly or indirectly influence starch granules, resulting in various modifications to their external and internal structures that exhibit desirable properties. These alterations are attributed to the dielectric characteristics of the system and the input energy from the microwaves. Consequently, the effective utilization of MW energy for heating or drying any material is contingent upon the dielectric properties, such as the dielectric constant and dielectric loss factor of the food material, which determine MW power absorption and penetration of MW energy (Abdel-Hay 2022; S. P. Singh 2018; Tao et al. 2020).

The heating method for a material using MW energy depends on the material's dissipation factor (tan δ). The dissipation factor or loss tangent is determined by the relationship between a material's dielectric loss or loss factor (ε_r'') and its dielectric constant (ε_r'). The dielectric constant quantifies a material's capacity to permit MW energy to traverse through it, while the loss factor indicates the material's propensity to dissipate energy, such as the amount of input MW energy converted into heat. The dissipation factor is constrained for absorbent materials and varies among different substances. A higher dissipation factor indicates a reduced penetration of MW energy at a specified frequency, leading to an increased conversion of MW energy into heat (S. Wang 2009). Consequently, the transformation of the electric field into heat can be expressed as follows:

$$\varepsilon_r'' = \varepsilon_r' \left(\tan\delta\right)$$

The dielectric properties comprise the dielectric constant (ε_r') and the dielectric loss factor (ε_r''), which represent the real and imaginary components, respectively, of the relative complex permittivity (ε_r). This relationship is expressed by the following equation:

$$\varepsilon_r = \varepsilon_r' - j\varepsilon_r''$$

Where $j = \sqrt{-1}$.

These properties determine the behavior of a food material when exposed to a MW field (dielectric heating); they indicate the ability of the material to reflect, transmit, or absorb energy from microwaves. ε_r', a dimensionless number, is related to the ability of the food to store electromagnetic energy (energy penetration rate of food), and ε_r'' also a dimensionless number is the measure of the ability of food to dissipate electrical energy into heat (S. Wang 2009). These characteristics impact how a food substance reacts to MW radiation (dielectric heating); they show how well the material can either reflect, transmit, or absorb energy from microwaves. ε0, a unitless figure, reflects the food's capacity to retain electromagnetic energy (penetration rate of energy in food), while ε, also unitless, measures the food's capability to convert electrical energy into heat.

6.3 Impact of Microwaves on Millet and Their Effects

MW has shown many beneficial effects on millet grains and its flour, which makes it suitable for various product development. In the study, kodo millet (KM) grains were microwaved at 360, 720, and 900 W for 3 and 5 min, then pan-roasted at 195 °C for 5 min. Physical, functional, pasting, rheological, antioxidant, morphological, and structural features were used to describe kodo millet flours (KMFs). Roasting reduced the bulk density (BD) of all treated samples while increasing the tap density, Hausner ratio, and compressibility index. Higher microwave power (720 and 900 W) and duration time (5 min) led to lower $L*$ values and higher $a*$ and $b*$ values. Roasting in a pan and microwave dramatically increased flour's water absorption capacity (WAC). However, the oil absorption capacity (OAC) was significantly decreased as a result of the hydrophobic functional groups being destroyed. The rheological and pasting curves demonstrated that thick pastes may be produced with minimum power, improving pasting qualities. However, the starch was broken down by the pan and microwave roasting (MWR) at 900 W for 5 min, which produced a flat pasting curve (D. K. Singh et al. 2024a). This was further supported by scanning electron microscope (SEM) micrographs, which revealed the creation of pore spaces, which began at 720 W and became more apparent until 900 W when the structure was destroyed. Except for the 900 W for 5 min and the pan-roasted KMF, all samples' X-ray diffraction (XRD) patterns indicated a type A crystalline structure, and the degree of crystallinity rose with microwave power. All roasted samples, except for 900 W for 5 min, had stronger antioxidant activity (AOA) than the control sample. All KMFs, except for 360 W for 3 min, had lower total flavonoid content (TFC) than the control. Additionally, compared to control and pan-roasted KMF, all microwave-roasted KMF displayed substantially greater total phenolic content (TPC). These results offer useful insight into how to best roast KMF to enhance its nutritional and functional qualities (D. K. Singh et al. 2024a).

In another study, the quinoa was treated under MW roasting and studied its impact on various properties. MW roasting considerably ($p < 0.05$) altered the WAC (1.04–1.41 g/g), OAC (1.09–0.96 g/g), foaming capacity (6.66–4.62%), emulsion capacity (57.06–43.18%), and pasting characteristics of quinoa. Roasted quinoa samples are high in TPC (78.54–105.9 mg GAE/100 g) and AOA (68.95–81.84%). In contrast, heavily roasted samples showed a larger drop in gelatinization temperature as well as loss of crystalline structure (amylopectin content: 85.23–80.78%) in starch molecules. Microwave roasting significantly reduced ordered structure (β-sheet and α-helix) and resulted in an unstructured and flexible structure (Kheto et al. 2022). In another study, microwave roasting affects sorghum's physical, thermal, functional, pasting, and antioxidant characteristics. Roasting significantly reduces the bulk density of sorghum (778–546 kg/m^3) ($p < 0.05$). Roasting raised the overall color change and browning index from 14.8 to 48.2 and 17.9 to 64.8, respectively, while decreasing the 'L*' brightness from 83.1 to 50.6. The amylose-lipid complex's onset, peak, and end temperatures fell by 21.00%, 19.01%, and 17.03%, respectively. Similarly, peak viscosity, pasting viscosity, breakdown viscosity, and setback from the trough were reduced from 2295 to 35.42, 26.15 to 16.67, 984.8 to 8.38, and 2013 to 17.97, respectively (Sharanagat et al. 2019).

In a different study, barley flour was treated using MW, and studies showed anticancerous and antioxidant effects using three different solvents, i.e., methanol, ethanol, and water. Among the several solvents, methanol was determined to be the best suited for extraction, as the extracts demonstrated the strongest antioxidant and anticancer activity, as well as the greatest protection against DNA damage. Microwave roasting had a mixed effect on barley's antioxidant capability, as the percentage inhibition of DPPH and reducing power rose but all other measures, such as OH radical scavenging activity, H_2O_2 radical scavenging activity, and total phenolic content decreased. In the instance of DPPH, a rise of 13.24% in methanol, 8.14% in ethanol, and 6% in water was recorded, whereas lowering power resulted in an increase of 21.37% in methanol, 21.43% in ethanol, and 15.23% in water (Baba et al. 2016). Microwave roasting was found to reduce total phenol content (TPC) values for methanolic, ethanolic, and aqueous extracts by 34.39%, 121.477%, and 50%, respectively. However, when the extract concentration rose, every test revealed an overall rise in a dose-dependent way. Regardless of the solvents utilized, roasting also reduced the anticancerous potential and prevented DNA damage from the barley extracts. Colo-205, T47D, and MCF7 cell lines were used to assess the antiproliferative properties of the native and microwave-roasted barley flour extracts. More so for native barley flour (39.81%) than roasted barley flour (22.91%), barley flour extracts reduced the development of cancer cells (Baba et al. 2016). The study aims to regulate the enzymatic and oxidative deterioration of pearl millet flour using microwave treatment. The three independent variables are moisture content (14, 18, 22%) and microwave power (600, 800, 1000 W). Microwave treatment of pearl millet grains was optimized based on exposure period (75, 85, 95 s) and various rancidity-producing chemicals. The optimal microwave processing conditions for pearl millet grains were 1000 W power, 18% moisture content, and 93 s exposure time. The dependent variables of lipase, peroxidase, free fatty acids (FFA), and total phenol were found to be 12.018 μM FFA/h/g, 32.843 μM/min/g, 0.29%, and 0.569%, respectively (Vidhya et al. 2024).

In a similar study, the pearl millet grains that underwent MW treatment resulted in decreased lipase activity by increase in moisture content, and it was observed that 18% moisture content for a 100 s treatment resulted best reduction. But, based on both pasting and lipase activity, 18% moisture and 80 s treatment are considered as optimal treatment. The flour of grains treated for 80 s at 18% moisture level showed a significantly ($p < 0.05$) reduced change in free fatty acid (FFA, % oleic acid) value (20.80–22.25) during storage up to 30 days compared to control flour (20.11–32.43). A subjective assessment of the flour samples' overall acceptability revealed that although control flour had an unpleasant off-odor and a bitter taste by the tenth day of storage, microwave-treated flour (18% moisture content, 80 s) was acceptable for up to 30 days of storage at room temperature (D. N. Yadav et al. 2012). Using MW for pretreatment of foxtail millet flour and the study of its impact on physical and functional qualities. Physical parameters such as tapped density, bulk density, Carr's index, and Hausner ratio showed no change in the obtained values. Although the color of microwave-treated flour changed noticeably, it did not affect its visual appeal. Oil absorption capacity reduced following microwave treatment; however, water absorption capacity rose. Swelling power is reduced with increasing power level and time. However, the solubility of flour improved with microwave treatment, with the maximum solubility observed at 840 W and 2-min treatment settings (Rao et al. 2021). These studies highlight the potential advantages of MW in food processing by showing notable changes in a variety of characteristics. Although MW has several advantages for improving the characteristics of different grains, its particular use is in millets.

6.4 Applications of Microwave in Millet Processing

MW technology is an important aspect of millet processing, providing several benefits that improve the quality and functionality of millet-based products. By enabling quick and uniform heating, microwaves effectively change the structural and functional properties of millet grains and flour, enhancing water absorption, gelatinization, and thermal stability. Specifically, MW processing is employed to precook millet for instant food products, improve the puffing of millet grains for snacks, and deactivate enzymes to increase shelf life. It also reduces anti-nutritional factors, such as phytic acid, while boosting the bioavailability of nutrients. The energy efficiency and shorter processing times make MW technology a favorable option for both industrial and small-scale millet processing. Various possible applications of MW technology in millet processing are explained as follows (Fig. 6.3).

6.4.1 Microwave-Assisted Drying or Microwave Pretreatment Before Drying

Millets are generally harvested at a moisture content between 15% and 20% w.b. (wet basis) to get maximum yield and minimum losses. This can be further brought down during storage up to 10–12% w.b. or lower, by drying at optimal temperatures of 40–60 °C, as exceeding 60 °C can lead to quality degradation, such as loss of germination potential, reduced nutritional value, and discoloration. This practice plays a crucial role in ensuring grain storage without the risk of germination, breakage, or infestations. Key factors such as grain moisture content, ambient

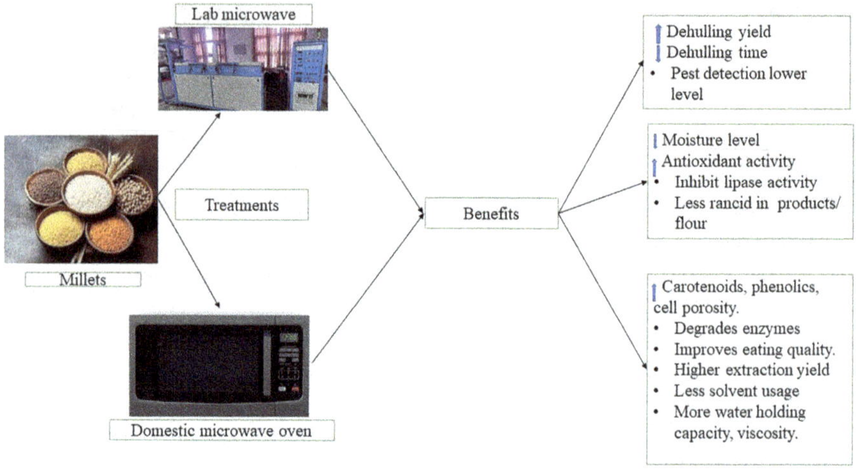

Fig. 6.3 Applications of microwave in the processing of millet

temperature, relative humidity, and other environmental conditions significantly influence the quality of stored grains. Inadequate control of temperature and humidity can result in the absorption or loss of moisture, leading to potential grain spoilage. Traditionally, millet drying is carried out both in the field and with the aid of mechanical drying equipment (Naik et al. 2022).

In thin-layer drying or sun or hot air drying, sometimes undesirable events such as uneven or inconsistent drying, increased risk of external contamination, weather dependency, discoloration due to over drying, and many others happen. One of the ways to reduce or counter these undesirable effects is reducing the drying time (at the same to get a better-quality product after drying). This can be achieved by pretreating the millets to enhance the drying rate or drying using alternative thermal technologies such as either MW drying or radiofrequency-based drying as these methods are technically superior to conventional drying methods as mentioned elsewhere (Radhika et al. 2013). In the conventional methods, energy is transferred (from surface to center) through conduction, convection, and radiation for heat and mass transfer, whereas in the case of MW technology, volumetric heating happens, i.e., heat is generated throughout the material irrespective of surface or center based on the basic phenomenon called dielectric heating (polar molecule interaction—microwaves cause polar molecules, such as water, to oscillate rapidly. As the molecules attempt to align themselves with the alternating electric field of the microwaves, they generate heat through friction. Ionic conduction—in materials containing free ions, microwaves cause these ions to move back and forth, also generating heat. This dual effect results in internal heating of the material.) and subsequent pressure gradient formation (Wray and Ramaswamy 2015). Various factors influence the efficacy of the treatment; they are (a) sample type and composition, (b) density of the material (available material readily interacts with microwaves), (c) temperature (at higher MW frequencies, loss factor decrease with temperature and vice versa at lower frequencies), (d) frequency of the MW field, and (e) storage time of the product (Sosa-Morales et al. 2010).

MW is rarely used in the convective drying process as a combination; in general, MW energy is combined with the flow of hot air in a continuous or static treatment chamber. Here, the hot air dries the surface or free moisture, and MW energy facilitates the internal or unbound and bound moisture to come toward the surface where the hot air takes care of the rest of the process. Moreover, the MW energy is given to the sample in two basic ways, i.e., (a) right from the start of the drying with hot air (interior is quickly heated and the product starts to turn into puffed form with a more porous structure which facilitates the quicker transfer of water vapor from inner to outer parts of grains) and (b) when the drying rates start to fall (once free or outer moisture is completely dried, application of MW energy facilitates inner moisture to move out toward surface) (Andrés et al. 2004; M. Zhang et al. 2006). Moreover, MW vacuum drying was also proven a useful drying technique by applying a minimum vacuum pressure of 7 kPa (at this pressure, boiling point of water is 39 °C); therefore, lesser heat is required for water removal which preserves volatiles, reduces heat transfer rates to solid material, and enhances moisture removal rate as there is no convection (González-Cavieres et al. 2021; Huang et al. 2021). Wray and Ramaswamy (2015) reviewed and reported a 20–90% reduction in drying times and significantly higher quality product output from MW-assisted convective drying MW vacuum drying and compared with convective drying alone across various fruits and vegetables such as carrots, bananas, cranberries, mushrooms, etc.

Very few studies were carried out on the drying of millets using MW energy; they are as follows. In the study, Radhika et al. (2013) attempted drying two minor millets (finger millet and foxtail millet) in a MW oven during a falling rate period; they found that the drying time decreased with the MW power and the effective diffusivity was also increased with increasing MW power for both millets. In another study, Y. Wang et al. (2022b) compared four different methods such as MW hot air drying (1.25 W/g power intensity at 60 °C at 0.5 m/s air velocity), MW vacuum drying (2.5 W/g power intensity, 5 kPa (kilopascals) pressure and 60 °C), hot air drying (60 °C at 0.5 m/s air velocity), and roasting (150 °C with 5 mm thickness) of foxtail millet. They found that the sample dried 100 and 160 min quicker during MW vacuum drying and MW hot air drying compared to normal hot air drying; at the same time, MW vacuum drying preserved nutritional parameters such as total phenol content (TPC) and yellow pigment content highest among all followed by MW hot air drying.

MW-based drying techniques face significant challenges in achieving temperature uniformity, as the standing wave pattern within the MW cavity can create hot spots, leading to rapid and uneven heating that may cause charring and other damage. To manage these issues, it's crucial to adjust power density by modifying the magnetron's duty cycle, decreasing output, increasing sample or cavity size, and ensuring proper airflow and sample mixing. While microwaves offer potential in drying applications, they are hindered by poor accuracy and repeatability, requiring careful optimization of input parameters. Problems like localized overheating, pressure buildup, and uneven drying, often due to sample heterogeneity, can be mitigated by using lower power density or integrating techniques like vacuum or airflow (Chahbani et al. 2018; Chang et al. 2011; Kaur and Suri 2023; Wray and Ramaswamy 2015).

6.4.2 Dehulling and Milling Efficiency Enhancement

Dehulling, also known as decortication, is one of the important unit operations in millet processing, which removes 12–30% of the outer seed coat/pericarp or hull covering the edible endosperm. There is difficulty in removing the hulls from the millets due to their smaller size and availability in different varieties (Chikkaballapur Krishnappa et al. 2022). In addition, the hulls are tightly attached to the endosperm because when force is applied, it causes damage to the endosperm, leading to an increase in the breakage of the endosperm, resulting in limiting consumer acceptance. For instance, kodo millet holds more hulls, resulting in difficulty in dehulling among the millets. The seed coat is high in fiber, which is indigestible in the digestible tract and contains fewer nutrients (Nayak et al. 2019). In general, major millets are ground directly into flour after cleaning, whereas minor millets are subjected to dehulling before being milled into flour. Millets are traditionally dehulled by using a mortar and pestle at the domestic level, which leads to labor-intensive, uneven hulling due to smaller size and more damage. Notably, polishing of millets after dehulling is essential to remove the bran and germ that contain antinutrients hindering the digestibility and absorption of nutrients (Singh et al. 2024b). These obstacles increase the research attention on employing pretreatments before dehulling and developing specific hullers for particular millet. Developing suitable processing equipment is essential to enhance the effectiveness and efficiency of dehulling and polishing millets. Different methods have been employed with slight modifications from the technologies of paddy processing to increase the efficiency of dehulling of the millets.

Millet dehulling is typically performed like that of pulses, involving two primary stages: loosening the hull through pretreatments and removing the hull using milling machines. Some of the pretreatment methods are used to facilitate the dehulling of millets such as soaking, germination, drying, thermal and nonthermal treatments, enzyme treatment, chemical treatment, etc. (Sunil and Chacko 2019). It also depends on factors like grain harvesting condition, moisture content (Vidhya et al. 2024), drying method, grain quality, machine efficiencies, etc. (Rathore and Singh 2017). Traditionally, dry and wet methods are employed to loosen the hull, after which the treated grains are milled using a wooden pestle in a wooden or stone mortar or under stone runners. Currently, hydrothermal treatments similar to those used in paddy processing are applied to millet processing, resulting in the hardening of the endosperm (Poongodi Vijayakumar et al. 2010). Adding about 10% water to moisten the grain helps remove fibrous bran and, if necessary, separates the germ from the endosperm. Although this method produces slightly moist flour, which can reduce shelf life, it enhances dehulling efficiency and eliminates stickiness. Millet dehulling is commonly carried out using processing equipment such as rice dehullers (rubber roll shellers) and abrasive dehullers. Thermal treatments can also be applied to loosen the husk. Traditional thermal methods, like hot air drying and LSU (Louisiana State University) drying, take longer time and can alter properties such as color and cooking time. Microwaves are considered more suitable pretreatment for dehulling due to their low operating costs, precise control, faster drying rates, uniform heating, shorter processing times, and the absence of chemical residues, making the product safer for consumption (Sunil and Chacko 2019).

For instance, Rao et al. (2021) found that drying of conditioned foxtail millet in a MW increased dehulling yield as power level and exposure time increased. Similarly, Sunil and Chacko conducted a study on the dehulling of foxtail millet using MW treatment at various power levels (900 W, 720 W, and 540 W) and exposure times (120 s, 100 s, and 80 s). They observed that as the MW power and exposure time increased, the dehulling time decreased from 191 to 88 s, with the grain surface temperature rising from 64.68 to 81.66 °C. Initially, the dehulling yield improved with higher MW power and longer exposure, reaching a peak yield of 85.64% at 720 W for 120 s. The lowest dehulling yield, 72.12%, was recorded at 900 W for 100 s. Additionally, an increase in broken content from 0.20% to 19.61% and a reduction in head yield from 99.80% to 85.40% were noted with increased power and exposure time. They predicted the optimal dehulling yield of 91.97% at 732 W and 80 s of exposure. Moreover, the cooking time of the dehulled grain decreased from 544 to 442 s as MW power and exposure time increased (Sunil and Chacko 2019). The heat generated during the drying process denatures proteins, exposing polar side chains and improving dehulling efficiency. This innovative pretreatment method focused on maximizing efficiency, reducing environmental impact, and achieving desired results such as higher yields or lower energy consumption. However, these approaches may require further research and development, and the initial investment costs could be higher than traditional methods.

However, the dehulling and polishing processes cause significant nutrient loss, exclusively protein and amino acids (Nanje Gowda et al. 2022). Another study reported that lipid and zinc content is reduced in pearl and sorghum millet by polishing with 10% abrasion, leading to the removal of germ and loss of half the amount of iron (Hama et al. 2011). Tatah et al. (2021) observed a reduction in nutrient content by 2–7% after dehulling, specifically noting decreases in zinc (Zn) from 38.53 to 35.97 mg/kg, magnesium (Mg) from 311.06 to 306.45 mg/kg, and iron (Fe) from 40.3 to 39.3 mg/kg. These reductions indicate that these micronutrients are primarily located in the outer layers of millet grains. Similarly, Babiker et al. (2018) noted a decrease in protein content after dehulling, which is attributed to the removal of the protein-rich aleurone layer, and fat content decreases due to the removal of the germ. The nutrient losses depend on factors like types and levels of dehulling, pretreatments used, grain size, and varieties. Hence, these factors should also be considered during the dehulling and polishing of millets. The design and processing conditions need to be standardized, which will enhance the utilization and milling efficiency of the millets. Thus, a suitable pretreatment with optimal conditions and a specific dehuller is necessary for a specific millet with higher dehulling efficiency and lesser millet loss.

6.4.3 Starch Modification and Effect on Other Functional Properties

Starch is the major constituent of the carbohydrate content (approximately 70%) in millets; it defines the processing and functional properties of the millet and millet-based products. Moreover, starch plays a crucial role in the final texture, nutritional composition, and other qualities like sensory, digestive, etc. of processed foods (Amarnath et al. 2024; Li et al. 2019a). The functional properties of millet starch, such as swelling capacity and oil/water absorption, are influenced by both the starch itself and the impurities it contains. Polar phospholipids in millet starch form complexes with amylose, reducing flowability and swelling capacity, while fiber reduces oil binding capacity. Native starches have limited uses due to issues like low paste clarity and sensitivity to heat, shear, and pH. However, modifying starches through physical, enzymatic, or chemical methods can enhance their characteristics, making them suitable for broader applications in food and other industries. For example, while native millet and corn starches have similar mechanical properties in tablets, modified starches result in slower dissolution and disintegration, highlighting the benefits of starch modification (Amarnath et al. 2024). Moreover, the starch modification in millets has essential limitations like low paste clarity, retrogradation, and sensitivity to heat, shear, and pH. These modifications improve functional properties like swelling, gelatinization, stability, and oil/water absorption, making the starch more suitable for a wide range of food and industrial applications. Additionally, processing can alter the digestibility of starch and enhance its mechanical properties, expanding its utility in products such as food formulations and pharmaceuticals (Karma et al. 2022; Maniglia et al. 2021). Advanced thermal methods such as MW processing help in doing so. MW processing modifies millet starch by rapidly gelatinizing it, altering its molecular structure, and enhancing its functional properties like water absorption, stability, and emulsification. It also improves thermal stability and reduces retrogradation, making the starch more suitable for diverse food applications while being energy-efficient and time-saving (Balakumaran et al. 2023; Lian et al. 2022; Zheng et al. 2020).

Li et al. (2019a) treated millet starch (30–50% moisture on w.b.) using MW at 700 W for 60 s, and it was observed that a significant decrease in the final viscosity, peak viscosity, breakdown value, and setback value of the starch along with both MW treatment and increase in moisture content. This could be due to the rapid friction and collision between the water and starch granules and the instantaneous heat generated, resulting in hydrolysis and degradation of starch structures. Other than that, the gelatinization temperature range was increased considerably and consistently with a combination of moisture increase and MW treatment (11.48 @ 35% m.c. (moisture content) to 15.94 °C @ 50% m.c.) and the similar trends reflected in transparency and in vitro digestibility of starch. In the other study, 24% (wb) m.c. foxtail millet grains were treated at 720–900 W for 2–6 min of MW treatment; it was reported that the oil absorption capacity of millet flour was decreased with treatment (840 W and 900 W) time (2–6 min) combinations, whereas water absorption capacity was increased at the same time. These changes are mainly due to heat-induced protein denaturation, which causes protein molecules to unfold, exposing their polar side chains. Similarly, the solubility of the flour was enhanced and swelling power was decreased with MW treatment where the latter was in negative correlation concerning the moisture content of the flour (Rao et al. 2021). Similarly, in another study, Li et al. (2019a, b) compared the effects of ultrasound and MW at the same power levels and treatment times on millet starch; they observed that MW effectively decreased the relative crystallinity of starch (100%) compared to ultrasound (23.2%); at the same time, MW decreased swelling power, transparency, and original appearance of starch granules. Therefore, it can be concluded that MW caused more serious damage to the starch particles with far less treatment time.

6.4.4 Enzyme Inactivation

Millets have an active enzyme called lipase in their bran that speeds up the lipid oxidation rate and can induce hydrolytic rancidity, resulting in off-odors and reducing the shelf life of millet and millet flours. MW heating has been used to inactivate lipase in millets and millet grain flours, which extends their shelf life. There is still debate over whether the inactivation of enzymes by MWs is due to thermal effects or the influence of electromagnetic fields. For instance, Rao et al. (2023) examined the effect of MW and hot air-assisted radiofrequency (HARF) technologies on lipase activity (LA) in foxtail millet protein (FMP). The study found that MW treatment resulted in lower LA, with 0.27% after 6 min and 8.17% after 4 min at a power level of 786.80 W. In contrast, HARF treatment showed 10.86% LA at 15% moisture with two passes for 28 min (Rao et al. 2023). A study reported a 92.9% reduction in lipase activity in pearl millet flour with 18% moisture content within 100 s at a constant MW power of 900 W (Yadav et al. 2012). This inactivation is likely due to the high temperature reached during the treatment. Optimal grain moisture content enhances the availability of polar water molecules, which convert MW energy into thermal energy. On the other hand, lesser polar water molecules are available at lower moisture content, while higher moisture content can lead to evaporative cooling, reducing the temperature increase. Furthermore, MW-treated pearl millet flour could be stored for up to 30 days with only a slight increase in free fatty acids (FFA), whereas untreated flour remained acceptable for only 10 days (Mohajerkhorasani et al. 2019). Also, a higher lipase inactivation rate in millet grain was achieved at 900 W within 90 s, with a similar trend observed for FFA levels during the 30-day storage period. Vidhya et al. (2024) optimized the MW treatment conditions for lipase inactivation in pearl millet flour by considering factors such as MW power (600, 800, 1000 W), exposure time (75, 85, 95 s), and moisture content (14%, 18%, 22%). The study identified the optimal conditions as a power level of 1000 W, 18% moisture content, and 93 s of exposure time to inactivate lipase and peroxidase, resulting in 12.018 μM FFA/h/g and 32.843 μM/min/g, respectively (Vidhya et al. 2024). A work suggested that MW at 90 kJ/100 g stabilized wholegrain sorghum flour than 36 kJ/100 g with partial inactivation of the flour lipases and free fatty acids (Adebowale et al. 2020).

6.4.5 Antinutritional Factor Reduction

Millets contain various antinutritional factors (ANFs), such as trypsin inhibitors, phytic acid, saponins, tannins, and oxalates that can inhibit the absorption and bioavailability of nutrients (Joshi et al. 2023). To enhance the safety and quality of edible grains, both conventional and novel processing methods have been employed to remove these ANFs. Among these methods, MW processing is fast, consistent, safe, efficient, and environmentally friendly (Singh et al. 2017). During MW treatment, the power level and exposure time are critical factors in effectively reducing ANFs, and these parameters are carefully evaluated during processing. Heating with microwaves forms insoluble complexes in millets, which help reduce heat-sensitive ANFs. Examples of millets and millet flours containing ANFs include barnyard millet with phytic acid (3.3–3.7 mg/g), finger millet flour with tannic acid (3.5 mg/g) and phytic acid (6.1 mg/g), and pearl millet flour with tannic acid (2.2 mg/g) and phytic acid (9.2 mg/g) (Singh et al. 2024b). Additionally, indigestible fibers and ANFs found in the hulls of millets are reduced during dehulling and polishing.

Water, a major polar compound in millets, plays a crucial role in MW heating. As microwaves cause the electric field to oscillate rapidly, water molecules align themselves within the magnetic field by dipole interaction, producing frictional heat. This heat is transferred to neighboring molecules via conduction and convection (Suhag et al. 2021). MW heating also affects ionic compounds like salts in millets, with charged ions generating heat through kinetic energy. Millet's specific heat and dielectric properties, such as dielectric constant and dielectric loss, influence their thermal behavior during MW processing. Microwaves reduce ANFs due to heat-induced changes, such as peptide bond hydrolysis and disulfide bond breakdown (Rahate et al. 2021). For example, MW treatment can reduce phytic acid and tannins due to their heat-labile nature, while trypsin inhibitors are inactivated by protein denaturation. Saponins, being thermolabile, also decrease during MW processing. Kala and Mohan reported that MW processing produces free radicals that reduce the levels of inositol and inositol phosphates, leading to a decrease in phytic acid concentration in millet (Kala and Mohan 2012). For instance, MW treatment at 900 W for 40–100 s reduced tannin levels in finger millet, pearl millet, and sorghum by 45.18%, 45.53%, and 77.54%, respectively (Singh et al. 2017). It also reduces the phytic acid content in pearl millet, finger millet, and sorghum to 70%, 58.14%, and 11.76% at MW treatment at 900 W, frequency 2450 MHz for 40–100 s (Singh et al. 2017). In addition, MW heating facilitates the dehulling of millet, which helps reduce ANFs by removing the hull.

6.4.6 Roasting/Flavor Development and Enhancing Sensory Properties/Consumer Acceptance

Generally, the traditional or primary methods that are used on millets for cooking are roasting and boiling. Boiling of millet foods are common and popular foods from ages such as porridges, soups, and instant millet rice powders from dried or extruded grains. However, the problem during these processes is the low processing rate and fewer instant products. Using both traditional and modern technology will help in increasing value-added products. For example, instant rice is the best example for the processed dried grains (Y. Wang et al. 2022b). Instant rice is a type of granular α-rice that is gelatinized and dehydrated before being rehydrated in hot water for a few minutes. To make instant rice, freshly cooked rice must be dried to a moisture level of 6–12 g/100 g dry matter. Therefore, drying is a crucial phase in the preparation process.

In a study, the use of MW vacuum drying, MW hot air drying, traditional roasting, and hot air drying was done on foxtail millet at power levels of 5 kPa, 2.5 W/g at 60 °C; 1.25 W/g at 60 °C; 60 °C; and 150 °C, respectively, and studied its sensory, physicochemical properties; it was concluded that MW vacuum drying and MW hot air drying are the best compared to the roasting and hot air drying. Finally, the sensory of the all samples was done, and it was found that MW-treated samples had an 8.3 overall acceptability than the traditional roasting process, i.e., 7.00. Majorly, the eating quality of the porridge for MW-treated samples was 7.4 for vacuum and 7.8 for MW hot air-assisted drying, whereas for traditional roasting and hot air drying alone, it was 3.9 and 7.1, respectively (Y. Wang et al. 2022b).

In a similar study, sorghum grains were treated with a pilot-scale commercial MW oven at a power of 36 and 90 kJ/100 g since it is sensitive to the rancid flavor. After treating the whole grains with MW, the grains were pulverized using a hammer mill and sieved with a 500 μm mesh size. The porridge was made and sensory was done with the selected panelist. It was found that oily, painty, rancid, and fermented aromas were more common in untreated porridge followed by 36 and 90 kJ/100 g treatment. It was also found that lipase inactivation was found more in 90 kJ/100 g made porridge than in 36 kJ/100 g; after a certain period, it was activated again; this is due to the improper deactivation of lipases by MW treatment, which means it needs more power to inactivate the lipase enzyme. On the other hand, it was also meant that porridges from treated flours are less rancid than the untreated flour, i.e., control.

Even after the 6 weeks of storage period, porridges made from these samples didn't show any changes, except for the increase in the painty aroma in 36 kJ/100 g treatment porridges (Adebowale et al. 2020).

Teff grains are one of the minor millets that are rarely utilized and recognized and belong to the origin of Ethiopia; its scientific name is *[Eragrostis tef (Zucc.) Trotter]*. These can be used in many food products such as varieties of beers, pancakes, flatbreads, etc. Subjecting these grains to traditional treatments such as fermentation, cooking, and baking will affect the nutritional and technological properties in both positive and negative ways. However, the impact of thermal treatments on teff is much less. So, in this very particular study, authors treated the grains using MW and oven roasting and studied their changes. Teff grains were treated with conventional roasting at 120 °C for 40 min, whereas MW oven treatment was done at 900 W for 5 min. It shows that a decrease in moisture content up to 2.4% was found in MW-treated grains than for unroasted grains (7.38%). Also, there is increase in oil and carotenoid content in MW-treated samples compared to the untreated grains, i.e., 2.85% and 0.53 μg/g and 2.20% and 0.26 μg/g respectively. This boost in oil and carotenoid content might be attributed to heat changes in cell wall construction, which increase cell porosity and liquefy cellular fluids, hence improving the extractability of these components. Because carotenoids are soluble in oil, the rise in their concentration is likely owing to the enhanced oil's accessibility to lipid-soluble carotenoids. Additionally, heat suppression of carotenoid-degrading enzymes might contribute to elevated carotenoids in roasted grains. Furthermore, roasting can cause carotenoids to attach to proteins during unfolding and refolding, resulting in thermostable complexes.

Also, flavonoid (627.50 mg/100 g), antioxidant (70.54%), and total phenolic content (111.88 mg/ 100 g) of MW-treated grains increased in soluble fractions compared to the untreated (620.50 mg/100 g, 93.06 mg/100 g, 68.75%) and conventional roasting (60.35 mg/100 g and 56.42%), respectively; this might due to the breakdown of cellular constituent and the release of free phenolic and flavonoid compounds. This decrease in contents during conventional roasting could be attributed to the degradation of heat-liable phenolic compounds during a long time (40 min) of roasting treatment. On the other hand, decrease in phenolics and antioxidants could be attributed to the formation of Maillard reaction brown products which can react with Folin-Ciocalteu reagent (Ahmed et al. 2021). The conclusion of this study indicates that teff grains are a suitable source of nutrients and bioactive compounds which can be enhanced via roasting especially MW roasting.

In another similar study, the barley flour was roasted using MW without any burning at the conditions of 900 W at 120 s. Later, the roasted barley flour was subjected to the extraction of phenolics and antioxidants using three different solvents, i.e., methanol, ethanol, and water. It was found that methanol showed the highest extraction compared to the other solvents. Using these extracts, further tests such as total phenolics and antioxidants were done, and it was found that there is a decrease in total phenolic content in barley extract by MW cooking this could be due to the thermal degradation of the heat susceptible phenolic compounds that leads to decrease in the content it also attributes to that the decrease in content might be due to the iteration in the molecular structure of phenolic compounds, which may lead to the decrease in extractability due to the degree of polymerization. Longer heat treatment in the MW oven leads to a greater decrease in the TPC. Initially, the total phenolic content for barley flour using methanol, ethanol, and water without any treatment was 16.92 mg gallic acid equivalent (GAE)/g d.b. (dry basis), 33 mg GAE/g d.b., and 14.1 mg GAE/g d.b., respectively, while after MW roasting of barley, TPC of methanolic, ethanolic, and aqueous solvents decreased to 12.59 mg GAE/g d.b., 14.9 mg GAE/g d.b., and 9.4 mg GAE/g d.b., respectively. The overall phenolic content of the three extracts varies due to the polarity of the solvents. MW roasting resulted in a substantial reduction of 34.39%, 121.477%, and 50% for methanolic, ethanolic, and aqueous extracts, respectively. A lower percentage drop was detected in the case of methanolic solvent, indicating that methanol had a better protective impact on the antioxidant chemicals recovered throughout the procedure (Baba et al. 2016).

Methanol may be a more stabilizing solvent for antioxidants, as evidenced by the methanolic extract's noticeably larger rise in the percentage suppression of DPPH free radicals. Methanolic extracts' correlation coefficients both before and after MW roasting point to a more easily released antioxidant. MW roasting increased the methanolic, ethanolic, and aqueous extracts by an average of 13.24%, 8.14%, and 6%, respectively. After being MW roasted, the IC 50 values for methanol, ethanol, and aqueous solvents decreased from 39.853 μg/mL to 36.182 μg/mL, 40.597 μg/mL to 38.712 μg/mL, and 41.866 μg/mL to 40.691 μg/mL, respectively. Therefore, methanol was the most effective solvent out of the three in terms of percentage suppression of free radicals generated by DPPH, while MW roasting increased the antioxidant potential, possibly because more antioxidants were released. Melanoidins, among other chemicals, are produced during MW roasting and have the potential to provide antioxidant action to the resulting extract (Baba et al. 2016). After being treated for 24 h, the extracts showed antiproliferative activity against Colo-205, T47D, and MCF7. They also dramatically slowed down the growth rate of cancer cells at an initial concentration of 50 μl/mL. However, Colo-205 experienced the most impact. It's interesting to note that native barley extract significantly inhibited Colo-205 growth at 50 μg/ml, demonstrating the strongest antiproliferative activity. This could be brought on by a decrease in phytic acids, saponins, and polyphenols; during the process of MW roasting, new compounds like melanoidins are produced. These compounds have the potential to significantly impact the study's parameters and require additional chemical characterization to better understand the mechanism underlying their antioxidant and antiproliferative activities and to inform applications in the fields of pharmacy, cosmetology, and medicine (Baba et al. 2016).

Another work is to formulate (Muala et al. 2024) a suitable weaning diet made from red millet and yellow maize enhanced with pretreatment *Parkia biglobosa* (PB) and appropriate protein and mineral contents. Dehulled seeds were subjected to fermentation (3–6 days), oven roasting (175 °C for 20 and 40 min), and MW roasting (450 W for 15 and 30 min). 1:1 w/w (weight/weight) as the ratio of maize and millet flour composition was taken as a control sample along with the pretreated PB seeds further divided into 14 formulations; all these trials were analyzed for antioxidant activity, proximate, and total phenolic contents. It was found that there was a significant increase in proximate contents made with PB roasted at 175 °C at 40 min and in MW-treated samples. It was mentioned that protein content for all formulated powders was above 15 g/100 g recommended values for infant foods by WHO (World Health Organization). It also shows that pretreated fermented PB flour exhibited the highest swelling, water-holding capacity, TPC, and antioxidant activity, i.e., 173.43%, 230.85%, 48.32 mg/GAE/g, and 95.78% respectively. All these results show that pretreated PB dehulled seeds can be recommended to enrich the foods made from maize and millet.

6.4.7 Microwave-Assisted Extraction

Sorghum has been recognized as a source of phenolic acids with potential health-promoting antioxidant action; most of the phenolics in sorghum are in bound form, i.e., attached to the cell wall and a small portion is in free form. Extraction of them is critical since they influence grain hardness. Ferulates in plant cell walls are theorized to etherify with previously esterified phenolic acids, creating ester-ether bridges between lignin and polysaccharides. Esterified ferulic acid forms ether bonds with lignin quinone methide intermediates seen in lignified cell walls. This cross-linking is anticipated to strengthen cell walls and impact grain mechanical characteristics like hardness. Since there are no studies on how bound phenolics affect grain hardness, there is a study in which the extraction of bound phenolics was extracted using MW-assisted extraction (Chiremba et al. 2012). Sorghum flour extract was extracted using 5 mL of 2 M NaOH using a polytetrafluoroethylene (PTFE) bomb vessel in a detailed procedure given by Chiremba et al. (2012). The extract was treated at 1400 W for 90 s in a domestic oven and ferulics were analyzed using HPLC-MS/MS compared to conventional alkaline extraction which took 2 h. This method reduced the extraction and showed complete solubility.

In another study, tartary buckwheat seeds (TBS) were treated using MW at different powers and times. It is said that TBS is a potential source of natural nutrient and plant protein ingredients. Initially, seeds were disinfected, washed, MW treated, incubated for germination, and powdered to grains. Protein was isolated using acid alkaline procedure. It was found that the germinated tartary buckwheat protein (TBP) exhibited a substantial rise in random coil contents and a decrease in α-helix content, along with an increase in the content of free sulfhydryl groups. Additionally, the intensity of the UV absorption peak increased ($p < 0.05$). Furthermore, the solubility (24.37%), water-holding capacity (38.95%), emulsifying activity index (17.21%), emulsifying stability index (11.22%), foaming capacity (71.43%), and foaming stability (33.60%) of germinated TBP were all significantly improved by MW pretreatment ($p < 0.05$). Additionally, the in vitro protein digestibility (5.56%) and antioxidant activities (2,2-diphenyl-1-picrylhydrazyl (DPPH) 32.35%, 2,2′-azino-bis (3-ethylbenzothiazoline-6-sulfonic acid (ABTS) 41.95%, and fluorescence recovery after photobleaching (FRAP) 41.46%) of germinated TBP were also enhanced. A MW level of 300 W/50 s produced the greatest outcomes for the parameters under study when compared to other treatment levels (S. Wang et al. 2022a).

In a different study, the extraction of β-glucan from hull-less barley was done using MW-assisted pressurized carbon dioxide/water extraction (MWAPCD/W). It was given that barley has good food processing properties such as emulsifying, thickening, and water holding. Also, they are water soluble and consist of β 1–4,1–3 glycosidic linkages; it also helps in decreasing cholesterol levels and blood pressure, and it was found that the 10.10:1 water/barley flour ratio was finalized along with the time and temperature of 19.92 min and 47.74 °C, respectively, were found to be the optimized conditions for the extraction (Özen et al. 2024). The maximum yield was found to be 62.43% with the help of these optimum conditions with less solvent and time. This might be due to the increase in the amount of starch in the extract with increasing temperature. The beta-glucan extract obtained from (MWAPCD/W) has a greater water-holding capacity (8.05%), which qualifies it for use as an industrial thickener as a functional ingredient. On the other hand, the results of DSC revealed that 122.34 and 145.28 °C results are related to glass transition and melting temperatures for extracted glucans. This means they can add to foods to be heated and able to withstand higher temperature processes such as sterilization, pasteurization, and cooking (Özen et al. 2024) (Table 6.2).

6.5 Optimization of Microwave Heating Parameters

At present, the application of MW treatment in food processing is showing an increasing trend due to their efficient heating with the benefits of quicker start-up time, rapid and selective heating, energy efficiency, space saving, and final products with improved nutritive quality (Sutherland Mishra et al. 2015). MW technology is widely used in millet processing for various applications such as the removal of anti-nutritional factors, increase in nutritional content, enzyme inactivation, cooking, dehusking, drying, roasting, baking, sterilization, and disinfection (Joshi et al. 2023; Sharanagat et al. 2019). As an example, Singh et al. (2017) studied that the MW treatment (900 W for 40–100 s) reduced the tannin content in finger millet, pearl millet, and sorghum by 45.18, 45.53, and 77.54%, respectively (A. Singh et al. 2017).

MW treatment can improve the physicochemical, functional, and antioxidant properties of millets. For instance, Sharanagat et al. (2019) examined the MW roasting of sorghum [*Sorghum bicolor* (L.) Moench] which decreased bulk density from 778 to 546 kg/m^3 and lightness' L*' from 83.1 to 50.6. It also reduced the onset, peak, and final temperatures and peak, pasting, and breakdown viscosity and setback value while increasing total color change and browning index from 14.8 to 48.2 and 17.9 to 64.8, respectively. The MW roasting does not affect the starch gelatinization and functional properties; however, roasting significantly affected TPC (2.64 to 4.67 mg GAE/100 g), total flavonoid content (3.89 to 2.73 mg QE/100 g), and antioxidant properties (92.50% to 83.95% DPPH), respectively. These studies suggested that the MW-treated millet flour with enhanced functional, pasting, and antioxidant properties offers for developing various functional and healthy bakery products, particularly for individuals with celiac disease (Sharanagat et al. 2019). In kodo millet, the MW treatment increased the biochemical and functional properties at high power and longer exposure time, resulting in higher TPC from 2.64 to 4.67 mg GAE/100 g and a reduction in free radical scavenging ability from 91.12% to 87.94% (Gopal and Bhuvana 2021). In addition, Rao et al. (2021) examined the impact of MW treatment on the functional and physical properties of

Table 6.2 Effects of microwave processing in different applications of millet processing

Millet	Treatment methodology	Major outcome(s)	References
Drying/dehydration/pretreatment to drying			
Foxtail millet (13% m.c. on w.b.)	MW drying (350–700 W) Hot air drying (95 °C for 2 h)	MW enhanced rehydration rates and shorter rehydration times Peak viscosities of millet pastes treated with MW were decreased by 50%	Wang et al. (2017)
Foxtail millet	MW vacuum drying (MVD) (2.5 W/g at 5 kPa) MW hot air drying (MHD) (1.25 W/g at 5 kPa and 60 °C) Hot air drying (HAD) (60 °C)	Drying time was reduced by 100 min during MVD and 160 min during MHD Moisture diffusion coefficients were enhanced during MVD and MHD	Wang et al. (2022b)
Dehulling and milling efficiency enhancement			
Sorghum	Microwave treatment followed by milling	TPC content was increased by 6.3% 74% retention of total carotenoids 50.8% net increase of vitamin E 11.01% increase in antioxidant activity	Cardoso et al. (2014)
Starch modification and effect on other functional properties			
Finger millet starch	High-pressure modification of starch at 400 MPa/10 min MW treatment of modified starch at 900 W for 50 s for preparing starch hydrogels Evaluation of the stability of hydrogels during storage (4 °C).	MW-induced synthesis of gels had lesser relaxation time MW-induced gels had greater wall thickness and finer pores MW effectively helped in the gelatinization of starch for hydrogels	Jamna et al. (2024)
Millet starch	The moisture content of starch 30, 35, 40, 45 and 50% w.b. MW treatment at 700 W for 60 s	MW treatment decreased peak viscosity Swelling power was reduced from 16 to 14.8–12.2 (at 30–50% m.c.) Transparency was also reduced by a minimum of 25% In vitro enzyme digestibility was increased from ~20% to 80–100% after treatment	Li et al. (2019a, b)
Proso millet starch	Heat moisture treatment (100 °C for 30 min in a sealed container) Autoclave treatment (121 °C for 30 min at 0.1 MPa) MW treatment (500 W for 10 min)	Water-holding capacity was increased during MW treatment by 65% compared to control Autoclave and MW treatment were more effective in reducing the structural crystallization MW treatment induced a thinner stromal wall, and the network structure was neater and more regular	Zheng et al. (2020)

(continued)

Table 6.2 (continued)

Millet	Treatment methodology	Major outcome(s)	References
White finger millet starch	MW treatment (510 W for 10 min)	Water activity was reduced significantly Slight reduction in oil absorption capacity and a significant increase in the water absorption index Least gelation concentration was up by 3%	Balakumaran et al. (2023)
Enzyme inactivation and antinutritional factor reduction			
Barnyard millet Proso millet Foxtail millet	MW treatment (700 W for 3 min)	44.4 and 17.56% increase of TPC in foxtail and proso millets after MW treatment Slight decrease of flavonoids with treatment Significant reduction in the activity of α-amylase and α-glucosidase	Pradeep and Sreerama (2015)
Pearl millet flour	MW treatment (900 W for 40 to 100 s)	92.9% of lipase activity was reduced after 100 s of treatment at 18% m.c. The shelf life was extended by 30 days	Yadav et al. (2012)
Other applications			
Kodo millet flour	Microwave roasting and pan roasting @ 195 °C	MW power: 360, 720, and 900 W Treatment time: 3 and 5 min Significant improvement in water absorption capacity of flour Significant reduction of oil absorption capacity of flour MW roasted flour had significantly higher TPC than pan-roasted one	Singh et al. (2024b)

foxtail millet flour, observing a decrease in oil absorption capacity (OAC) and an increase in water absorption capacity (WAC). With increasing MW power and exposure time, the swelling power of the millet flour decreased (Rao et al. 2021). MW treatment is also applied to stabilize sorghum flour and significantly reduce the acidity of flour lipid and anisidine levels during storage. The treatment partially inactivated lipases in the flour, slowing the oxidation of free fatty acids. Additionally, the thermal effects of MW heating increased starch damage in the flour, interfering with starch and protein interactions and reducing the peak viscosity of flour paste in treated samples. They also assessed the sensory properties of porridge made from treated flour, finding no negative impact on sensory attributes and significantly lower rancidity during storage compared to control samples (Adebowale et al. 2020). Likewise, Zhi et al. also reported a decrease in peak viscosity with increased MW treatment time and initial moisture content of millet kernels (Zhi et al. 2022). Yadav et al. found that MW heating of pearl millet grains significantly reduced lipase activity. They also conducted a sensory evaluation of flour samples, showing that MW-treated flour could be stored at room temperature for up to 30 days, whereas control flour developed an unpleasant off-odor and bitter taste by the tenth day (S. Yadav et al. 2021). Formation of hot spots is a major drawback of MW pretreatment due to nonuniform temperature distribution, which can reduce grain germination rates (I. Das et al. 2013).

Apart from that, the effect of MW pretreatment on foxtail millet dehusking using power levels (900 W, 720 W, and 540 W) and exposure time (120 s, 100 s, and 80 s) was analyzed, and it was found that increased grain surface temperature (64.68–81.66 °C) reduced dehusking time (191–88 s). It is indicated that higher power levels and exposure times increased broken content (0.20–19.61%) and reduced head yield (99.80–85.40%) and cooking time (544–442 s). At the same time, the MW treatment does not affect the millet's color and suggested that the 732 W for 80 s is more suitable for dehusking of foxtail millet with a maximum yield of 91.97% (Sunil and Chacko 2019).

Some studies showed the MW treatment is used to enhance the yield of millet starch as well as the obtained starch quality. Kanagaraj et al. (2019) worked on increasing the yield of resistant starch (RS) in both selected barnyard millet and rice by comparing the dry heat methods of pan roasting, fluidized bed drying, tray drying, and convective MW heating. Among the methods, the convective MW heating provided a higher yield of RS in both barnyard millet (15.08 ± 0.20 g/100 g at 170 °C, 20 min) and rice (26.25 ± 0.14 g/100 g at 180 °C, 10 min). The MW treatment is also combined with other treatments to enhance millet and millet-based products' nutritional and functional properties. For instance, the MW treatment combines with hydrothermal treatment and modifies white finger millet starch. The hybrid modification methods significantly alter the properties of finger millet starch, such as color, gelation, viscosity, and crystallinity, resulting in extending their applications in the food industry (Balakumaran et al. 2023). Furthermore, Rao et al. (2023) found that combining MW treatment with hot air-assisted radiofrequency (HARF) technologies reduced residual relative lipase activity (LA) to 10.86% at 15% moisture (w.b.) after a 28-min treatment in HARF. In contrast, MW heating alone resulted in lower LA at 786.80 W, benefiting 0.27% in 6 min and 8.17% in 4 min. The peak viscosities of MW and HARF-treated flour decreased and increased, respectively, for emulsifying and oil/water absorption properties, while foaming properties and protein purity showed the opposite trend compared to untreated samples. These changes may be due to decreased relative crystallinity and modifications in secondary structural elements in millet flour (Rao et al. 2023).

The MW treatment is extensively used as a pretreatment for making various millet-based products (particularly puffed and extruded products) to increase palatability and sensory characteristics. For instance, the optimized formulations of composite flour from sorghum (31.96%), finger millet (13.04%), and gluten (3.40%) are hydrated with water, and the dough is treated in the MW for 1.5 min to improve texture and cooking properties followed by extrusion. MW-treated pasta's in vitro protein digestibility (76.51%) was significantly altered as compared to control pasta (without MW treatment) (89.69%) (Kamble et al. 2019). Sutherland Mishra et al. (2015) conducted a study on optimizing the pretreatments (moisture content,

12–20%; salt addition, 0–2%; oil, 0–10% w/w) before popping and MW process parameters (power density, 9–18 W/g and residence time, 100–180 s) for producing higher-quality pop sorghum using a domestic MW oven based on popping yield, volume expansion ratio, and sensory score. The study found that the optimized condition for sorghum popping is moisture (16.62% (w.b.)), salt (0.55%), and oil (10%) with an 82.228% popping yield, 14.564 volume expansion, and 8.495 overall acceptability. They found that MW power density and residence time required for the production of superior quality pop sorghum are 18 W/g and 140 s (Sutherland Mishra et al. 2015).

The pasta obtained from finger millet flour, soy flour, and potato mash is developed using a cold extruder followed by MW puffing using an MW oven. The process parameters such as convective heating temperature (CT), convective heating time (Ct), MW power, % of 1350 W (MP), and MW puffing time (Mt) based on the product moisture content (MC), expansion ratio (ER), hardness (HD), crispness (CSP), color (L-value), and color difference (ΔE) are optimized using RSM (central composite rotatable design). The study reported that the optimal conditions of cold extrudate could be 200 °C and 300-s convective heating and time followed by MW puffing with 90% of 1350 W for 90 s. MC, ER, HD, CSP, L-value, and ΔE value of MW puffed product are found to be 0.088 kg/kg DM. 2.58,1424.93 g, 20.21 + ve peaks, 49.93, and 16.98 respectively (Dhurve et al. 2015). Similarly, Pawar et al. (2015) developed a puffed product from composite millet flour-based RTE snack using MW puffing optimal conditions of convective heating at 196 °C for 132 s followed by MW power of 520 W for 173 s with 0.0304 kg/kg dm of average moisture content, 1315.4 g of hardness, 15.125 (+ve peaks) of crispness,1.96 of expansion ratio, and 43.97 of color (L-value) (Pawar et al. 2015). MW puffing is used to make novel ready-to-eat (RTE) oil-free snack products by combining barnyard, finger millet, and tubers like potato and sweet potato. The process parameters (heating temperature and time) of MW puffing are optimized RSM. The study optimized the MW puffing process parameters: convective heating temperature and time is 220 °C and 274 s, 80% of a total power of 1350 W, and MW heating time 60 s followed by MW oven toasting at 125 °C for 12.5 min. The developed product can be stored in 150-gauge metalized polyester for 6 months at 30 °C and 35% RH (Dhumal et al. 2014).

These studies explained that the application of MW treatment in millet processing offers numerous benefits, including enhanced nutritional quality, improved functional properties, and efficient processing methods. These advancements promote the development of healthier bakery and snack products, particularly beneficial for individuals with dietary limits like celiac disease. However, the existing research has not studied the effect of MW treatment on all the millet quality. Hence, more research is required on optimizing MW treatment parameters for specific millet varieties to enhance nutritional content and functional properties. Also, the effect of hybrid technologies of MW with other thermal and nonthermal methods needs to be studied to improve processing efficiency with product innovations. The drawback of nonuniform heating and maintaining sensory quality over extended storage periods will be addressed to scale up the MW treatment in millet processing.

6.6 Challenges, Future Prospects, and Research Directions in Microwave Processing

Despite the numerous advantages of microwave processing, there are a few problems in the microwave processing of pseudocereals, which need to be resolved for full exploitation. Among many, one of the major concerns is that of achieving uniform heating, since on occasion, microwave energy can lead to a distributed temperature inhomogeneity among grains. This variability can negatively affect product quality, such as influencing fundamental properties like texture, nutrient retention, and functional aspects. Also, microwave processing, though very effective at the small-scale level, faces significant scaling-up problems for industrial applications. Maintaining consistent quality and efficiency on a larger scale involves close attention to energy consumption and cost-effectiveness, which is considered fairly challenging. Other concerns encompass the possible devastation of sensitive nutrients, including some vitamins and bioactive compounds, by high temperatures and a very rapid rate of heating during microwave treatment. This may affect the nutritional quality of the final product. In addition, consumer acceptance of microwave-treated foods can be slow due to safety and quality considerations, which will need education among consumers. Ultimately, the relatively scant research on microwave treatment of pseudocereals compared to other types of grains creates substantial knowledge gaps in understanding its overall impact on the functionality and nutritional aspects of such products, thus limiting the potential for optimization and more general use of this technology.

Notwithstanding these challenges, microwave processing holds tremendous potential for novel applications in the future, especially with concerted research efforts along with technological development. Optimization of the processing variables such as power input, duration of processing, and moisture content can result in improved uniform heating and enhancement of the quality of pseudocereal products. Combining microwave processing with complementary technologies, such as infrared heating or extrusion, may further improve nutrient retention and textural properties and also extend the shelf life of products. Special microwave apparatus, designed specifically for pseudocereals, could overcome the problems of uneven heating and scalability, thus allowing industrial applications. Moreover, microwave treatment may improve the nutritional bioavailability and functional properties of these pseudocereals, thus opening their use to the creation of new functional foods, including gluten-free or low-glycemic ones that can meet specific health needs. Investigating methods to enhance the energy efficiency and sustainability of microwave processing, potentially through the incorporation of renewable energy resources, is gaining prominence due to escalating environmental apprehensions. Furthermore, performing comprehensive analyses on the nutritional effects, preservation of bioactive compounds, and consumer attitudes toward microwave-processed pseudocereals will be crucial for the progression of this technology and for fostering consumer confidence. This will surely expand the market for these healthy grains when the innovation of new value-added products from pseudocereals-snacks, convenience meals, and nutritionally enhanced foods are matched with long-term storage capabilities that improve their availability and enhance benefits for more people.

References

Abdel-Hay MM (2022) Principles of microwave heating for the food industry. In: Emerging thermal processes in the food industry: unit operations and processing equipment in the food industry. Elsevier Inc. https://doi.org/10.1016/B978-0-12-822107-5.00006-4

Adebowale OJ, Taylor JRN, de Kock HL (2020) Stabilization of wholegrain sorghum flour and consequent potential improvement of food product sensory quality by microwave treatment of the kernels. LWT 132:109827

Adedeji AA, Gachovska TK, Ngadi MO, Raghavan GSV (2008) Effect of pretreatments on drying characteristics of okra. Dry Technol 26(10):1251–1256. https://doi.org/10.1080/07373930802307209

Ahmed IAM, Uslu N, Al Juhaimi F, Özcan MM, Osman MA, Alqah HAS, Babiker EE, Ghafoor K (2021) Effect of roasting treatments on total phenol, antioxidant activity, fatty acid compositions, and phenolic compounds of teff grains. Cereal Chem 98(5):1027–1037. https://doi.org/10.1002/cche.10442

Amarnath MS, Muhammed A, Antony AK, Yashini M, Sherin AJ, Sunil CK (2024) Millet starch modification: effect of thermal and non-thermal methods on its properties and their application – a review. Food Human 3(July):100356. https://doi.org/10.1016/j.foohum.2024.100356

Andrés A, Bilbao C, Fito P (2004) Drying kinetics of apple cylinders under combined hot air-microwave dehydration. J Food Eng 63(1):71–78. https://doi.org/10.1016/S0260-8774(03)00284-X

Baba WN, Rashid I, Shah A, Ahmad M, Gani A, Masoodi FA, Wani IA, Wani SM (2016) Effect of microwave roasting on antioxidant and anticancerous activities of barley flour. J Saudi Soc Agric Sci 15(1):12–19. https://doi.org/10.1016/j.jssas.2014.06.003

Babiker E, Abdelseed B, Hassan H, Adiamo O (2018) Effect of decortication methods on the chemical composition, antinutrients, Ca, P and Fe contents of two pearl millet cultivars during storage. World J Sci Technol Sustain Dev 15(3):278–286. https://doi.org/10.1108/wjstsd-01-2018-0005

Balakumaran M, Gokul Nath K, Giridharan B, Dhinesh K, Dharunbalaji AKK, Malini B, Sunil CKK (2023) White finger millet starch: hydrothermal and microwave modification and its characterisation. Int J Biol Macromol 242(March):124619. https://doi.org/10.1016/j.ijbiomac.2023.124619

Banerjee P, Maitra S (2020) The role of small millets as functional food to combat malnutrition in developing countries cropping system view project plant nutrient management by organic inputs view project. Ind J Nat Sci. https://www.researchgate.net/publication/342153966

Bunkar DS (2021) Nutritional, functional role of Kodo millet and its processing: a review. Int J Curr Microbiol Appl Sci 10(01):1972–1985. https://doi.org/10.20546/ijcmas.2021.1001.229

Cardoso L d M, Montini TA, Pinheiro SS, Pinheiro-Sant'Ana HM, Martino HSD, Moreira AVB (2014) Effects of processing with dry heat and wet heat on the antioxidant profile of sorghum. Food Chem 152:210–217. https://doi.org/10.1016/j.foodchem.2013.11.106

Chahbani A, Fakhfakh N, Balti MA, Mabrouk M, El-Hatmi H, Zouari N, Kechaou N (2018) Microwave drying effects on drying kinetics, bioactive compounds and antioxidant activity of green peas (Pisum sativum L.). Food Biosci 25:32–38. https://doi.org/10.1016/j.fbio.2018.07.004

Chang YK, Steel CJ, Clerici MTPS (2011) Use of microwave radiation to process cereal-based products. In: Advances in induction and microwave heating of mineral and organic materials, pp 531–552. http://www.intechopen.com/books/

advances-in-induction-and-microwave-heating-of-mineral-and-organic-materials/the-use-of-microwave-radiation-energy-to-process-cereal-root-and-tuber-based-products

Chikkaballapur Krishnappa S, Saravanan S, Natarajan V (2022) Development and performance evaluation of foxtail millet (Setaria italica L.) dehuller. J Food Process Eng 45(6):e13937. https://doi.org/10.1111/jfpe.13937

Chiremba C, Rooney LW, Beta T (2012) Microwave-assisted extraction of bound phenolic acids in bran and flour fractions from sorghum and maize cultivars varying in hardness. J Agric Food Chem 60(18):4735–4742. https://doi.org/10.1021/jf300279t

Crane CA, Pantoya ML, Weeks BL (2013) Spatial observation and quantification of microwave heating in materials. Rev Sci Instrum 84(8):084705. https://doi.org/10.1063/1.4818139

Dahlberg JA, Wilson JP, Snyder T (2004) Sorghum and pearl millet: health foods and industrial products in developed countries. In: Alternative uses of sorghum and pearl millet in Asia. Proceedings of an Expert Meeting, ICRISAT, Patancheru, Andhra Pradesh, India, 1–4 July, 2003, pp 42–59

Das I, Kumar G, Shah NG (2013) Microwave heating as an alternative quarantine method for disinfestation of stored food grains. Int J Food Sci 2013(1):926468. https://doi.org/10.1155/2013/926468

Das S, Khound R, Santra M, Santra DK (2019) Beyond bird feed: proso millet for human health and environment. Agriculture (Switzerland) 9(3). https://doi.org/10.3390/agriculture9030064

Devi PB, Vijayabharathi R, Sathyabama S, Malleshi NG, Priyadarisini VB (2014) Health benefits of finger millet (Eleusine coracana L.) polyphenols and dietary fiber: a review. J Food Sci Technol 51(6):1021–1040. https://doi.org/10.1007/s13197-011-0584-9

Dhumal CV, Pardeshi IL, Sutar PP, Babar OA (2014) Optimization of process parameters for development of microwave puffed product. J Ready Eat Food 1(3):111–119

Dhurve P, Bakane PH, Agrawal S (2015) Optimization of process parameters of finger millet–soy pasta using response surface methodology. Int J Eng Res Gen Sci 3(2):415–427

González-Cavieres L, Pérez-Won M, Tabilo-Munizaga G, Jara-Quijada E, Díaz-Álvarez R, Lemus-Mondaca R (2021) Advances in vacuum microwave drying (VMD) systems for food products. Trends Food Sci Technol 116:626–638. https://doi.org/10.1016/j.tifs.2021.08.005

Gopal DS, Bhuvana S (2021) Effect of microwave treatment on phytochemical, functional and rheological property of Kodo millet (Paspalum scrobiculatum). Pharma Innov J 10(11):397–401

Guo Q, Sun D-W, Cheng J-H, Han Z (2017) Microwave processing techniques and their recent applications in the food industry. Trends Food Sci Technol 67(2):236–247. https://doi.org/10.1016/j.tifs.2017.07.007

Hama F, Icard-Vernière C, Guyot JP, Picq C, Diawara B, Mouquet-Rivier C (2011) Changes in micro- and macronutrient composition of pearl millet and white sorghum during in field versus laboratory decortication. J Cereal Sci 54(3):425–433. https://doi.org/10.1016/j.jcs.2011.08.007

Hariprasanna (2016) Foxtail millet—nutritional importance and cultivation aspects. Indian Farm 65(July):25–29

Hong X, Wang J (2015) Use of electronic nose and tongue to track freshness of cherry tomatoes squeezed for juice consumption: comparison of different sensor fusion approaches. Food Bioprocess Technol 8(1):158–170. https://doi.org/10.1007/s11947-014-1390-y

Huang LL, Lian MM, Duan X, Li B, Yang SZ (2021) Studies on the quality and moisture distribution of kiwifruit dried by freeze drying combined with microwave vacuum drying. J Food Process Eng 44(1). https://doi.org/10.1111/jfpe.13581

India Science Technology and Innovation Portal (2023) Millets: the future food. https://www.indiascienceandtechnology.gov.in/listingpage/millets-future-food#:~:text=In%20April%202018%2C%20Millets%20were,extensive%20promotion%20and%20demand%20generation

Jamna A, Sunil CK, Venkatachalapathy N (2024) Effect of microwave irradiation and conventional heating on high pressure modified white finger millet starch hydrogel properties. Food Human 3(July):100380. https://doi.org/10.1016/j.foohum.2024.100380

Joshi TJ, Singh SM, Rao PS (2023) Novel thermal and non-thermal millet processing technologies: advances and research trends. Eur Food Res Technol 249(5):1149–1160. https://doi.org/10.1007/s00217-023-04227-8

Kala BK, Mohan VR (2012) Effect of microwave treatment on the antinutritional factors of two accessions of velvet bean, Mucuna pruriens (L.) DC. Var. Utilis (wall. Ex wight) bak. Ex burck. Int Food Res J 19(3):961–969

Kamble DB, Singh R, Rani S, Kaur BP, Upadhyay A, Kumar N (2019) Optimization and characterization of antioxidant potential, in vitro protein digestion and structural attributes of microwave processed multigrain pasta. J Food Proces Preserv 43(10):e14125. https://doi.org/10.1111/jfpp.14125

Kanagaraj SP, Ponnambalam D, Antony U (2019) Effect of dry heat treatment on the development of resistant starch in rice (Oryza sativa) and barnyard millet (Echinochloa furmantacea). J Food Process Preserv 43(7):1–7. https://doi.org/10.1111/jfpp.13965

Karma V, Gupta AD, Yadav DK, Singh AA, Verma M, Singh H (2022) Recent developments in starch modification by organic acids: a review. Starch/Staerke 74(9–10). https://doi.org/10.1002/STAR.202200025

Kaur H, Sharma S (2020) An overview of barnyard millet (Echinochloa frumentacea). J Pharmacogn Phytochem 9(4):819–822

Kaur S, Suri K (2023) A review on the impact of microwave processing on physicochemical properties of different cereal grains and flours. J Food Agric Res 03(01):45–83

Kheto A, Joseph D, Islam M, Dhua S, Das R, Kumar Y, Vashishth R, Sharanagat VS, Kumar K, Nema PK (2022) Microwave roasting induced structural, morphological, antioxidant, and functional attributes of Quinoa (Chenopodium quinoa Willd). J Food Process Preserv 46(5). https://doi.org/10.1111/jfpp.16595

Li Y, Hu A, Wang X, Zheng J (2019a) Physicochemical and in vitro digestion of millet starch: effect of moisture content in microwave. Int J Biol Macromol 134:308–315. https://doi.org/10.1016/j.ijbiomac.2019.05.046

Li Y, Hu A, Zheng J, Wang X (2019b) Comparative studies on structure and physiochemical changes of millet starch under microwave and ultrasound at the same power. Int J Biol Macromol 141:76–84. https://doi.org/10.1016/j.ijbiomac.2019.08.218

Lian F, Sun DW, Cheng JH, Ma J (2022) Improving modification of structures and functionalities of food macromolecules by novel thermal technologies. Trends Food Sci Technol 129:327–338. https://doi.org/10.1016/j.tifs.2022.10.001

Maniglia BC, Castanha N, Le-Bail P, Le-Bail A, Augusto PED (2021) Starch modification through environmentally friendly alternatives: a review. Crit Rev Food Sci Nutr 61(15):2482–2505. https://doi.org/10.1080/10408398.2020.1778633

Malik, S (2015) Pearl Millet-Nutritional Value and Medicinal Uses! Ijariie-Issn(O) 1(3); 414–418

Mohajerkhorasani S, Alami M, Kashaninejad M, Shahiritabarestani H (2019) Increasing the shelf life of millet flour by using heat-moisture and microwave treatments. J Food Sci Technol (Iran) 16(86):83–93

Mounika A, Shanmugam A, Rao TJ, Sinija VR, Rawson A, Baskaran N (2024) Evaluation of quality characteristics of ultrasound–treated browntop millet grains. Chem Eng Process Process Intensif 203:109919. https://doi.org/10.1016/j.cep.2024.109919

Muala WCB, Charnelle TK, Fabrice TD, Bernard T, Ghislain MN, Eric Serge N (2024) Formulation of weaning food from yellow maize (Zea mays L.). and red millet (Eleusine coracana L.), enriched with pretreated African locust beans (Parkia biglobosa Jacq.) flour. J Agric Food Res 16. https://doi.org/10.1016/j.jafr.2024.101080

Muthukumarappan K, Swamy GJ (2019) Microwave processing of foods. In: Handbook of farm, dairy and food machinery engineering, pp 417–438. https://doi.org/10.1016/B978-0-12-814803-7.00016-6

Naik M, Modupalli N, Sunil CK, Rawson A, Natarajan V (2022) Major millet processing. In: Handbook of millets – processing, quality, and nutrition status. Springer, Singapore, pp 63–80. https://doi.org/10.1007/978-981-16-7224-8_4

Nanje Gowda NA, Siliveru K, Vara Prasad PV, Bhatt Y, Netravati BP, Gurikar C (2022) Modern processing of Indian millets: a perspective on changes in nutritional properties. Foods 11(4):1–18. https://doi.org/10.3390/foods11040499

Nayak P, Gupta AK, Jain P, Pandey S (2019) Effect of moisture and machine parameters on dehusking efficiency of Kodo Millet. Int J Curr Microbiol Appl Sci 8(02):1792–1804. https://doi.org/10.20546/ijcmas.2019.802.211

Nguyen VP, Le Trung H, Nguyen TH, Hoang D, Tran TH (2021) Advancement of microwave-assisted biosynthesis for preparing au nanoparticles using Ganoderma lucidum extract and evaluation of their catalytic reduction of 4-nitrophenol. ACS Omega 6(47):32198–32207. https://doi.org/10.1021/acsomega.1c05033

Nithiyanantham S, Kalaiselvi P, Mahomoodally MF, Zengin G, Abirami A, Srinivasan G (2019) Nutritional and functional roles of millets—a review. J Food Biochem 43(7):e12859. https://doi.org/10.1111/JFBC.12859

Orsat V, Raghavan GSV, Krishnaswamy K (2017) Microwave technology for food processing: an overview of current and future applications. In: The microwave processing of foods, 2nd edn. Elsevier. https://doi.org/10.1016/B978-0-08-100528-6.00005-X

Özen S, Ünlü A, Özbek HN, Göğüş F (2024) β-Glucan extraction from hull-less barley by a novel approach: microwave-assisted pressurized CO2/H2O. Food Bioprocess Technol 17:4781. https://doi.org/10.1007/s11947-024-03420-1

Paschapur AU, Joshi D, Mishra KK, Kant L, Kumar V, Kumar A (2021) Millets for life: a brief introduction. In: Millets and millet technology, pp 1–32. https://doi.org/10.1007/978-981-16-0676-2_1

Pawar SG, Pardeshi IL, Mohril MP, Borkar PA, Khodke SU (2015) Optimization of microwave puffing process for preparing ready-to-eat snack using composite millet flours. J Ready Eat Food 2(4):101–111

Poongodi Vijayakumar T, Mohankumar JB, Jaganmohan R, Janaki A (2010) Effect of pre-milling treatments on characteristics of barnyard millet (Echinochloa Colona). Int J Curr Res 2:5–11

Pradeep PM, Sreerama YN (2015) Impact of processing on the phenolic profiles of small millets: evaluation of their antioxidant and enzyme inhibitory properties associated with hyperglycemia. Food Chem 169:455–463. https://doi.org/10.1016/j.foodchem.2014.08.010

Radhika GB, Satyanarayana SV, Rao DG (2013) An investigation on drying of millets in A microwave oven. Int J Emerg Technol Adv Eng 3(6):583–589

Rahate KA, Madhumita M, Prabhakar PK (2021) Nutritional composition, anti-nutritional factors, pretreatments-cum-processing impact and food formulation potential of faba bean (Vicia faba L.): a comprehensive review. In: LWT, vol 138. Elsevier, p 110796. https://doi.org/10.1016/j.lwt.2020.110796

Rao MV, Akhil K, Sunil C, Venkatachalapathy N, Jaganmohan R (2021) Effect of microwave treatment on physical and functional properties of foxtail millet flour. Int J Chem Stud 9(1):2762–2767. https://doi.org/10.22271/chemi.2021.v9.i1am.11641

Rao MV, Sunil CK, Venkatachalapathy N (2023) Effect of microwave and hot air radiofrequency treatments on physicochemical and functional properties of foxtail millet flour and its protein isolate. J Cereal Sci 114:103774. https://doi.org/10.1016/j.jcs.2023.103774

Rathore S, Singh K (2017) A critical optimization study on hydrothermal treatment for decortication of pearl millet to improve its consumption efficiency. J Food Meas Charact 11(3):1501–1515. https://doi.org/10.1007/s11694-017-9529-y

Shaheen MS, El-Massry KF, El-Ghorab AH, Anjum FM (2012) Microwave applications in thermal food processing. In: The development and application of microwave heating, vol 11. https://doi.org/10.5772/48716

Sharanagat VS, Suhag R, Anand P, Deswal G, Kumar R, Chaudhary A, Singh L, Kushwah OS, Mani S, Kumar Y (2019) Physico-functional, thermo-pasting and antioxidant properties of microwave roasted sorghum [Sorghum bicolor (L.) Moench]. J Cereal Sci 85:111–119. https://doi.org/10.1016/j.jcs.2018.11.013

Shirsat B (2020) A review of finger millet (Eleusine coracana L.) on nutrition, health benefits and value addition. Int J Chem Stud 8(2):2764–2769. https://doi.org/10.22271/chemi.2020.v8.i2ap.12405

Singh A, Gupta S, Kaur R, Gupta HR (2017) Process optimization for anti-nutrient minimization of millets. Asian J Dairy Food Res 36(4):322–326. https://doi.org/10.18805/ajdfr.dr-1215

Singh DK, Kurichh R, Bist Y, Kheto A, Kumar Y, Sharma R, Shikha D, Saxena DC (2024a) Effect of microwave roasting on the physicochemical, functional, rheological, and antioxidant properties of Kodo millet flour. J Food Process Preserv 2024(1). https://doi.org/10.1155/2024/6861190

Singh SM, Joshi TJ, Rao PS (2024b) Technological advancements in millet dehulling and polishing process: an insight into pretreatment methods, machineries and impact on nutritional quality. Grain Oil Sci Technol 7:186. https://doi.org/10.1016/j.gaost.2024.05.007

Singh SP (2018) An analysis of dielectric parameters and penetration depth of tomato sauces. J Food Process Technol 09(07):7–10. https://doi.org/10.4172/2157-7110.1000742

Singh S, Suri S, Singh R (2022) Potential and unrealized future possibilities of browntop millet in the food sector. Front Sustain Food Syst 6. https://doi.org/10.3389/fsufs.2022.974126

Sosa-Morales ME, Valerio-Junco L, López-Malo A, García HS (2010) Dielectric properties of foods: reported data in the 21st century and their potential applications. LWT 43(8):1169–1179. https://doi.org/10.1016/j.lwt.2010.03.017

Suhag R, Dhiman A, Deswal G, Thakur D, Sharanagat VS, Kumar K, Kumar V (2021) Microwave processing: a way to reduce the anti-nutritional factors (ANFs) in food grains. LWT Food Sci Technol 150(November 2020):111960. https://doi.org/10.1016/j.lwt.2021.111960

Sun J, Wang W, Yue Q (2016a) Review on microwave-matter interaction fundamentals and efficient microwave-associated heating strategies. Materials 9(4). https://doi.org/10.3390/ma9040231

Sun J, Wang W, Yue Q, Ma C, Zhang J, Zhao X, Song Z (2016b) Review on microwave-metal discharges and their applications in energy and industrial processes. Appl Energy 175:141–157. https://doi.org/10.1016/j.apenergy.2016.04.091

Sunil CK, Chacko N (2019) Microwave assisted de-husking of foxtail millet. J Agric Eng 55(4):21–33

Sutherland Mishra G, Joshi DC, Mohapatra D (2015) Optimization of pretreatments and process parameters for sorghum popping in microwave oven using response surface methodology. J Food Sci Technol 52:7839–7849. https://doi.org/10.1007/s13197-015-1898-9

Tao Y, Yan B, Fan D, Zhang N, Ma S, Wang L, Wu Y, Wang M, Zhao J, Zhang H (2020) Structural changes of starch subjected to microwave heating: a review from the perspective of dielectric properties. Trends Food Sci Technol 99(January):593–607. https://doi.org/10.1016/j.tifs.2020.02.020

Tatah SV, Ikwebe J, Mayel MH, Tatah SV, Mamman E, Sabo DJ (2021) Effects of dehulling on the levels of micronutrients in maize, milled and sorghum grains. FUW Trends Sci Technol J 6(1):220–222. http://www.iita.org/crops/maize/

Tiwari S, Caiola A, Bai X, Lalsare A, Hu J (2020) Microwave plasma-enhanced and microwave heated chemical reactions. Plasma Chem Plasma Process 40(1):1. https://doi.org/10.1007/s11090-019-10040-7

Tripathi T, Vyas S (2023) From ancient grains to modern solutions: a history of millets and their significance in agriculture and food security. Int J Home Sci 9(2):72–78

United Nations Population Division (UNPD) (2019) Our growing population. World Population Dashboard, p 1

Varith J, Dijkanarukkul P, Achariyaviriya A, Achariyaviriya S (2007) Combined microwave-hot air drying of peeled longan. J Food Eng 81(2):459–468. https://doi.org/10.1016/j.jfoodeng.2006.11.023

Vidhya V, Sangani V, Davara P, Diwate P, Chocha D (2024) Optimization of microwave treatment for storage life enhancement of pearl millet flour. Int J Adv Biochem Res 8(8):96–103. https://doi.org/10.33545/26174693.2024.v8.i8b.1706

Wang R, Chen C, Guo S (2017) Effects of drying methods on starch crystallinity of gelatinized foxtail millet (α-millet) and its eating quality. J Food Eng 207:81–89. https://doi.org/10.1016/j.jfoodeng.2017.03.018

Wang S (2009) Emerging technologies for food processing. In: Sun D-W (ed) Trends in food science, vol 20. Elsevier, San Diego, CA, pp 105–106. 2005, 792 pages, Price GBP 100.00, ISBN: 0-12-676757-2, 978-0-12-676757-5

Wang S, Xu X, Wang S, Wang J, Peng W (2022a) Effects of microwave treatment on structure, functional properties and antioxidant activities of germinated Tartary buckwheat protein. Foods 11(10). https://doi.org/10.3390/foods11101373

Wang Y, Zhao H, Song X, Zhang W, Yang F (2022b) Drying kinetics, physicochemical properties and sensory quality of the instant foxtail millet as affected by drying methods. Polish J Food Nutr Sci 72(1):69–78. https://doi.org/10.31883/pjfns/146175

Wei W, Shao Z, Qiao R, Chen W, Zhou H, Yuan Y (2021) Recent development of microwave applications for concrete treatment. Constr Build Mater 269(October):121224. https://doi.org/10.1016/j.conbuildmat.2020.121224

Wray D, Ramaswamy HS (2015) Novel concepts in microwave drying of foods. Dry Technol 33(7):769–783. https://doi.org/10.1080/07373937.2014.985793

Xiong Y, Zhang P, Warner RD, Fang Z (2019) Sorghum grain: from genotype, nutrition, and phenolic profile to its health benefits and food applications. Compr Rev Food Sci Food Saf 18(6):2025–2046. https://doi.org/10.1111/1541-4337.12506

Yadav DN, Anand T, Kaur J, Singh AK (2012) Improved storage stability of pearl millet flour through microwave treatment. Agric Res 1(4):399–404. https://doi.org/10.1007/s40003-012-0040-8

Yadav S, Mishra S, Pradhan RC (2021) Ultrasound-assisted hydration of finger millet (Eleusine Coracana) and its effects on starch isolates and antinutrients. Ultrason Sonochem 73:105542. https://doi.org/10.1016/j.ultsonch.2021.105542

Zamri AA, Ong MY, Nomanbhay S, Show PL (2021) Microwave plasma technology for sustainable energy production and the electromagnetic interaction within the plasma system: a review. Environ Res 197(February):111204. https://doi.org/10.1016/j.envres.2021.111204

Zhang M, Tang J, Mujumdar AS, Wang S (2006) Trends in microwave-related drying of fruits and vegetables. Trends Food Sci Technol 17(10):524–534. https://doi.org/10.1016/j.tifs.2006.04.011

Zheng MZ, Xiao Y, Yang S, Liu HM, Liu MH, Yaqoob S, Xu XY, Liu JS (2020) Effects of heat–moisture, autoclaving, and microwave treatments on physicochemical properties of proso millet starch. Food Sci Nutr 8(2):735–743. https://doi.org/10.1002/fsn3.1295

Zhi W, Zhou Y, Wang R, Wang M, Wang W, Hu A, Zheng J (2022) Effect of microwave treatment on the properties of starch in millet kernels. Starch-Stärke 74(9–10):2200063. https://doi.org/10.1002/star.202200063

Waves of Change: Radiofrequency in Millet Processing (Electromagnetic Solutions for Improved Millet Characteristics)

7

K. Sandhya and Venkatachalapathy Natarajan

7.1 Introduction

Ensuring the shelf life of food commodities is crucial for maintaining food security. Traditional thermal processing methods like pasteurization, sterilization, and ultra-high-temperature (UHT) treatment use high heat to eliminate pathogenic microorganisms and extend the safety of food throughout its shelf life. However, these methods can lead to partial destruction of heat-sensitive nutrients, especially with long processing times required for large quantities thereby changing the sensory attributes of the food product (Pathaw et al. 2022). To address these issues, innovative novel thermal processing techniques have been developed that employ EM waves for faster and more efficient treatment. The EM spectrum includes both ionizing and nonionizing radiations. Gamma rays and X-rays are classified as ionizing radiations, while ultraviolet (UV) rays, visible light, infrared (IR) rays, microwaves (MW), and RF waves fall into the nonionizing category. RF waves have longer wavelengths and lower energy than MW, allowing them to penetrate more deeply. They heat food products through dielectric heating (Bhatt et al. 2020).

MW and RF heating are two prominent novel thermal technologies in food processing. They work by targeting the polar molecules in the food, causing them to rotate and generate heat. RF and MW heating are commonly known as dielectric heating and are used to heat dielectric materials, including food products. RF technology generates heat throughout the entire volume of the food, resulting in quicker, more uniform heating within the food matrix with a greater depth of penetration of RF waves (Jiao et al. 2018).

K. Sandhya · V. Natarajan (✉)
Department of Food Process Engineering, National Institute of Food Technology, Entrepreneurship and Management – Thanjavur, Thanjavur, Tamil Nadu, India

Ministry of Food Processing Industries, Government of India, Thanjavur, Tamil Nadu, India
e-mail: venkat@iifpt.edu.in

S. Jaddu, R. Chandra Pradhan (eds.), *Innovative Millet Processing*,
https://doi.org/10.1007/978-3-031-99864-5_7

RF processing in the food sector has gained popularity recently due to its high heating efficiency. Stabilizing millets for extended shelf life is a major challenge in the millet processing industry. Researchers have extensively studied the use of various novel technologies to stabilize millets, and each of the processing methods is limited in its application due to limitations such as ineffectiveness in uniform heating, reduced efficiency, traces of grain damage, higher energy requirements, and long processing times in the case of bulk materials (Rao et al. 2023). As traditional millet processing technologies encounter issues due to inefficient uniform heating, RF heating emerges as a possible option, providing precise, controlled, and quick heat application.

This chapter provides an overview of the basics of RF heating technology, examining how it works and its benefits specifically for millet processing. It also covers how RF processing affects the disinfestation and storage stability of millets, as well as its impact on their nutritional content, functional properties, cooking qualities, and texture. Additionally, the chapter outlines the advantages and limitations of RF processing and discusses potential future developments in this area.

7.2 Fundamentals of RF Processing Technology

RF waves cover a broad range in the EM spectrum, from 3 kHz to 300 MHz. MW and RF frequencies fall within the radar range and can interfere with communication and navigation signals, which is why only specific frequencies are used based on international agreements. Frequencies commonly utilized in various sectors, especially for industrial, scientific, and medical purposes, include 6.78 MHz, 13.56 MHz, 27.12 MHz, and 40.68 MHz (Wang et al. 2021b). Industrial RF applications are categorized into two types: self-oscillator RF generators and 50 Ω RF amplifiers. A detailed explanation of the oscillators used in RF processing is given elsewhere (Wang et al. 2021b). RF heating occurs when EM field energy is converted to thermal energy through dielectric absorption. Since most food products are poor conductors of heat and electricity, dielectric heating is a more effective option. The food is placed between the electrodes transmitting EM energy. Traditional heating methods require more time to heat food from the surface to the interior due to a temperature gradient, resulting in longer overall heating times. In contrast, RF heating using dielectric methods heats the food rapidly without warming the surrounding environment (Jiao et al. 2015). RF processing produces heat throughout the entire volume of the food as the waves penetrate and are absorbed, leading to quick and even heating. This efficiency is due to the mechanisms of dipole rotation and ionic conduction (Fellows 2016).

7.2.1 Dipole Rotation and Ionic Conductivity

Water represents a major portion of the food products. The molecular structure of water consists of a negatively charged oxygen atom with positively charged hydrogen atoms which creates an electronic dipole. With the application of an RF electric field, the water molecules in foods, in some cases, ionic components, such as salt or sugar, reorient themselves due to the change in dipole with the field. With the rapid oscillation in the electric field, from positive to negative and vice versa, for a million times, the dipoles try to reorient themselves with the changing electric field causing molecular vibration which ultimately heats the product (Fellows 2016). Dielectric heating is also energy-efficient, which is crucial from both an industrial and environmental standpoint. In certain instances, combining dielectric heating with traditional methods can reduce processing time and energy consumption while preserving product quality (Demeczky 1985). In certain cases, dielectric sterilization can produce high-quality results because EM waves heat the product 3–5 times faster than traditional sterilization methods (Wang et al. 2003) (Fig. 7.1).

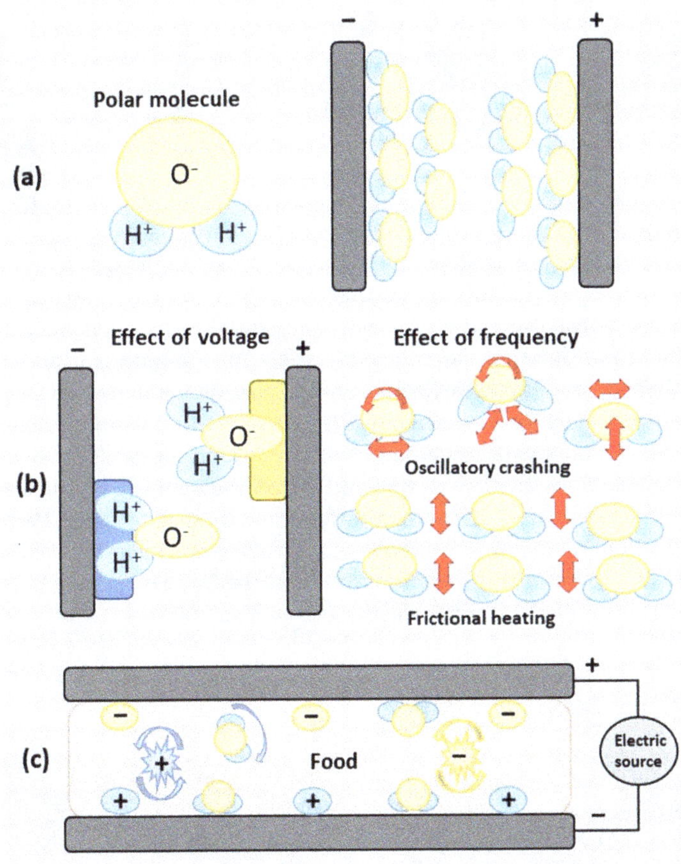

Fig. 7.1 Mechanism of RF heating ((**a**) alignment of polar molecules with electrons, (**b**) effect of voltage and friction on heating, and (**c**) ionic collision causing dielectric heating). (Obtained with permission from Costa and Marra 2024)

In the case of heating using RF, the energy is transferred to the food product through the electric field developed between the electrodes. RF is effective for large quantities of materials with high ionic conductivity, while MW technology is more suitable for smaller quantities of products with a dipolar nature (Orsat and Jusoh 2014). In an RF heating system, a generator creates a fluctuating electric field between two electrodes. The material to be heated is placed between these electrodes, where the alternating energy causes polarization. This process forces the molecules within the material to constantly adjust their alignment to match the shifting electric field. At radio frequencies like 27.12 MHz, the electric field reverses direction 27,120,000 times per second. The heat is produced by the friction resulting from the molecules' rotational movement and displacement of space charge within the material. The heat produced depends on factors such as the frequency, the applied voltage raised to the power of two, the product size, and the material's dielectric loss factor, which indicates how easily can radio waves heat the material. To assess the feasibility of heating a material, its dielectric properties must be studied. The addition of ionic compounds to low-loss material helps in enhancing its effective loss and makes it suitable for dielectric processing. In the case of grains, the addition of salt, in some cases, had significantly affected impedance, with electrode voltage decreasing with time with a proportional increase in salt concentration (Orsat 1999).

When the electric field reverses at such high rates, it causes the charged ions to align with the electric field, where the charged ions move in the direction opposite to their polarity. Due to a change in polarity, a reversal in the acceleration of ions takes place. These strong collisions convert the frictional energy into heat (Dragoev et al. 2024). A critical review by Piyasena et al. (2003) gives a detailed explanation of the influence of different properties that affect the RF heating. Similarly, work done by Maloney and Harrison (2016) provides the interpretation of loss tangent, power dissipation, and the generation of RF energy and its control mechanisms. RF heating works without depending on conductive heating. It produces internal heat by causing the polar molecules to spin and the movement of ions. Dielectric heating, with its superior wave penetration, offers benefits such as rapid, uniform, and volumetric heating, resulting in high energy efficiency (Costa and Marra 2024). Comprehensive details on different RF generation methods and power-heat calculation formulas are covered in other sources (Tang et al. 2004).

7.3 RF Processing of Millets

RF processing is utilized to achieve maximum moisture removal in low-fat baked goods like crackers, cookies, and snacks. When the final 1–2% of moisture is removed using MW or other methods, it often results in surface hardening, crumbling, and breaking. However, RF heating, having a longer wavelength of EM energy than microwave, which heats from the inside out, helps draw moisture from the interior to the outer surface or crust (Orsat and Jusoh 2014). Inside to outward heating equalizes the moisture throughout the product by retarding over-dehydration and overheating effects of the product surface. This method is particularly effective in preserving the texture of millet-based products such as breads, cookies, and biscuits. In some cases, RF drying is also used in the removal of moisture in pasta.

RF drying of high-moisture foods is another key area of interest, where RF is initially used to rapidly raise the sample temperature to the desired drying level. This approach helps avoid the prolonged heating times needed to reach the target drying temperature. In a study by Peng et al. (2019), apple slices were dried using a two-stage process: first, air jet impingement, followed by hot air-assisted RF drying. The study found that 100 min of RF drying with a 12 cm electrode gap and hot air at 60 °C significantly reduced the moisture content to the target level. This method of drying using a hybrid method reduced the processing time significantly and at the same time maintained the product quality. Millet-based food products, such as millet porridge, pasta, or millet-based food analogs, can also benefit from RF technology for effective drying. A major challenge in using RF for high-moisture foods is the development of cold spots caused by uneven heating. This issue can be mitigated by incorporating hot air assistance (Fig. 7.2).

Processing of buckwheat with hot air-assisted RF shows the presence of cracks on the flour due to the release of water. RF treatment of buckwheat at 100 °C for 10 min caused numerous minute pieces to appear and scatter around the clustered particles, indicating reduced compactness of structure and easier to break. The force released by water evaporation during RF heating created cracks between the molecules, leading to these structural changes (Xu et al. 2023). RF heating also results in slight agglomeration of protein particles due to the effects of superheated steam. This agglomeration occurs due to increased cross-linking between denatured protein molecules. The interaction between protein and starch in their native state during heating forms a protein-starch network, which contributes to the agglomeration.

Another important application of RF is on pearl millet as it undergoes rancidity within 10 days after milling due to poor stability which causes early rancidity and

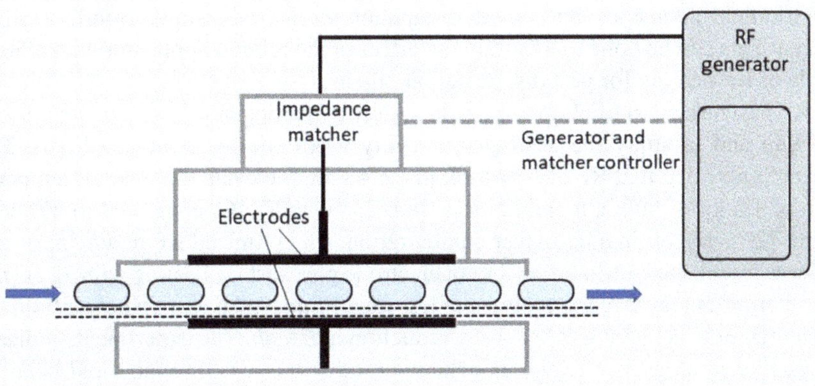

Fig. 7.2 Schematic of RF system. (Obtained with permission from Costa and Marra 2024)

unpleasant off-flavor development on storage. Thus, hot air-assisted RF heating helps in the inactivation of enzymes that act as precursors of rancidity to extend the storage period (Yarrakula et al. 2024).

7.3.1 RF on Millet Disinfestation

This section provides insights into the effect of RF heating technology in the application of pest control. RF heating acts as a replacement to chemical fumigation to manage pest infestation in grains. Moreover, environmental issues, the growth of the organic food market, and the insecticide resistance have pushed the industry to create environmentally friendly methods for disinfestation. Additionally, synthetic chemicals pose risks to human health due to potential product contamination (Costa and Marra 2024).

RF systems can be categorized into two types: the free-running oscillator system, also known as the conventional system, and the 50 Ω system. In a conventional system, RF generators send oscillating EM waves to the material positioned between the anode and ground electrodes. The 50 Ω system, on the other hand, uses an RF generator (RFG) to produce RF energy, which is delivered to a matching network through a 50 Ω cable. The matching network connects to the electrodes, through which the radio waves are emitted and delivered to the materials being processed (Moirangthem et al. 2020). Matching networks are specialized transformers featuring a network of series and parallel impedances, unlike conventional wire-wound transformers. In research on RF heating of bulk canola seeds, a 50 Ω cable was employed to study the maintaining of uniform temperature throughout the bulk without temperature gradients. Canola, a major oilseed crop in Canada, is treated with RF to achieve disinfestation and extend its storage life. The canola seeds were packed in polypropylene bags in a tubular form to be placed inside the tubular applicator of the RF unit. The study indicated that despite placing temperature sensors in nine different locations within the canola seed bags, temperature inconsistencies were observed. These inconsistencies can be reduced by optimizing power levels, electric field intensities, and power dissipation within the sample. Similar studies with millets can be done to ascertain the effect of the tubular applicator with a 50 Ω cable of RF heating for pest disinfestation effects.

RF pasteurization of almond kernels to destroy pathogenic *E. coli* using a 6 kW RF unit and to study the heating uniformity was done by Li et al. (2018). The research revealed that RF energy took just 6.4 min to elevate the internal temperature of 1.5 kg of bulk almond kernels from 25 °C to 72 °C, compared to 276 min using 72 °C hot air in a five-layer configuration. An optimized RF treatment protocol was established, utilizing an 11 cm electrode gap, surface heating with 72 °C hot air, intermittent layer rearrangement in less than 1 min, and a 15-min hot air hold for pasteurization, achieving over a 4-log reduction in *E. coli*. The experiments indicate that RF technology is effective in achieving log reductions through rapid heating. Another study on the use of RF heating for disinfestation was done on 6 Kg of the bulk of mung beans with a 12.5 cm electrode gap for 6 min to reach a target temperature of 54 °C (Song et al. 2020). Although forced air was used in the experiment to attain a uniform heating index, there was an insignificant difference in the physicochemical properties of the grain, thus suggesting the efficacy of RF in disinfestation. A detailed study on the use of RF heating in the disinfestation of food grains is given elsewhere (Sirohi et al. 2021). Due to the nonuniform heating of grain bed in RF, synergistic effects of forced hot air circulation throughout the sample bed can be done to achieve greater mortality of pests in the sample.

7.4 Effects of RF Processing on the Nutritional Quality of Millets

This section details the changes in the nutritional quality of millets following RF processing. Numerous studies examine the impact of RF processing on starch and protein, which are crucial to the nutritional attributes of millet.

7.4.1 Effect on Protein

The study on RF heating of buckwheat presents the following changes in protein after treatment (Xu et al. 2023). The deconvoluted IR spectra of the amide I band were used to predict changes in secondary structures of proteins, which are primarily associated with C=O stretching vibrations and H–O–H bending vibrations. After RF heating, the percentage of β-sheets increased, while random coils and α-helix content decreased. Thermal disruption of hydrogen bonds between neighboring peptide chains in buckwheat protein triggered the unfolding of the α-helix structure, resulting in a decrease in α-helix content and a corresponding rise in β-sheet content. Additionally, the soluble aggregate formation and hydrophobic interactions contributed to changes in the protein's secondary structure. This explains the initiation of protein-starch agglomerates, as denatured proteins may expose previously hidden reactive groups, increasing their tendency to adhere to starch particles. The purity of protein percentage reduces with RF treatment compared to the untreated samples due to the denaturation of protein that occurs during treatment. In a study by Rao et al. (2023), a 5% reduction in the purity of protein was observed with hot air-assisted RF drying. Enzyme inactivation is another influence of RF treatment where the exposure to RF causes changes in thermal stability of enzymes leading to protein rearrangement or denaturation due to the non-covalent bond breaking leading to unfolding of the enzymes (Chen et al. 2024). The effect of protein influence is also reflected on the gelling time, least gelation concentration, and emulsifying properties of the flour treated using RF technology (Yarrakula et al. 2024). The heat generation during RF processing of millets causes heat denaturation of proteins. During prolonged exposure to such high temperatures, degradation of heat-sensitive amino acids occurs which reduces the nutritional value of the grains. However, moderate RF treatment can enhance protein digestibility by breaking down anti-nutritional factors while also inducing structural modifications such as protein aggregation or unfolding, affecting millet flour properties like water absorption and foaming capacity (Jahan and Gnananethri 2024) (Table 7.1).

7.4.2 Effect on Starch

RF technology significantly impacts starch modification. Changes in starch granule morphology can be observed using FESEM (field emission scanning electron microscope). In some cases, RF treatment did not affect starch, while in others, it caused surface cracks in the granules, altering their surface structure and reducing their rigidity. Research on the impact of RF on potato, maize, and pea starch indicates that RF modifies starch by converting moisture into steam during heating. This process creates low pressure outside the starch granules while maintaining high pressure inside. As EM energy is absorbed and converted to kinetic energy, the rapid movement and collisions of starch molecules lead to increased moisture absorption, swelling, and heightened internal hydration. This ultimately results in the leaching of amylose from the starch and adhesion between the granules (Zhou et al. 2023).

Table 7.1 Effect of RF techniques on the quality of millets

Type of millet	Treatment condition	Significant result	References
Pearl millet	HARF (10 kW, 40.68 MHz ± 0.5%), 10.50 ± 0.50%, 12%, and 15% moisture levels; 5-, 10-, 15-, 20-, and 25-min exposure; 180 mm electrode position	Maximum lipase inactivation (97.30%) achieved with better quality characteristic retention at 15-min exposure	Yarrakula et al. (2022a, b)
Pearl millet	HARF treatment at different exposure periods (5, 8, 11, and 14 min)	Moisture content and water activity reduces with increase in exposure time; water absorption capacity and water solubility index reduced; oil absorption capacity increased	Yarrakula et al. (2022a, b)
Foxtail millet	Electrode distance = 18 cm, conveyor speed = 2.5 m/h, one pass (14 min) and 2 passes (28 min)	Changes in secondary structural elements (increase in β-turn, decrease in α-helix and β-sheets) and crystallinity led to increase in emulsifying and oil and water absorption properties	Rao et al. (2023)
Brown millet	500 g millet bed with thickness 6 cm in PP container, electrode gap = 9.5 cm, 80 °C hot air temperature	Less damage to the structure of grains, structural integrity of proteins and starch intact, lipase activity and peroxidase values reduced	Chen et al. (2024)
Buckwheat	1 kg grains with heating rate 5 °C/min, with RF temperature at 80, 90, and 100 °C and exposure time 5 and 10 min	Reaggregation of protein-starch complexes, crystallinity, random coiling and led to increase in β-sheet with reduction in α-helix structure	Xu et al. (2023)

Another study of RF treatment on buckwheat concluded the ineffectiveness of RF treatment on starch (Xu et al. 2023). The changes in starch are quantified using the FTIR (Fourier transform infrared) spectra. Typically, the bands at 995 cm^{-1} and 1047 cm^{-1} correspond to the hydrated and crystalline starch structures, while the band at 1022 cm^{-1} is linked to the amorphous fraction of starch. As a result, the ratios of absorbances at 1047 and 1022 cm^{-1} ($R_{1047/1022}$) and 995 and 1022 cm^{-1} ($R_{995/1022}$) were calculated to serve as indicators for quantifying the degree of short-range order in the starch molecules. As RF heating time and temperature increased, the $R_{1047/1022}$ values dropped from 0.947 to 0.927, indicating damage to the starch conformation after the heating treatment due to the disruption of intra- and intermolecular hydrogen bonds within the starch granules during RF heating, leading to the unfolding of the double helix structure in the starch molecules. The $R_{995/1022}$ ratio for the sample treated with RF heating at 100 °C for 10 min is generally lower than that of the untreated samples, suggesting increased disruption of the double helix and crystalline structures after the heating process.

7.5 Methods to Enhance Heating Uniformity in RF Processing

To enhance the uniformity of heating in samples subjected to RF heating, modifications to the equipment are necessary to achieve consistent heating by adding some auxiliary devices. PVC (polyvinyl chloride)-based screw conveyors are such devices to induce heating uniformity in granular materials. In pest disinfestation of soybean, with an electrode gap of 13 cm and 1.2 rpm of the screw conveyor, the best heating uniformity was achieved (H. Zhou and Wang 2019). Various factors such as electric motors, speed of rotation of screw shaft, and direction of rotation can be controlled by a speed controller. The optimization of process parameters for uniformity depends on conveying directions, rate of filling of grains, and speed of screw. Excessive increases in screw speed can negatively impact heating uniformity. Higher grain filling rates in the conveyor improve the heating rate during RF heating. In laboratory settings, materials like PVC, acrylic, nylon, and polyamide offer good heating efficiency without becoming heated themselves (Gao et al. 2023). However, in industrial setups, prolonged use of these sheet materials can lead to heating loss. To prevent the absorption of RF energy, which causes heating and distortion of the material, low-heating-loss materials such as PP (polypropylene) can be used instead. Further development of the screw conveyor led to the production of a double screw conveyor to enhance the heating uniformity of the samples. This design effectively minimized heat accumulation at the edges or corners of the sample trough, thereby reducing the risk of product overheating. The screw conveyor's rpm can be adjusted according to the granular size of the samples to improve heating uniformity. However, in some instances, when the rpm exceeds 2, there is a decline in the slope of heating uniformity (Li et al. 2022).

Another study utilized a novel rotator to improve the heating uniformity of raisins for microbial control. When operated at speeds of 20, 40, and 60 rpm, the rotator reduced temperature differences between the surface and interior of the sample to 12.2 °C, 8.24 °C, and 7.92 °C, respectively. This decrease in temperature variation helped maintain the sample at the required time and temperature, resulting in a greater log reduction of microbial colonies (Cui et al. 2021). Similar rotators could be applied in millet processing to enhance heating uniformity and achieve higher log reductions, ensuring a longer and safer storage life.

To expand the use of RF heating at an industrial scale, polypropylene blocks are utilized to address the complexities of material geometry during the processing of millets. The study concentrated on convex-concave sample shapes rather than the conventional rectangular blocks (Wang et al. 2021a). The findings showed that the uniformity index of RF heating for millets declined as the convexity of the sample's top surface increased, while a decrease in concavity had a similar effect. Moreover, placing PP blocks at the base of the container significantly improved RF heating uniformity, as it better accommodated varying sample sizes and promoted more even temperature distribution. This research introduced an innovative method to enhance RF heating uniformity by modifying the shape of samples on either the top or bottom surface.

7.6 RF Processing on Millets to Improve Storage Stability

This section explores how RF treatment affects the storage stability of millet products. To ensure proper storage and prevent pest infestations, it is essential to treat the grains to eliminate adult pests and weaken the eggs within. During storage, two key factors to monitor are peroxide value and free fatty acid value, as these indicate the freshness of the grains. Even minor changes in these properties during storage can signal the beginning of rancidity, making the grains unsuitable for consumption. Once the millet kernel loses its protective husk and is taken for milling into flour, its storage stability declines rapidly, even when the water activity is reduced below 0.65. This results in rancidification and bitterness during storage, attributed to an increase in free fatty acids (FFA) (Liang et al. 2021). The rise in FFA is primarily driven by two interconnected factors: the production of FFA from triglycerides through lipolysis, facilitated by highly active lipase enzymes found in the pericarp, aleurone layer, and germ of millets, and oxidative rancidity, where unsaturated fatty acids react with molecular structures, leading to the formation of volatile secondary metabolites that produce off-odors (Adebowale et al. 2020). To achieve longer storage periods, inactivation of lipase enzymes is a primary approach.

Many studies suggest that grains or other food products prone to rancidity should be processed with RF treatment, using a minimal distance between the electrodes. Reducing the gap between the electrodes increases the temperature, which can make the grains more susceptible to rancidity during storage. The high temperature between the minimal electrode gap accelerates the oxidation of lipids and hydrolyzation. In a study examining RF treatment on walnuts, both peroxide value and free fatty acid levels remained within acceptable limits after 2 years at 4 °C. This suggests that RF treatment is a promising method for extending shelf life and meeting industrial standards (Mao and Wang 2021). This suggests that RF processing could be beneficial for millets with higher fat content, as it may enhance their storage stability over longer periods.

Another study focused on the effects of RF heating on improving the storage stability of brown millet by inhibiting rancidity (Chen et al. 2024). RF treatment significantly reduced lipase and peroxidase activity while preserving the structural integrity of millet, even under high-temperature conditions. Although the structure of starch and protein remained intact, high-temperature RF treatments led to a reduction in phenolics and flavonoids, whereas lower RF temperatures had a much smaller impact on these compounds. Hot air-assisted RF treatment also minimized the increase in fatty acid values and malondialdehyde content, thereby enhancing storage stability. The storage analysis, conducted using hydrolytic rancidity kinetic models, indicates the stability of millets throughout the storage period.

7.6.1 Lipase Activity

The lipase activity of RF-processed millet flour can be estimated by the addition of 2 mL of olive oil and 0.3 mL of distilled water to 2 g of defatted RF-processed millet flour. The mixture is incubated at 45 °C for 4 h, following which the released free fatty acids are measured at 715 nm using a UV spectrophotometer (Rao et al. 2023). The study on foxtail millet treated with hot air-assisted RF revealed that lipase activity decreased with increased heat generation. Millet samples with 15% moisture showed a greater reduction in lipase activity compared to those with 24% moisture. This indicates that higher moisture content during processing leads to less temperature rise due to the cooling effect of evaporation:

$$\text{Residual relative enzyme activity}(\%) = \frac{\text{Value of residual enzyme activity}}{\text{Value of enzyme activity of raw sample}} \times 100 \tag{7.1}$$

The observed phenomenon suggests that polar groups accelerate the hydrolysis of peptide bonds, leading to disulfide interchange and the deamination of glutamine and asparagine. This process resulted in reduced enzyme thermal stability and increased moisture (Liao et al. 2020). A greater heat generation and greater heating uniformity index during RF processing leads to increased lipase activity inactivation. The increase in the number of passes through the RF heating system also helps in reducing lipase activity.

RF treatment has proven effective in inactivating enzymes by affecting their thermal stability and causing protein rearrangement or denaturation through the non-covalent bond destruction, leading to the unfolding of enzymes. In comparison to control samples, the residual activity of lipase and peroxidase enzymes varies inversely with temperature and holding time. Brown millet exhibited greater sensitivity to hot air-assisted RF heating compared to millet bran, demonstrating a more significant reduction in enzyme activity. This also indicates that lipase has enhanced thermal resistance than peroxidase. The duration of treatment is crucial; in some cases, enzyme activity ceases to decrease even with extended treatment times, indicating a maximum RF treatment duration. For brown millet, a 10-min treatment at 105 °C for millet and 115 °C for bran effectively reduces lipase activity, extending shelf life. Conversely, a 20-min treatment at lower temperatures—95 °C for millet and 105 °C for bran—is sufficient to inactivate peroxidase (Chen et al. 2024). The detailed procedure to evaluate the free fatty acid content and lipase activity is given in Yarrakula et al. (2024).

7.7 Effect of RF Processing Functional, Structural, Cooking, and Textural Attributes of Millets

This portion of the chapter discusses the functional, structural, cooking, and textural properties of millet grains or flours treated with RF technology. This section also sheds insight into the changes that occur in samples in terms of dependent and independent variables, as well as the reasons for such changes.

7.7.1 Functional Properties

The functional features include emulsifying ability, foaming capacity, oil and water absorption, bulk density, and the lowest gelation concentration. Functional property changes occur as a result of molecular changes in the sample during treatment, such as protein denaturation and starch depolymerization. The next sections provide comprehensive explanations for such phenomena.

7.7.1.1 Emulsifying Property

The emulsifying properties of millet grains processed using RF technology are assessed based on the protein extracted from defatted flour. This defatted flour, rich in protein, is analyzed for its emulsifying capabilities. The emulsifying activity is measured by the emulsion stability index, where the millet protein is combined with vegetable oil to create an emulsion at high mixing speeds. The stability of the emulsion is then evaluated after incubating the mixture at 80 °C for 30 min by measuring the volume of the emulsion. RF treatment improves emulsion stability and the emulsifying activity index due to the partial unfolding of proteins, increased exposure of hydrophobic groups, and changes in the protein's secondary structure at elevated temperatures, which enhance the protein's ability to interact with the water-oil interface (Rao et al. 2023). With the increase in treatment temperature in RF, the emulsifying capacity and emulsion stability reduce due to protein denaturation (Xu et al. 2023). Emulsion capacity increases with the interference of interface absorption of proteins thus preventing the coalescence flocculation of the emulsion due to increasing of hydrophobic moieties and induction of flexibility on thermal denaturation. Under more intense conditions, the unfolded protein rearrangement and the reduced solubility caused by heat-induced aggregation resulted in a discontinuous phase at the water-oil interface, negatively impacting emulsification properties.

7.7.1.2 Foaming Property

To assess the functional properties of millet protein, its foaming ability, similar to its emulsifying properties, is evaluated. Contrary to the emulsifying property, the foaming stability or capacity of millet flour treated with RF was found to be lower than that of untreated samples. This reduction may indicate protein denaturation due to the high temperatures experienced between the electrodes in the RF system. This thermal denaturation can cause proteins to form insoluble aggregates, which hinders their ability to migrate and stabilize air-water interfaces (Ling et al. 2019). To measure foaming capacity, millet protein is dispersed by adjusting the pH to 7.0 and then whipped with a rotor-stator homogenizer. The foaming capacity is determined by calculating the initial foaming volume and the final foaming volume after 30 min. A study on RF treatment on buckwheat flour exhibited a decrease in foaming capacity compared to the native flour, while the foaming stability increased abundantly from 1.74% in native flour to 22.37% in treated flour (Xu et al. 2023). Foam capacity primarily relies on the dispersion of protein at the air-water surface tension, while foam stability relies on the formation of an opaque, cohesive layer entrapping the air bubble. The reduction in foam capacity could be attributed to the insoluble aggregate formation and protein denaturation following RF treatment, which makes it strenuous for proteins to disperse and adsorb at the air-aqueous interface. Nevertheless, the expansion of peptide chains induced by RF heating can form a relatively stable network structure at the interface, which traps air and improves foam stability. This stability is crucial for maintaining the soft texture of food products.

7.7.1.3 Water and Oil Absorption Capacity

The water and oil absorption capacities are evaluated by mixing the millet protein in distilled water and vegetable oil, respectively. The water absorption capacity of millet flour increases after hot air-assisted RF treatment. This paves the way for the utilization of hot air-assisted treated millet flour in dough preparation of cookies and pasta due to better water hydration properties. OAC (oil absorption capacity) refers to the flour's capacity to absorb oil, which is beneficial because oil helps retain flavor and enhances the texture of the product. With the increase in RF treatment time, the water absorption capacity decreases, while the OAC increases (Yarrakula et al. 2022b). In a study examining foxtail millet protein, an increase in both oil and water absorption was noted when the millet was treated with RF (Rao et al. 2023). This enhancement is attributed to the increased accessibility of both nonpolar and polar groups of the proteins, which allows them to bind more oil and water. Compared to microwave treatment, RF treatment results in higher oil and water absorption due to the greater exposure of hydrated carboxyl and hydroxyl groups and improved capillary and interfacial actions of the proteins (Jiang et al. 2021).

In some instances, water absorption halts because denatured protein forms a barrier around the starch molecules, preventing them from diffusing into the aqueous phase and limiting both diffusion and solubility. Conversely, when starch granules are damaged, increased heating enhances the interaction between water and hydrogen, leading to greater solubility (Xu et al. 2023). Similarly, the swelling power also increases as the incubation temperature rises. The formation of a gel network structure at higher temperatures limits starch swelling. Additionally, the breakdown of amylose and amylopectin chains in starch molecules caused by RF treatment may further inhibit starch swelling at elevated temperatures.

7.7.1.4 Least Gelation Concentration

The millet samples mixed in an aqueous phase at an ascending concentration from 2 g to 20 g per 100 mL are prepared in test tubes. The tubes are then subjected to heating in a water bath at 80 °C for an hour with rapid cooling using tap water. The tubes are then kept in the refrigerator for 2 h and inverted afterward. The least gelation is determined by the concentration above which the millets ceased to flow when placed in an upright position (Yarrakula et al. 2024). A lower least gelation concentration value indicates better gelation ability of protein and greater swelling capacity of the flour. The hot air-assisted RF treatment did not impair the protein component's gelation capability. Gel formation mainly depends on the hydration and swelling of the amorphous regions within starch molecules. Carbohydrates such as lactose, maltose, and sucrose are recognized for decreasing the protein groups' thermodynamic affinity for water, which enhances interactions between protein units and thus improves their gelling properties.

7.7.1.5 Bulk Density

The bulk and tapped density of grains reflect their shape. When grains undergo specific heating treatments, their moisture content decreases, which alters their density. A reduction in both bulk and tapped densities was observed in pearl millet flour treated with hot air-assisted RF (Yarrakula et al. 2022b).

7.7.2 Structural Property

This section gives information on the changes in structural properties of the samples following RF treatment. These experiments aid in measuring the effect of treatment on molecular rearrangement in the samples. FTIR provides the molecular fingerprint of the sample, whereas XRD analysis provides the crystallinity of the sample.

7.7.2.1 FTIR

The FTIR spectra reveal the molecular structure of starch processed with RF technology. Some studies indicate that RF has minimal impact on the chemical properties of starch, with only physical changes being noted (Zhou et al. 2023). Protein denaturation caused by the high temperatures of RF treatment, as discussed in previous sections, can be confirmed using FTIR spectra. In this analysis, the changes in the intensity of secondary structural elements of proteins are directly related to the heating temperatures induced by RF treatment, which leads to a higher rate of protein denaturation (Rao et al. 2023). The presence of cellulose in the millet samples can be confirmed by observing peak intensities between 1380 and 1420 cm^{-1}, which indicate hydroxide in-plane bending vibrations. Functional changes in lipids, specifically in the v(C=O) bond due to the ester linkage connecting fatty acids to the glycerol backbone of triglycerides, can be observed in the wavelength range of 1700–1800 cm^{-1}. Protein configurations are typically assessed by analyzing changes in the peak position, particularly in the amide I region (1700–1600 cm^{-1}). This region is primarily focused on due to the C-O stretch vibrations of protein linkages and is the most sensitive to modifications in the secondary structure of the protein, more so than the amide II band region (1200–1400 cm^{-1}) (Naik et al. 2022). Hot air-assisted RF heating enhances the peak intensity in the amide III region, while the amide II region reflects intramolecular or intermolecular hydrogen bonding. Additionally, RF heating leads to a reduction in the β-sheet and α-helix content of proteins, accompanied by a shift in the amide I band peak. This occurs due to the disruption of certain hydrogen bonds, hydrophobic interactions, and Van der Waals forces, driven by the carbon-centered free radical reaction and the thermal nature of RF heating.

7.7.2.2 XRD (X-Ray Diffraction)

The XRD spectra reveal the crystalline nature of the samples. In the case of hot air-assisted RF drying of millet protein isolate, the identified peak in the spectra indicates that the protein exhibits an amorphous halo nature. Changes in protein crystal size can be linked to the diffraction intensity observed in XRD spectra. Higher diffraction peak intensity indicates a greater percentage of crystallinity in protein concentrate or isolate. In the case of RF heating, the untreated samples showed higher diffraction peak intensity compared to the treated samples, indicating that RF treatment reduces the crystallinity size of millet protein (Rao et al. 2023). Additionally, as treatment times increase, the relative crystallinity of the samples decreases. This is due to mild protein denaturation caused by thermal treatment and the nonthermal effects of carbon-centered free radicals. The denatured protein portions form interlinked protein molecules instead of aggregates, where the interlinking results in the formation of new chemical bonds. This disrupts the protein configurations, leading to a reduction in overall rigidity and a subsequent decrease in crystalline size (Malik and Saini 2019).

The XRD pattern of the study on RF treatment on buckwheat concluded a decrease in crystallinity with an increase in RF heating conditions (temperature and time) (Xu et al. 2023). The reduction in crystallinity suggests that the long-range ordered structure of buckwheat starch was disrupted by RF treatment. This could be attributed to the breakdown of H_2 bonds during RF heating, resulting in the destruction and melting of the crystalline regions.

7.7.3 Cooking Properties

This section discusses the effect of RF treatment on cooking attributes such as cooking time, texture of cooked millets, and color of cooked millets.

7.7.3.1 Cooking Time

Approximately 10 grams of decorticated grains were placed in a glass beaker with 100 mL of boiling water and boiled. At 10-min intervals, a few grains were removed from the beaker and pressed between glass plates. The cooking time was determined by the translucency and constant diameter of the spread of the grains. Hot air-assisted RF treatment does not significantly affect cooking quality, as studies have reported no differences in cooking quality between treated and untreated millet samples.

7.7.3.2 Color of Cooked Millets

The color of cooked RF-treated grains differs significantly from that of untreated pearl millet grains, primarily due to the decortication process, which reduces the whiteness index. Although there is a noticeable change in the color of cooked millet, it is subtle and not easily detected by the human eye, suggesting minimal change in color attributes. However, hot air-assisted RF treatment significantly increases the redness and yellowness of the grains.

7.7.3.3 Textural Attributes

Texture profile analysis was conducted to assess the quality characteristics of cooked grains using a compression load cell. Although hot air-assisted RF treatment increased the hardness of the samples, the changes were not highly significant. The springiness remained consistent before and after treatment. The decortication of pearl millet led to notable changes in the hardness and cohesiveness of the RF-treated samples due to the rigid cell wall structure (Yarrakula et al. 2024).

7.7.3.4 Pasting Property

The grains treated with RF are initially milled to obtain flour. This flour is further taken for peak viscosity analysis. The peak viscosity and pasting properties of treated millet flour are estimated using a Rapid Visco Analyzer. RF treatment causes a decrease in the peak viscosity of samples, but peak viscosity increases with longer residence times. This increase is attributed to the formation of simple sugars and water resulting from starch depolymerization. Moisture content in millet flour samples also significantly affects peak viscosity. A study by Rao et al. (2023) investigated how the moisture content of foxtail millet grains influences the RF treatment. The results showed that peak viscosity rises with increased moisture content. Foxtail millet flour samples with 12% and 15% moisture content followed this trend, while samples with moisture levels of 18% and 24% exhibited a reverse trend. This reversal is due to the impact of higher moisture on the size distribution of starch granules, which affects their behavior during processing.

RF treatment reduces the peak viscosity, trough viscosity, breakdown, and final viscosity of buckwheat flour as treatment time and temperature increase (Xu et al. 2023). This decrease results from reduced starch granule swelling, leading to weak gel formation. The reduction is likely attributable to the breakdown of starch chains and the weakening of the interactive forces between starch units due to the heating process. The reduced breakdown can be attributed to the strengthening of the amorphous regions through RF treatment, which enhances the starch's resistance to damage during water absorption. Setback, reflecting amylose retrogradation during cooling, was decreased, probably due to the new interactions formed between amylose and amylopectin chains. Conversely, the pasting temperature, which indicates the minimum temperature required for starch gelatinization, increased due to the formation of an amylose-lipid complex from the degraded starch, which impedes typical gelatinization.

7.8 Advantages and Disadvantages of RF Processing of Millets

RF technology offers several benefits over traditional heating methods. Unlike ohmic heating, where electrodes are submerged in food products like slurries or liquids, RF processing generates alternating electric fields between its electrodes. This results in volumetric heating, which heats the product quickly and evenly from the inside out, thereby minimizing the risk of overheating. The total cost for implementing an RF technology is given in detail by Costa and Marra (2024). Because of their higher power capacity, RF waves can penetrate deeper into the material than conventional MW. As a result, heat is generated internally within the sample material, paving the way to a more uniform distribution. Additionally, the design of large-scale RF units is more practical and helps enhance the final product's quality. Other significant benefits of this sustainable technology include higher energy efficiency, better moisture control, contactless heating, and quicker drying and curing times (Sirohi et al. 2021).

Another advantage of RF heating systems is their ability to reach the desired processing time more quickly, and they can be easily turned on and off as needed. However, despite these benefits, RF heating also has a few disadvantages. These include high installation costs for the source electronics and the fact that heating efficiency is influenced by several factors, such as the product's dielectric properties, shape, size, position between electrodes, and electrode configuration, all of which significantly affect temperature and heating uniformity (Mahendran and Jojo 2013).

7.9 Mathematical Modeling of RF Processing

This section deals with the mathematical modeling for RF heating. To ensure uniform RF heating, temperature measurements from various locations are analyzed, and a model is created incorporating power dissipation density and a simulated heating uniformity index using Maxwell's equations. This model can be simplified to the Laplace equation using a quasi-static approximation. A study applied this mathematical approach to assess heating uniformity for RF-treated eggshells using Maxwell's method (Palazoğlu and Miran 2019). The study used the Joule heating interface of COMSOL (Computer Simulation Object Laboratory) Multiphysics to simulate the RF heating process. The governing equations discussed in the study include a simplified Laplace equation from Maxwell's equation, given by

$$-\nabla.\left(\left(\sigma + j2\pi f\varepsilon_0\varepsilon'\right)\nabla V\right) = 0 \tag{7.2}$$

where σ is the electrical conductivity of the material (S/m), $j = (-1)^{0.5}$, f is the frequency (Hz), ε_0 is the permittivity of free space (8.854×10^{-12} F/m), ε' is the dielectric constant of the material, and V is the potential difference across the electrodes (V). The quasi-static electric field (EF) strength, (E) (a vector quantity) measured in (V/m), between the RF electrodes can be derived from the negative gradient of the voltage (V).

The EM energy is converted to thermal energy and dissipated as heat inside the dielectric material and is obtained as total power dissipation density (Q) in W/m^3 by

$$Q = 2\pi f\varepsilon_0\varepsilon'' E^2 \tag{7.3}$$

where ε'' is the dielectric loss factor of the material. The transient heat conduction equation to obtain the temperature gradient is given by

$$\rho C_p \frac{dT}{dt} = k\nabla^2 T + Q \tag{7.4}$$

where (ρ) is the density (kg/m^3), (C_p) is the specific heat (J/kg·K), and (k) is the thermal conductivity (W/mK) of the material.

Huang et al. (2018) provide a comprehensive review of the mathematical modeling of various parameters influencing heating uniformity. The paper covers heat and mass transfer models, as well as uniformity equations for different RF processing applications such as pasteurization, sterilization, disinfestation, and thawing. Additionally, it highlights the factors that contribute to the nonuniform heating of samples.

7.10 Conclusion

Applying RF treatment to millets offers a promising solution for enhancing their characteristics. A major issue with storing millets is rancidity, which causes unpleasant flavors and shortens shelf life. RF treatment helps mitigate this problem by reducing the activity of enzymes like lipase and peroxidase, which are precursors to rancidity. Despite its benefits in improving storage stability, RF treatment has drawbacks, such as a reduction in phenolics and flavonoids at high temperatures. Another significant issue is nonuniform heating, which can result in localized overheating and diminished millet quality. To address these challenges, innovations in RF generators, such as the use of screw conveyors, appropriate coverings, and polypropylene heating blocks, have been developed to enhance heating uniformity. These improvements also optimize energy use and facilitate scaling up for industrial applications. Additionally, advancements in artificial intelligence and neural networks can further address the issue of nonuniform heating, thereby improving the quality of millet grains.

References

Adebowale OJ, Taylor JRN, de Kock HL (2020) Stabilization of wholegrain sorghum flour and consequent potential improvement of food product sensory quality by microwave treatment of the kernels. LWT 132:109827

Bhatt K, Vaidya D, Kaushal M, Gupta A, Soni P, Arya P, Gautam A, Sharma C (2020) Microwaves and Radiowaves: in food processing and preservation. Int J Curr Microbiol Appl Sci 9(9):118–131. https://doi.org/10.20546/ijcmas.2020.909.015

Chen C, Yin J, Jing P, Jiao S (2024) Effects of hot air-assisted radio frequency stabilization of brown millet and millet bran: enzyme inactivation, microstructure, nutritional quality, and storage stability. Innov Food Sci Emerg Technol 91(September 2023):103529. https://doi.org/10.1016/j.ifset.2023.103529

Costa JM, Marra F (2024) Advances in food processing through radio frequency technology: applications in Pest control, microbial and enzymatic inactivation. Food Eng Rev:0123456789. https://doi.org/10.1007/s12393-024-09372-8

Cui B, Fan R, Ran C, Yao Y, Wang K, Wang Y, Fu H, Chen X, Wang Y (2021) Improving radio frequency heating uniformity using a novel rotator for microorganism control and its effect on physiochemical properties of raisins. Innov Food Sci Emerg Technol 67(November 2020):102564. https://doi.org/10.1016/j.ifset.2020.102564

Demeczky M (1985) Application of dielectric techniques in food production and preservation. Developments in Food Preservation 3(133):127–181.

Dragoev S, Kolev N, Vlahova-Vangelova D, Balev D (2024) Radio frequency food processing. Current status and perspectives—a review. Food Sci Appl Biotechnol 7(1):103–121. https://doi.org/10.30721/fsab2024.v7.i1.304

Fellows P (2016) Dielectric, ohmic and infrared heating. In: Food processing technology: principles and practices, 3rd edn. Woodhead Publishing Ltd, Oxford/Cambridge, pp 581–612

Gao J, Wu M, Du S, Zhang H, Wang S, Ling B (2023) Recent advances in food processing by radio frequency heating techniques: a review of equipment aspects. J Food Eng 357(May):111609. https://doi.org/10.1016/j.jfoodeng.2023.111609

Huang Z, Marra F, Subbiah J, Wang S (2018) Computer simulation for improving radio frequency (RF) heating uniformity of food products: a review. Crit Rev Food Sci Nutr Comput 58(6):1033–1057. https://doi.org/10.1080/10408398.2016.1253000

Jahan A, Gnananethri CJ (2024) Novel thermal technologies on the storage of pearl millet grain and flour. August, 1–6

Jiang Y, Zhou X, Zheng Y, Wang D, Deng Y, Zhao Y (2021) Impact of ultrasonication/shear emulsifying/microwave-assisted enzymatic extraction on rheological, structural, and functional properties of Akebia trifoliata (Thunb.) Koidz. seed protein isolates. Food Hydrocoll 112:106355

Jiao Y, Shi H, Tang J, Li F, Wang S (2015) Improvement of radio frequency (RF) heating uniformity on low moisture foods with Polyetherimide (PEI) blocks. Food Res Int 74:106–114. https://doi.org/10.1016/j.foodres.2015.04.016

Jiao Y, Tang J, Wang Y, Koral TL (2018) Radio-frequency applications for food processing and safety. Annu Rev Food Sci Technol 9:105–127. https://doi.org/10.1146/annurev-food-041715-033038

Li R, Kou X, Hou L, Ling B, Wang S (2018) Developing and validating radio frequency pasteurisation processes for almond kernels. Biosyst Eng 169:217–225. https://doi.org/10.1016/j.biosystemseng.2018.03.001

Li H, Wang J, Wang S, Ling B (2022) Performance evaluation of the double screw conveyor in radio frequency systems: heating uniformity and quality of granular foods. Innov Food Sci Emerg Technol 77(April):102990. https://doi.org/10.1016/j.ifset.2022.102990

Liang K, Liu Y, Liang S (2021) Analysis of the characteristics of foxtail millet during storage under different light environments. J Cereal Sci 101:103302

Liao M, Damayanti W, Zhao Y, Xu X, Zheng Y, Wu J, Jiao S (2020) Hot air-assisted radio frequency stabilizing treatment effects on physicochemical properties, enzyme activities and nutritional quality of wheat germ. Food Bioprocess Technol 13:901–910

Ling B, Ouyang S, Wang S (2019) Effect of radio frequency treatment on functional, structural and thermal behaviors of protein isolates in rice bran. Food Chem 289:537–544

Mahendran R, Jojo S (2013) Radio frequency heating and its application in food processing: a review. Int J Curr Agric Res 1(9):42–46. https://www.researchgate.net/publication/313041438_Radio_Frequency_heating_and_its_application_in_food_processing_A_Review

Malik MA, Saini CS (2019) Heat treatment of sunflower protein isolates near isoelectric point: effect on rheological and structural properties. Food Chem 276:554–561

Maloney N, Harrison M (2016) Advanced heating technologies for food processing. In: Innovation and future trends in food manufacturing and supply chain technologies. Elsevier Ltd. https://doi.org/10.1016/B978-1-78242-447-5.00008-3

Mao Y, Wang S (2021) Simultaneous hot-air assisted radio frequency drying and disinfestation for in-shell walnuts using a two-stage strategy. LWT 151(April):112134. https://doi.org/10.1016/j.lwt.2021.112134

Moirangthem TT, Macana R, Baik OD (2020) Characterization of 50-ohm radio frequency heating of bulk canola seeds (Brassica napus. L) in a tubular applicator with parallel electrodes and post-treatment quality. Innov Food Sci Emerg Technol 64(February):102409. https://doi.org/10.1016/j.ifset.2020.102409

Naik M, Natarajan V, Modupalli N, Thangaraj S, Rawson A (2022) Pulsed ultrasound assisted extraction of protein from defatted bitter melon seeds (Momardica charantia L.) meal: kinetics and quality measurements. LWT 155:112997

Orsat V (1999) Radio-frequency thermal treatments for agrifood products. McGill University, Montreal

Orsat V, Jusoh YMM (2014) Electrical conductivity effect on dielectric properties and radio-frequency heating. In: Radio-frequency heating in food processing: principles and applications. https://doi.org/10.1201/b17740

Palazoğlu TK, Miran W (2019) Computational investigation of the effect of orientation and rotation of shell egg on radio frequency heating rate and uniformity. Innov Food Sci Emerg Technol 58(September):102238. https://doi.org/10.1016/j.ifset.2019.102238

Pathaw MS, Mukhim, Rani S, Kamble D, Swer TL (2022) Principles of radiofrequency processing in the food industry. In: Emerging thermal processes in the food industry: unit operations and processing equipment in the food industry. Elsevier. https://doi.org/10.1016/B978-0-12-822107-5.00008-8

Peng J, Yin X, Jiao S, Wei K, Tu K, Pan L (2019) Air jet impingement and hot air-assisted radio frequency hybrid drying of apple slices. LWT 116(August):108517. https://doi.org/10.1016/j.lwt.2019.108517

Piyasena P, Dussault C, Koutchma T, Ramaswamy HS, Awuah GB (2003) Radio frequency heating of foods: principles, applications and related properties – a review. Crit Rev Food Sci Nutr 43(6):587–606. https://doi.org/10.1080/10408690390251129

Rao MV, Sunil CK, Venkatachalapathy N (2023) Effect of microwave and hot air radiofrequency treatments on physicochemical and functional properties of foxtail millet flour and its protein isolate. J Cereal Sci 114(September):103774. https://doi.org/10.1016/j.jcs.2023.103774

Sirohi R, Tarafdar A, Kumar Gaur V, Singh S, Sindhu R, Rajasekharan R, Madhavan A, Binod P, Kumar S, Pandey A (2021) Technologies for disinfection of food grains: advances and way forward. Food Res Int 145(May):110396. https://doi.org/10.1016/j.foodres.2021.110396

Song X, Ma B, Kou X, Li R, Wang S (2020) Developing radio frequency heating treatments to control insects in mung beans. J Stored Prod Res 88:101651. https://doi.org/10.1016/j.jspr.2020.101651

Tang J, Wang Y, Chow TV, Chan T (2004). Radio-frequency heating in food processing. In Novel food processing technologies (pp. 501–524). CRC press.

Wang Y, Wig TD, Tang J, Hallberg LM (2003) Sterilization of foodstuffs using radio frequency heating. J Food Sci 68(2):539–544

Wang P, Liu J, Mao Y, Guan X, Wang S (2021a) Improvement of radio frequency heating uniformity for millets by changing shape and adding polypropylene blocks. Innov Food Sci Emerg Technol 74(October):102856. https://doi.org/10.1016/j.ifset.2021.102856

Wang S, Llave Y, Kong F, Marra F, Erdoğdu F (2021b) Update on emerging technologies including novel applications: radio frequency. In: Food engineering innovations across the food supply chain, pp 163–186. https://doi.org/10.1016/B978-0-12-821292-9.00013-3

Xu J, Yang G, Zhou D, Fan L, Xu Y, Guan X, Li R, Wang S (2023) Effect of radio frequency energy on buckwheat quality: an insight into structure and physicochemical properties of protein and starch. Int J Biol Macromol 251(July):126428. https://doi.org/10.1016/j.ijbiomac.2023.126428

Yarrakula S, Mummaleti G, Pare A, Vincent H, Saravanan S (2022a) Hot air–assisted radio frequency hybrid technology for inactivating lipase in pearl millet. J Food Process Preserv 46(10):e16178. https://doi.org/10.1111/jfpp.16178

Yarrakula S, Suraj AB, Rehaman A, Saravanan S (2022b) Effect of hot air assisted radio frequency technology on physical and functional properties of pearl millet. Pharma Innov 11(March):633–637. http://www.thepharmajournal.com

Yarrakula S, Mummaleti G, Toleti KS, Saravanan S (2024) Functional characteristics and storage stability of hot air assisted radio frequency treated pearl millet. Food Prod Process Nutr 6(1). https://doi.org/10.1186/s43014-024-00228-6

Zhou H, Wang S (2019) Developing a screw conveyor in radio frequency systems to improve heating uniformity in granular products. Int J Agric Biol Eng 12(3):174–179

Zhou D, Yang G, Tian Y, Kang J, Wang S (2023) Different effects of radio frequency and heat block treatments on multi-scale structure and pasting properties of maize, potato, and pea starches. Food Hydrocoll 136(PB):108306. https://doi.org/10.1016/j.foodhyd.2022.108306

Ohmic Heating: A New Dimension in Millet Processing (Examining the Electrical Approach for Efficient Heat Transfer)

8

Prabha Haldkar and Alka Mishra

8.1 Introduction

Millets are usually processed to remove inedible parts, extend shelf life, and improve nutritional and sensory qualities, before consumption. Various food processing methods have been explored to stabilize millet products while maintaining their nutritional value. Heating is a widely used traditional treatment for food processing and preservation. In the food industry, thermal processing plays a crucial role as one of the most important unit operations for destroying microbes and preserving food (Tadini and Gut 2022). Thermal processing fundamentally involves the heat transfer mechanisms of conduction, convection, and radiation. These thermal methods require a long time to transfer adequate heat from the surface to the center to ensure complete processing for solid foods. This extended process can result in overcooked food and degradation of quality in the outer layer (Piette et al. 2004). To address these issues and meet the growing consumer demand for high-quality food, several alternative processing technologies have been proposed. These technologies include high- intensity pulsed electric field, pulsed light field, high- and ultra-high-pressure processing, ohmic heating and radio frequency heating (Yogesh 2016), *etc.* Alternative technologies that employ electrical energy directly for food processing have captured the food industry's interest. Among these, the "Electro-pure Process" in the late 1910s is now referred to as electroconductive or ohmic heating (OH) which has particularly garnered significant attention from both researchers and industry professionals among these technologies (Tian et al. 2016). This is an electrical resistance heating method where alternating current is passed through the food to generate heat (Alkanan et al. 2021; Jan et al. 2021; dos Santos et al. 2024).

P. Haldkar
Jawaharlal Nehru Krishi Vishwavidyalaya, Jabalpur, Madhya Pradesh, India

A. Mishra (✉)
Sardar Vallabhbhai Patel University of Agriculture and Technology, Meerut, Uttar Pradesh, India

S. Jaddu, R. Chandra Pradhan (eds.), *Innovative Millet Processing*, https://doi.org/10.1007/978-3-031-99864-5_8

This method not only improves the efficiency of heat transfer but also preserves essential nutrients and enhances product quality by minimizing thermal degradation (Dornoush and Bagher 2022). In the context of millet processing, where maintaining the grain's nutritional profile is critical, ohmic heating represents a new dimension in food processing technology.

This chapter delves into the mechanics of ohmic heating, its application in millet processing, and the potential benefits it holds for improving efficiency, product quality, and sustainability. We explore the underlying principles of electrical heating, the technological advancements that have made ohmic heating viable for large-scale applications, and the future implications for the grain processing industry. This chapter aims to provide a comprehensive overview of how ohmic heating can revolutionize millet processing and contribute to the broader goal of innovative food technology solutions.

8.2 Working Principle of Ohmic Heating

Ohmic heating is also known as Joule heating, electrical resistance heating, electro-heating, or electroconductive heating. It is an advanced thermal processing technique that relies on the principle of electrical resistance to generate heat within food materials. In this process, food is placed in between two electrodes, and an alternating electrical current is passed directly through the food, which acts as a conductor (Leizerson and Shimoni 2005). As the electrical current moves through the food, it encounters resistance, leading to the conversion of electrical energy into heat. This internal heat generation allows for rapid, uniform heating throughout the material, as opposed to conventional methods like conduction or convection, which rely on external heat sources. The amount of heat generated in OH is directly proportional to the current induced by the voltage gradient in the applied electrical field and the material's electrical conductivity. This relationship is described by Joule's law, which connects heat generation to electrical current and resistance. The electric field strength can be modified by changing the electrode gap or adjusting the applied voltage (Kumar and Ramaswamy 2004). The key factor is the product's electrical conductivity and its dependence on temperature. In case the product consists of multiple phases, such as a liquid- particulate mixture, the electrical conductivity of each phase must be taken into consideration (Knirsch et al. 2010). As temperature increases, electrical conductivity rises, making OH more efficient and potentially leading to runway heating. Differences in electrical resistance and its temperature dependence between the two phases can complicate the system's heating characteristics (Sastry 2005). Since electrical conductivity is affected by ionic content, adjusting the ion levels (e.g., salt) in the product can modify its electrical conductivity in both phases thereby optimizing ohmic heating efficiency (Icier and Ilicali 2005). The principle of OH is shown in Fig. 8.1.

In millet processing, the use of ohmic heating presents several advantages. Millets are small, dense grains that can be difficult to heat uniformly using traditional methods. When external heat is applied, it often leads to uneven temperature

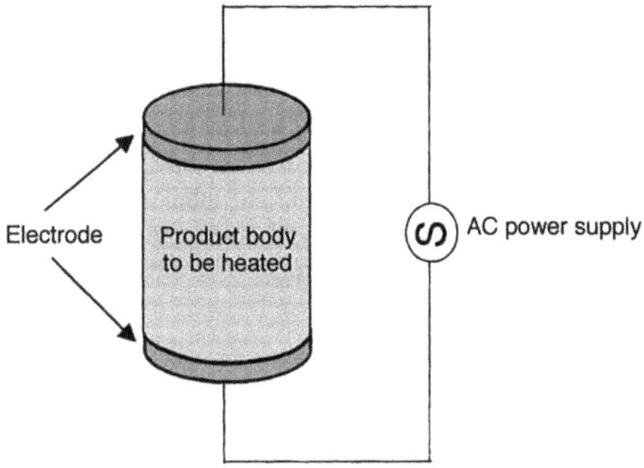

Fig. 8.1 Principle of ohmic heating

distribution, with the outer layers becoming overcooked while the core remains undercooked. This can result in degradation of the grain's nutritional value, texture, and flavor. Ohmic heating, by contrast, heats the entire millet mass uniformly because the electrical current passes through all parts of the grain. This uniformity ensures consistent heating, reducing the risk of overprocessing or underprocessing certain portions. Furthermore, ohmic heating is an energy-efficient method, as the electrical energy is converted directly into heat within the food product, reducing energy losses typically associated with heat transfer through external sources. In the case of millet, which can require substantial heat input during certain processing stages like cooking, drying, or sterilization, the efficiency of ohmic heating can lead to reduced energy consumption and operational costs.

8.2.1 Electrical Energy Conversion in Ohmic Heating: Joule's Law and Heat Generation

In ohmic heating, electrical energy is converted into heat based on Joule's law, which states that the heat generated is proportional to the square of the electric current and the resistance of the product, which can be expressed as

$$Q = I^2 R \tag{8.1}$$

where Q is the volumetric heat generated (W/m^3), I is the electrical current (A), and R is the electrical resistance (Ω).The rate of heat generation in an ohmic heater under constant voltage condition is expressed as

$$Q = (\Delta V)^2 \sigma \tag{8.2}$$

where ΔV is the voltage gradient (V/m) and

σ is the electrical conductivity (S/m).

In case of a constant current, the relationship can be expressed as

$$Q = \frac{|J|^2}{\sigma} \tag{8.3}$$

where | J | is the current density (A/m^2).

Most commercial ohmic heaters operate under constant voltage conditions. When convection, conduction, and other significant heat transfer mechanisms are negligible and heat losses to the surrounding are ignored, the heating rate can be calculated under constant voltage condition which is expressed as

$$\frac{dT}{dt} = \frac{(\Delta V)^2 \sigma}{\rho C_p} \tag{8.4}$$

where ρ is the product density (Kg/m^3) and *Cp* is the specific heat of the product (J/ Kg- °C).

According to Skudder and Bliss (1987), with a commercial residence time of 10–100 seconds, a temperature rise of 100 °C in the product can be achieved. Typically, heating rates of 1–10 °C/s are reported. The temperature increase in each section of a continuous ohmic heater tube can be expressed as

$$\Delta T = \frac{V^2 \sigma A}{L m C_p} \tag{8.5}$$

where *A* is the cross-section area of product (cm^2), *L* the distance between electrode (cm), *m* the mass flow rate (Kg/s), *V* the voltage (V), and *m* the mass flow rate (m/s).

The required electrical power, which determines the transformer size, is calculated as

$$P = mC_p \left(T_{out} - T_{in}\right) \tag{8.6}$$

where Tout and Tin are outlet and inlet temperature, respectively.

8.2.2 Power Requirement of Ohmic Heating

It has been demonstrated by Zhang et al. (2021) that the maximum power per unit volume (*P*v) delivered to a medium for a given electric field intensity can be expressed as

$$Pv = E^2 \omega \varepsilon'' \tag{8.7}$$

where *E* is the RMS value of the electric field intensity (V/m), ω the angular frequency (rad/s), and ε'' the permittivity loss factor (F/m).

For a given electric field intensity, increases in the permittivity loss factor (ε'') will lead to greater energy transfer to the product, resulting is more efficient and faster heating.

The heating time tn required for a given temperature increases (ΔT) can be expressed by the following equation (Zhang et al. 2021):

$$t_{\mathrm{h}} = \frac{\rho C_p \Delta T}{E 2\omega\ \varepsilon''} = \frac{\rho C_p \Delta T}{Pv} \tag{8.8}$$

The penetration depth (δ), where the field decreases to 37% of the surface level of the food product, can be approximated by Eq. (8.9) in the case of high conduction or dielectric losses:

$$\delta = \frac{\sqrt{2}}{\sqrt{\omega\mu\ (\sigma + \omega\ \varepsilon'')}} \tag{8.9}$$

where μ is the magnetic permeability (H/m).

8.3 Effect of Ohmic Heating on Physical Properties of Millets

Physical properties are characteristics of a substance that can be observed or measured without changing the substance's chemical identity. These properties are used to describe and identify materials and are essential for designing any millet processing equipment. Millet grains have variable size, shape, and color. They may be obovoid, cono-spherical, oblanceolate, elliptical, hexagonal, globular in shape and brown, ivory, purple, cream, light blue, yellow, grayish white, or white in color according to their variety (Taylor and Duodu 2017). Physical properties of millets are essential for the design of equipment for harvesting, handling, processing, and storing the grains. They affect conveying characteristics as well as cooling and heating loads of millets (Sahay and Singh 2010). Ohmic heating like other thermal processing technique significantly affects the physical characteristics of millets (Gavahian et al. 2019b). Physical properties include size, shape, cooking time, texture, color, water absorption capacity, viscosity (Dias-Martins et al. 2019), etc.

8.3.1 Texture

Millet grains have a naturally hard texture, which is beneficial for their stability during storage as it protects against mechanical damage and pest infestation, thus extending shelf life (Arendt and Zannini 2013). However, this hardness decreases significantly upon cooking, resulting in a softer texture ideal for dishes like porridge or as a component in multigrain products. The crispness of millet-based products, particularly those made through puffing or extrusion, is influenced by moisture content and processing conditions, such as temperature, pressure, and duration, with higher amylose content contributing to a firmer, crisper texture (Shukla and Srivastava 2014). Cooked millet often develops a sticky texture, especially when high in amylopectin, making it suitable for binding in dishes like dumplings and flatbreads (Saleh et al. 2013). Millet's chewiness, especially in its whole-grain form, is due to its intact bran and fiber content, which provides a firm texture desirable in products like breads and cereal bars. This chewiness can be modified through cooking methods or fermentation, making millet versatile in various culinary applications (McDonough et al. 2000). Finely ground millet flour is known for its smooth texture, essential for creating a pleasant mouthfeel in porridges, baked goods, and beverages, with the smoothness further enhanced by advanced milling techniques (Adekunle and Oluwole 2008). The granularity of millet flour also affects the final product's texture, with coarser flour adding a rustic feel to bread and finer flour producing a soft, uniform consistency in cakes and porridges (Chandrasekara and Shahidi 2011). Although millet flour lacks gluten, it can still achieve a slightly elastic texture in flatbreads when combined with binding agents or through fermentation, allowing for a soft and flexible dough despite the absence of gluten (Chitra et al. 2012). The process conditions in ohmic heating can impact the texture of the product and its overall acceptability to consumers (Apaiah et al. 2005).

The texture profile analysis includes the determination of the hardness, adhesiveness, cohesiveness, springiness, gumminess, and chewiness of the cooked grains (Dias-Martins et al. 2019). With ohmic technology, heating happens rapidly and uniformly, ensuring that finished products maintain their tissue firmness (Sain et al. 2024). In a study by Dias-Martins et al. (2019), the textural properties of whole and decorticated pearl millet cooked using ohmic heating were analyzed with a texture profile analyzer. The whole cooked pearl millet grains exhibited higher values of hardness, springiness, cohesiveness, and gumminess compared to the decorticated grains. Conversely, the whole cooked grains demonstrated lower adhesiveness than the decorticated grains. It is likely that the moderate electric field applied during ohmic heating did not significantly affect the texture parameters due to the lower electrical conductivity of whole grains. However, the electric field had a significant impact on the texture attributes, specifically the hardness and gumminess values, of the decorticated grains (Table 8.1).

Hardness, defined as the maximum force needed to compress the sample, is the most crucial parameter for assessing the quality of cooked cereal (Yu et al. 2009). Firmness is the degree of compactness or solidity of the millet grains. Dias-Martins et al. (2019) in a study on pearl millet found that hardness of ohmic-heated whole grain samples was similar to that of conventionally heated whole grain samples. However, when whole grain samples were compared with decorticated ones, whole grain samples found higher hardness values. The increased hardness of whole grains is attributed to the structure of the pericarp in whole pearl millet grains, which

Table 8.1 Textural attributes of whole and dehulled grains cooked by conventional and OH methods (Source: Dias-Martins et al. 2019)

Treatments	Hardness (N)	Adhesiveness (N.s)	Springiness	Cohesiveness	Gumminess
CONV-whole	34.57 ± 7.18^{a}	0.01 ± 0.004^{b}	0.37 ± 0.07^{a}	0.24 ± 0.03^{a}	9.77 ± 2.32^{a}
OH-whole	37.67 ± 9.48^{a}	0.01 ± 0.004^{b}	0.37 ± 0.06^{a}	0.24 ± 0.03^{a}	9.64 ± 2.09^{a}
CONV-dehulled	11.97 ± 1.18^{c}	0.03 ± 0.006^{a}	0.30 ± 0.05^{b}	0.18 ± 0.01^{b}	2.17 ± 0.64^{c}
OH-dehulled	21.21 ± 2.97^{b}	0.03 ± 0.011^{a}	0.26 ± 0.04^{b}	0.20 ± 0.03^{b}	4.86 ± 1.35^{b}

Note: Different letters (a, b, c) in the same column indicate significant differences between samples ($p < 0.05$)

contains fibers, proteins, and lipids that contribute to greater rigidity. Removing the pericarp through decortication improves water absorption, thereby reducing hardness.

In another study, the feasibility of ohmic baking for cakes and its impact on texture was studied and compared with conventional baking. Initially, the firmness of ohmic-baked crustless cakes was similar to conventionally baked cakes (around 2.5 N). However, after 12 days of storage, the firmness increased to 6.7 N for ohmic baking and 9.9 N for conventional baking. The authors suggested that greater moisture migration from crumb to crust in conventional cakes led to a higher rate of hardening compared to the crustless ohmic cakes. Additionally, the shorter duration of ohmic baking led to less moisture loss, resulting in a moister cake after storage (Luyts et al. 2013). Ohmic cooking has been explored for bread baking, with a study showing that voltage and frequency influence the final product's hardness. The authors noted that an optimized ohmic process can produce bread with the same hardness as oven-baked bread (Gavahian et al. 2019b). As per these studies, we can say that baking a millet flour using ohmic heating technology will be an innovative baking technology.

D.cruz et al. (2023) parboiled paddy using ohmic heating technology at various salt content. They observed that as salt concentration increases, hardness of paddy increases. The reason is that higher salt content has greater ionic concentration, allowing more electric current to pass through (Kaur and Singh 2016). This increased current causes more electroporation, as seen in the electric current values (Kaur and Singh 2016; Knirsch et al. 2010). The electric field exceeds the cell wall's dielectric strength, creating pores by disrupting the membranes. This enhances membrane permeability, making starch more accessible to water (Koral 2004) which enhances gelatinization of starch and improves hardness also. Since, parboiling of millets also occur that heals grain cracks, reduces breakage, improves de-husking, and boosts milling yield. It also alters the physical, chemical, and nutritional properties of millets, enhancing dry-milling, shelf life, insect and mold resistance, cooking, and overall processing qualities (Sai Prasanna 2022). Thus, ohmic heating can be an alternative in parboiling of millets.

A study conducted by Jittanit et al. (2017) showed that the hardness of rice cooked by ohmic heating was lower than that cooked using electric heater. This is because, during ohmic heating, the temperature distribution within the rice-solution mixture is more uniform due to internal heat generation. As a result, the rice kernels gelatinize evenly throughout. In contrast, an electric rice cooker relies on heat conduction and convection, causing gelatinization to occur more slowly and starting from the outer layer toward the center. This process leads to a firmer texture in conventionally cooked rice. Millets are rich in starch ranging from 60% to 75% of their dry weight, depending on the type of millet and its variety. Thus, starch gelatinization takes place evenly in millets also if cooked using ohmic heating technology. In other study conducted by Gavahian et al. (2019a), the effect of ohmic heating was compared with conventional and microwave heating methods. The initial hardness of the soaked rice was 5125±238 g, which decreased to 747±33 g, 1282±96 g, and 1117±80 g after six minutes of ohmic, microwave, and traditional cooking, respectively. However, extending the cooking time in the traditional and microwave methods further reduced the residual texture of their final products. Ohmic heating efficiently generates heat within electroconductive materials like salted water, thus allow for rapid heating, while microwave heating is influenced by the dielectric properties of the materials. Although higher salt concentrations can boost microwave heating rates, the low dielectric constants of lightly salted water and cereal grains result in slower heating and reduced textural softening of microwave-treated rice.

In a study by Gaytán-Martínez et al. (2011), the instant corn flour was obtained by ohmic heating for preparation of masa. Hardness, defined as the force required to deform masa under optimal moisture conditions without disintegration, showed significant differences with ohmic heating, ranging from 2.52 to 1.33 N, while traditionally processed masa had a hardness of 1.38 N. The textural properties of masa, such as hardness and adhesiveness, depend on water content. Small particles bind larger ones, while water serves as a plasticizer, influencing masa's rheology.

Farahnaky et al. (2012) conducted a study to examine the effects of various thermal processing methods, including ohmic heating (both high and low power), microwave heating, and conventional heating, on the textural properties of cylindrical pieces of root vegetables such as carrots, red beets, and golden carrots. The results revealed that the hardness of all samples decreased with heating time across all processing methods. The initial hardness of uncooked carrot, red beet, and golden carrot samples was approximately 270 N, 280 N, and 240 N, respectively. After cooking, the residual hardness values were 29.66 N, 27.32 N, and 22.68 N for samples heated conventionally, and 0.14 N, 5.74 N, and 7.40 N for those heated using ohmic methods.

Adhesiveness is defined as the negative force area during the first bite and is measured using a cylindrical probe. The probe is pressed onto the food's surface, and the force required to pull it off is then recorded (Bourne 2002). Adhesiveness measures how sticky or tacky the millet grains become when cooked. In a study on ohmic heating of whole and decorticated pearl millet, Dias-Martins et al. (2019) found that OH cooking did not cause significant changes in the adhesiveness of the millet grains. In rice also, the ohmic heating results were similar to that of electric heater, since adhesiveness is related to the surface of grains and ohmic heating is similar to conventional heating when surface is considered (Jittanit et al. 2017). However, due to the volumetric heating nature of ohmic cooking, differences arise within the interior of the grains, leading to variations in properties that involve the grain's interior compared to conventional methods.

In a study by Gaytán-Martínez et al. (2011), the instant corn flour was obtained by ohmic heating for preparation of masa. When masa has a good texture, it possesses sufficient adhesive force to be rolled and separated into sheets, which is essential for tortilla production (Bello-Pérez et al. 2002). The adhesiveness ranged from 0.44 to 0.21 N using ohmic heating, with traditional process exhibiting a lower adhesive force of 0.17 N (Gaytán-Martínez et al. 2011).

Springiness is an attribute related to the height that the grain recovers during the time between the end of the first bite and the beginning of the second bite (Bourne

2002). This property describes how well millet grains return to their original shape after being compressed. Springiness is an indicator of the grain's resilience and elasticity, contributing to the overall mouthfeel. A higher springiness value in whole grains, whether cooked using conventional or ohmic heating methods, indicates that decortication significantly altered the grain structure. However, no significant difference was observed between the two cooking processes (Dias-Martins et al. 2019). Rice also showed similar springiness values when compared with conventional cooking (Jittanit et al. 2017).

Cohesiveness can be measured by the rate at which food disintegrates under mechanical action, making it a useful indicator of how well the sample holds together during cooking (Yu et al. 2009). Whole grains require greater mechanical resistance to disintegrate compared to decorticated grains, as the pericarp, the outer layer, adds both hardness and cohesiveness (Dias-Martins et al. 2019). The cohesiveness values of rice is also found similar in both conventional and ohmic heating technology (Jittanit et al. 2017).

Gumminess is defined as the product of a grain's hardness and cohesiveness (Bourne 2002). The OH-decorticated grains exhibited greater gumminess than the conventional-decorticated grains (4.86 vs. 2.17), primarily due to the higher hardness of the OH- decorticated grains (21.21 N) (Dias-Martins et al. 2019). The gumminess and chewiness values of some varieties of OH-cooked rice are lower than those of conventionally cooked. This is likely due to the uniform cooking within the rice kernels and the less soggy outer layer in the case of ohmic cooking (Jittanit et al. 2017).

8.3.2 Color

Color is a key quality factor that is closely related to the product's acceptability. Millet color parameters are influenced by various factors and processing methods. The color of millet extrudates is significantly affected by legume content and die head temperature during extrusion (Chakraborty et al. 2014). Different millet varieties exhibit variations in color properties, with pearl millet showing higher L* values for grain compared to finger millet varieties (Dasa and Bình 2019). The color values of pearl millet flour vary among cultivars (Kulthe et al. 2016). Millet pigments, primarily C-glycosylflavones, are concentrated in the peripheral areas of the seed and contribute to its color (Chauhan et al. 2018; Sharma et al. 2021). Processing methods such as decortication and cooking can reduce the absorbance and flavanol content of millet flour. The appearance of millet grains is related to their flavanol content, with brown grains containing more flavanols than straw-colored grains. These color parameters play a crucial role in the acceptance and quality of millet-based products (Table 8.2).

The CIELAB color parameters a* (indicating redness or greenness), b* (indicating yellowness or blueness), and L* (indicating brightness) were generally measured to assess the color of the samples. From these values, chroma (C*) representing color intensity, hue angle (h°) indicating the color shade, and color variation (ΔE^*)

Table 8.2 Colour parameters of cooked pearl millet grains by conventional and OH cooking (Source: Dias-Martins et al. 2019)

Treatments	L*	a*	b*	C*	h°	ΔE*
CONV-whole	48.04±0.47[c]	2.75±0.42[a]	10.44±0.79[b]	10.80±0.86[ab]	75.32±1.26[d]	0.49±0.09
OH-whole	48.36±0.23[c]	2.43±0.12[b]	10.26±0.26[b]	10.55±0.27[ac]	76.68±0.40[c]	
CONV-dehulled	53.88±0.63[b]	1.50±0.02[c]	11.38±0.36[a]	11.48±0.35[a]	82.47±0.31[b]	3.07±0.79
OH-dehulled	56.70±0.33[a]	1.04±0.27[d]	10.31±1.12[b]	10.36±1.14[bc]	84.31±0.9[a]	

*Note: Different letters (a, b, c) in the same column indicate significant differences between samples ($p < 0.05$)

showing the overall color difference were calculated. The color of ohmic-heated whole pearl millet samples were found similar to that of conventional heated samples (Dias-Martins et al. 2019). However, in decorticated samples, ohmic heating gave higher lightness compared to conventional heating (ΔL* = 2.82). Additionally, the ohmic heating showed a reduction in both yellowness (b*) and redness (a*) when compared to the conventional heating. According to Khashayar et al. (2014), a color difference of ΔE* < 1 is imperceptible to the human eye, while a difference of ΔE* > 3 is noticeable. Therefore, ohmic heating causes significant changes in the color of decorticated pearl millet grains, with a ΔE* value of 3.05.

The other study on rice showed that the lightness of the rice decreased from 82.9 ± 0.1 to 78.6±0.1 upon a 7.5 min ohmic process which was similar to that of the microwave and conventional methods (Gavahian et al. 2019a). A similar trend was observed in the color intensity (chroma) of the samples, with all methods (ohmic, microwave, and conventional) decreasing chroma from 9.9±0.2 to approximately 5 after 7.5 minutes of cooking. Electrochemical reactions and electrode corrosion during ohmic heating may also influence the color of processed food (Mercali et al. 2014), which could explain the observed color changes in the rice grains cooked using this method.

8.3.3 Cooking Time

The cooking time is the time required to reduce 90 % of grain's hardness (Gavahian et al. 2019a). It is also recorded when at least 95 % of the grains were no longer opaque or did not have a translucent center (Dias-Martins et al. 2019). Ohmic heating is the method which reduces the cooking time of grains thus saves energy. Grain size directly affects physical properties like cooking time. Larger grains have a bigger endosperm, which requires more time for complete starch gelatinization compared to smaller grains. Pearl millet reaches optimal cooking when the starch granules absorb enough water, resulting in 95% of the grains having a soft texture and a translucent center, indicating complete starch gelatinization and readiness for consumption. The optimal cooking time for whole and decorticated pearl millet grains is 30 minutes and 20 minutes, respectively, using both conventional and ohmic heating methods. The cooking times for different millets vary considerably.

Whole pearl millet grains require 5 minutes (Eyzaguirre et al. 2006), 15 minutes (N'Dri et al. 2013; Hithamani and Srinivasan 2014), or 30 minutes (Suma Pushparaj and Urooj 2014) at 100°C. Decorticated pearl millet grains also take 30 minutes to cook (Chandrasekara et al. 2012). Whole finger millet requires 17 minutes, while decorticated finger millet cooks in just 5 minutes (Shobana and Malleshi 2007). For barnyard millet, decorticated grains need 8 to 12 minutes (Veena et al. 2005). These variations arise from differences in variety, cooking conditions, grain size, water-to-millet ratio, and composition.

In a study, the cooking time required for rice is 7.5 min for ohmic heating while 15 min for microwave and conventional heating (Gavahian et al. 2019a). The reason for shorter cooking time in ohmic heating compared to microwave heating is that the energy conversion efficiency of the magnetron in microwave generation was around 50%, which is quite low compared to the nearly 100% efficiency of ohmic heating (Ramaswamy et al. 2014). Lakshmi et al. (2007) also found that the microwave absorbance efficiency of the rice-water mixture could be as low as 68%. Also, the Pyrex vessel absorbed some energy during microwave or conventional heating, slowing the cooking process. In contrast, ohmic heating generated heat directly within the rice-water mixture. Therefore, ohmic heating proved to be superior to both microwave and conventional methods in terms of cooking time and energy consumption.

8.3.4 Viscosity

Millets, including pearl millet and browntop millet, have been studied for their physicochemical properties, particularly viscosity. The viscosity of whole browntop millet grain pastes ranged from 40,911 to 66,867 mPa-s, while flour pastes showed lower viscosity (Singh et al. 2024). Pearl millet grains exhibited an average viscosity of 1,769.22 mPa-s (Majumdar et al. 2022). Millet starches demonstrate unique gelatinization characteristics, with nonwaxy and waxy millet starches showing maximum viscosities of 300–460 and 800– 1080 B.U., respectively (Kim et al. 2009). The Brabender visco-amylograms of milled millet flour indicated gelatinization temperatures around 75°C, with peak viscosities ranging from 220 to 560 B.U. (Dharmaraj et al. 2012). These viscosity properties are influenced by factors such as amylose content, water binding capacity, and starch granule size. Understanding these characteristics is crucial for exploring the potential of millets in various food applications, including convenience foods and minimally processed products.

Whole and decorticated pearl millets when cooked with ohmic heating have shown viscosity different from conventional heating. The peak viscosity of decorticated pearl millet cooked with OH found lower than conventional heating. The differences in pasting properties between CONV and OH may be attributed to their distinct cooking mechanisms. Ohmic heating generates heat through electrical resistance when an alternating current passes through an electrically conductive food. However, as starch gelatinization increases, the electrical conductivity (σ)

decreases (Wong et al. 2011). The viscosity of all samples increased during the cooling cycle. Setback viscosity measures the retrogradation of starch molecules, primarily amylose, which tend to reassociate during cooling (Jittanit et al. 2017). Therefore, cooked pearl millet samples with higher setback viscosity demonstrate a greater tendency to retrograde into a thicker gel. Preprocessed millet flours with lower viscosity are ideal for weaning foods, as they suit infants' limited chewing ability. Their low viscosity also benefits formulations aimed at improving nutrition without thickening. The ability of modified starchy pastes to alter viscosity with temperature changes makes them versatile for industrial applications, while lower setback values suggest suitability for canned and frozen foods.

The effect of ohmic heating on viscosity of instant corn flour was found significant (Gaytán-Martínez et al. 2011). Maximum viscosity reflects the balance between granule swelling and polymer leaching in starch, influenced by processing conditions. The viscosity of ohmic-heated instant corn flour was significantly affected by temperature and moisture, with differences noted between ohmic heating and traditional processing due to heating methods. In ohmic heating, corn-water mixtures are heated internally by electric power, causing structural starch changes and reduced viscosity. Despite lower viscosity in ohmic-heated instant corn flour compared to traditional processed instant corn flour, the decrease was proportional to temperature and moisture, allowing control over starch gelatinization. Viscosity increases during cooling also called final viscosity, indicating that the components in the flour interact during the retrogradation phase, which is linked to the flour's quality during storage (Guzmán et al. 2009; Fernández-Muñoz et al. 2004). The ohmic-heated instant corn flour showed lower final viscosity than traditional processed instant corn flour, indicating greater starch stability and less retrogradation, leading to softer tortillas when reheated. This suggests that ohmic heating produces tortillas with better shelf life stability compared to the traditional process, where hardening is common.

8.4 Effect of Ohmic Heating on Nutritional Properties of Millet

Ohmic heating (OH) has demonstrated significant potential for preserving the nutritional content of millet, particularly in terms of retaining essential vitamins and minerals. This innovative technique provides uniform and rapid heating, which minimizes nutrient degradation typically seen with conventional cooking methods like boiling or steaming. The quick, even distribution of heat in OH prevents the formation of hot spots, reducing the risk of nutrient loss caused by prolonged exposure to high temperatures. Although research specifically focused on millet is still limited, findings from studies on other grains and food products suggest that OH could enhance the nutritional quality of millet.

8.4.1 Carbohydrates

Ohmic heating significantly enhances the gelatinization of carbohydrates in millet by facilitating rapid and uniform energy transfer through the application of electrical currents. This results in more efficient gelatinization of starch granules compared to conventional cooking methods, improving both the palatability and digestibility of millet- based products (Dias-Martins et al. 2019). Additionally, ohmic heating has been shown to alter starch composition by reducing the levels of both amylose and amylopectin in millets such as kodo millet. This is due to enhanced starch gelatinization and partial hydrolysis during heating. A study by Kabui et al. (2025) reported a maximum of 13.32% reduction in starch content after ohmic heating-assisted parboiling, likely due to the breakdown of crystalline starch into more amorphous forms (Table 8.3).

A study by Li et al. (2018) investigated the impact of ohmic heating on cornstarch and found that applying electric field strengths of 75 and 100 V/cm led to the formation of resistant starch. This increase was associated with the generation of hydroxyl radicals during electrolysis, which interacted with starch components, augmenting the percentage of resistant starch. The study noted that the nonthermal effects of ohmic heating, such as electrolysis, contributed to these changes.

Studies indicate that ohmic heating helps preserve the nutritional integrity of millet carbohydrates by minimizing the leaching of soluble carbohydrates, which are crucial for maintaining the grain's nutrient profile. Conventional methods often result in greater carbohydrate loss due to prolonged exposure to heat and water, whereas the rapid heating facilitated by ohmic technology reduces such nutrient losses.

8.4.2 Protein

Ohmic heating emerged as a novel thermal processing technique that offers unform, rapid heading through the internal generation of heat via electrical resistance within the food matrix. It's applications in cereal processing have prompted interest in evaluating how ohmic heating may impact the protein content and structural

Table 8.3 Effect of ohmic heating on nutritional properties of millets

Nutrient	Effect of ohmic heating	Mechanism	Reference
Protein	Slight increase or restructuring	Protein unfolding or aggregation	Sereechantarerk et al. (2021)
Fat	Minor decrease	Possible lipid migration or oxidation	Sain et al. (2024)
Carbohydrates (Starch)	Slight reduction, increased digestibility	Gelatinization and partial hydrolysis	Kabui et al. (2025)
Vitamins	Moderate loss, especially of heat-sensitive vitamins (e.g., vitamin C, some B- complex)	Heat degradation and leaching effects	Ottaway (2002)

attributes of gains. Specifically, in the context of millets, a study on kodo millet demonstrated that ohmic heating parboiling improved functional properties like water uptake ratio and reduced cooking time, suggesting alterations in the grain's protein matrix (Kabui and Athmaselvi 2024). Additionally, research on proso millet indicated that heat treatments, including ohmic heating, can lead to the formation of hydrophobic protein aggregates, thereby reducing protein digestibility (Gulati et al. 2017).

Ramírez-Jiménez et al. (2019) in a study have shown that ohmic heating can induce measurable alterations in protein composition. For instance, corn subjected to ohmic heating exhibited a marginal increase in protein content, rising from 7.98% to 8.01%. While this change appears modest, it suggests that ohmic heating may improve protein extractability or cause mild structural alterations that enhance protein availability.

Although specific research focused on millet protein under ohmic heating, insights from research on other plant proteins can provide valuable information. A study on soybean milk proteins demonstrated that ohmic heating induced structural modifications, such as increased free amino groups and decreased surface hydrophobicity, without causing significant denaturation. These structural changes suggest enhanced digestibility and functional properties of plant proteins subjected to ohmic heating (Li et al. 2018; Rocha et al. 2022).

Similarly, research on whey protein revealed that ohmic heating at temperature above 70°C led to conformational changes, producing small protein aggregates and increased clarity compared to conventional thermal treatment, indicating that ohmic heating may preserve nutritional and functional characteristics more effectively (Sereechantarerk et al. 2021).

A comprehensive review on the modification of plant-based proteins emphasized the versatility of ohmic heating in tailoring techno-functional properties. Specifically, the review identified improvement in protein solubility, emulsification, and foaming capacity, all achieved without the use of chemical additives (Avelar et al. 2021). This technology allows for precise customization of processing conditions, making it promising for enhancing the quality of plant protein.

The specific effects on protein content remain less defined, as many studies emphasize overall quality improvements, stabilization, and sensory aspects rather than explicitly detailing protein changes. While direct information on the effects of OH (ohmic heating) on protein content in millet is limited, it is known that heating techniques like this can impact protein structure in food products. These effects include protein aggregation, the formation of new molecular bonds, and conformational changes that influence protein functionality (dos Santos et al. 2024). These structural changes are crucial, particularly for how proteins behave during further processing and their nutritional quality.

Research on pearl millet cooked using OH has shown improvements in sensory properties and stability, although the specific effects on protein content have not been extensively studied (Dias-Martins et al. 2019). The studies suggest that OH treatment reduces water solubility and absorption indices, pointing to possible protein denaturation. However, the primary focus tends to be on overall quality stabilization rather than detailed protein measurements (Dias-Martins et al. 2019). The influence of heating parameters such as temperature and treatment time is critical in determining the extent of protein denaturation and functionality during OH treatment (Sain et al. 2024; dos Santos et al. 2024).

8.4.3 Fat

Ohmic heating (OH) can enhance the stability of fat content in millet during processing. Unlike traditional thermal treatments, which often involve high temperatures that accelerate fat oxidation, OH operates at lower, more controlled temperatures, reducing oxidation rates and preserving fat quality and functionality (Sain et al. 2024). This results in a longer shelf life for millet-based products and helps prevent the degradation of essential fatty acids.

8.4.4 Vitamins

Vitamins vary in their sensitivity to heat and moisture, and traditional heating methods can cause significant losses. Conventional thermal processes often degrade heat- sensitive vitamins such as vitamin C, B vitamins, and folate (Ottaway 2002). In contrast, OH's uniform heating may help mitigate such losses. OH is generally less damaging to food components, which could help preserve vitamins more effectively than traditional methods (Table 8.3). Although the benefits of OH in maintaining food quality and nutritional value have been well documented, there is limited research specifically on its effects on vitamin retention in millet. Most studies focus on the efficiency of OH in preserving the quality of other food products, like fruits, vegetables, and juices, with evidence of vitamin preservation (Sain et al. 2024). Further research on millet is needed to draw definitive conclusions about the impact of OH on its vitamin content.

8.4.5 Minerals

Research indicates that specific minerals in millet, such as calcium and magnesium, may benefit from the application of OH. However, more focused studies are needed to determine the comprehensive effects of OH specifically on millet's mineral retention and bioavailability. Such studies could provide clearer insights into how OH can be optimized for processing millet while maximizing its nutritional value (Kabui and Athmaselvi 2024).

8.5 Conclusion

Ohmic heating technology holds significant promise for processing millet-based products, enhancing both nutritional and functional qualities. OH provides rapid and uniform heating, which minimizes nutrient loss and preserves sensitive vitamins and minerals, offering an advantage over conventional method. By promoting effective gelatinization without nutrient leaching, OH improves millet's digestibility and texture, essential for consumer acceptability. This technology also optimizes texture, controlling qualities like hardness, springiness, and cohesiveness, especially in decorticated grains. Additionally, OH maintains or enhances millet's color properties, making it suitable for visually appealing, market-ready product.

Energy-efficient and adoptable to diverse processing condition, OH also reduces cooking time, making it ideal for developing ready-to-cook and shelf-stable millet products. Further research could refine OH applications to maximize nutritional retention and structural integrity in millet products, positioning OH as an innovative approach in the development of high-quality, nutrient-dense millet-based foods.

References

Adekunle AO, Oluwole OB (2008) Impact of particle size and soaking on functional properties of millet flour. Food Sci Technol Int 14(6):501–508

Alkanan ZT, Altemimi AB, Al-Hilphy AR, Watson DG, Pratap-Singh A (2021) Ohmic heating in the food industry: developments in concepts and applications during 2013–2020. Appl Sci 11(6):2507

Apaiah RK, Hendrix EM, Meerdink G, Linnemann AR (2005) Qualitative methodology for efficient food chain design. Trends Food Sci Technol 16(5):204–214

Arendt EK, Zannini E (2013) Cereal grains for the food and beverage industries. Elsevier

Avelar Z, Vicente AA, Saraiva JA, Rodrigues RM (2021) The role of emergent processing technologies in tailoring plant protein functionality: new insights. Trends Food Sci Technol 113:219–231

Bello-Pérez LA, Osorio-Díaz P, Agama-Acevedo E, Núñez-Santiago C, Paredes-López O (2002) Chemical, physicochemical and rheological properties of masas and nixtamalized corn flour. Agrociencia 33(3):319–328

Bourne M (2002) Food texture and viscosity: concept and measurement. Elsevier

Chakraborty SK, Singh DS, Kumbhar BK (2014) Influence of extrusion conditions on the colour of millet-legume extrudates using digital imagery. Irish J Agric Food Res:65–74

Chandrasekara A, Shahidi F (2011) Antiproliferative potential and DNA scission inhibitory activity of phenolics from whole millet grains. J Funct Foods 3(3):159–170

Chandrasekara A, Naczk M, Shahidi F (2012) Effect of processing on the antioxidant activity of millet grains. Food Chem 133(1):1–9

Chauhan M, Sonawane SK, Arya SS (2018) Nutritional and nutraceutical properties of millets: a review. Clin J Nutr Dietetics 1(1):1–10

Chitra M, Singh V, Ali SZ (2012) Effect of processing on the microstructure of finger millet by scanning electron microscopy and its implications for starch digestibility. Food Res Int 49(1):172–179

Dasa F, Binh LN (2019) A comparative study on rheological, functional and colour properties of improved millet varieties and Injera. J Agric Sci Food Res 10(3):1–8

D.cruz V, Chandran M, Athmaselvi KA, Rawson A, Natarajan V (2023) Ohmic heating using electrolytes for paddy parboiling: a study on thermal profile, electrical conductivity, milling quality, and nutritional attributes. J Food Process Eng 46(3):e14276

Dharmaraj U, Ravi R, Malleshi NG (2012) Physicochemical and textural characteristics of expanded finger millet. Int J Food Prop 15(2):336–349

Dias-Martins AM, Cappato LP, da Costa Mattos M, Rodrigues FN, Pacheco S, Carvalho CW (2019) Impacts of ohmic heating on decorticated and whole pearl millet grains compared to open-pan cooking. J Cereal Sci 85:120–129

Dornoush J, Bagher HSM (2022) Ohmic heating application in food processing: recent achievements and perspectives. Foods Raw Mater 10(2):216–223

dos Santos IF, Pimentel TC, da Cruz AG, Stringheta PC, Martins E, Campelo PH (2024) Ohmic heating in food processing: an overview of plant-based protein modification. PRO 12(9):1800

Eyzaguirre RZ, Nienaltowska K, De Jong LE, Hasenack BB, Nout MR (2006) Effect of food processing of pearl millet (Pennisetum glaucum) IKMP-5 on the level of phenolics, phytate, iron and zinc. J Sci Food Agric 86(9):1391–1398

Farahnaky A, Azizi R, Gavahian M (2012) Accelerated texture softening of some root vegetables by Ohmic heating. J Food Eng 113(2):275–280

Fernandez-Munoz JL, Rojas-Molina I, Gonzalez-Davalos ML, Leal M, Valtierra M E, Martin-Martinez ES, Rodriguez ME (2004) Study of calcium ion diffusion in components of maize kernels during traditional nixtamalization process. Cereal Chemistry 81(1): 65–69

Gavahian M, Chu YH, Farahnaky A (2019a) Effects of ohmic and microwave cooking on textural softening and physical properties of rice. J Food Eng 243:114–124

Gavahian M, Tiwari BK, Chu YH, Ting Y, Farahnaky A (2019b) Food texture as affected by ohmic heating: mechanisms involved, recent findings, benefits, and limitations. Trends Food Sci Technol 86:328–339

Gaytán-Martínez M, Figueroa JDC, Morales-Sanchez E, Vazquez-Landaverde PA, Martínez-Flores HE (2011) Physicochemical properties of masa and corn tortilla made by ohmic heating. Afr J Biotechnol 10(71):16028–16036

Gulati P, Li A, Holding D, Santra D, Zhang Y, Rose DJ (2017) Heating reduces proso millet protein digestibility via formation of hydrophobic aggregates. J Agric Food Chem 65(9):1952–1959

Guzman AQ, Flores MEJ, Escobedo RM, Guerrero LC, & Feria JS (2009) Changes on the structure, consistency, physicochemical and viscoelastic properties of corn (*Zea mays* sp.) under different nixtamalization conditions. Carbohydrate Polymers 78(4): 908–916

Hithamani G, Srinivasan K (2014) Effect of domestic processing on the polyphenol content and bioaccessibility in finger millet (Eleusine coracana) and pearl millet (Pennisetum glaucum). Food Chem 164:55–62

Icier F, Ilicali C (2005) Temperature dependent electrical conductivities of fruit purees during ohmic heating. Food Res Int 38(10):1135–1142

Jan B, Shams R, Rizvi QEH, Manzoor A (2021) Ohmic heating technology for food processing: a review of recent developments. J Postharvest Technol 9(1):20–34

Jittanit W, Khuenpet K, Kaewsri P, Dumrongpongpaiboon N, Hayamin P, Jantarangsri K (2017) Ohmic heating for cooking rice: electrical conductivity measurements, textural quality determination and energy analysis. Innovative Food Sci Emerg Technol 42:16–24

Kabui KK, Athmaselvi KA (2024) Ohmic heating parboiling as a novel pretreatment to enhance the milling properties of Kodo millet. J Cereal Sci 120:104049

Kabui KK, Venkatachalapathy NA, Baskaran N, Athmaselvi KA (2025) Ohmic heating-assisted parboiling: a modern approach to Kodo Millet processing and its effect on starch content and antinutritional factors. Available at SSRN: https://ssrn.com/abstract=5228248 or https://doi.org/10.2139/ssrn.5228248

Kaur N, Singh AK (2016) Ohmic heating: concept and applications-a review. Crit Rev Food Sci Nutr 56(14):2338–2351

Khashayar G, Bain PA, Salari S, Dozic A, Kleverlaan CJ, Feilzer AJ (2014) Perceptibility and acceptability thresholds for colour differences in dentistry. J Dent 42(6):637–644

Kim SK, Sohn EY, Lee IJ (2009) Starch properties of native foxtail millet, Setaria italica Beauv. J Crop Sci Biotechnol 12:59–62

Knirsch MC, Melo C, Martinelli JR (2010) Ohmic heating in food processing. Food Sci Technol 30(2):240–248

Koral T (2004) Radio frequency heating and post-baking. Biscuit World 7(4):1–7

Kulthe AA, Thorat SS, Lande SB (2016) Characterization of pearl millet cultivars for proximate composition, minerals and anti-nutritional contents. Adv Life Sci 5(11):4672–4675

Kumar C, Ramaswamy HS (2004) Application of ohmic heating for food processing. Food Eng Rev 6(1):45–59

Lakshmi S, Chakkaravarthi A, Subramanian R, Singh V (2007) Energy consumption in microwave cooking of rice and its comparison with other domestic appliances. J Food Eng 78(2):715–722

Leizerson S, Shimoni E (2005) Effect of ultrahigh-temperature continuous ohmic heating treatment on fresh orange juice. J Agric Food Chem 53(9):3519–3524

Li X, Ye C, Tian Y, Pan S, Wang L (2018) Effect of ohmic heating on fundamental properties of protein in soybean milk. J Food Process Eng 41(3):1–10

Luyts A, Wilderjans E, Van Haesendonck I, Brijs K, Courtin CM, Delcour JA (2013) Relative importance of moisture migration and amylopectin retrogradation for pound cake crumb firming. Food Chem 141(4):3960–3966

Majumdar AD, Munjal N, Kamboj U (2022, May) Characterization of Pennisetum Glaucum (Pearl Millet). In: J Phys 2267(1):012020. IOP Publishing

McDonough CM, Rooney LW, Serna-Saldivar SO (2000) The millets. In: Handbook of cereal science and technology revised and expanded. CRC Press, pp 177–201

Mercali GD, Schwartz S, Marczak LDF, Tessaro IC, Sastry S (2014) Ascorbic acid degradation and colour changes in acerola pulp during ohmic heating: effect of electric field frequency. J Food Eng 123:1–7

N.Dri D, Mazzeo T, Zaupa M, Ferracane R, Fogliano V, Pellegrini N (2013) Effect of cooking on the total antioxidant capacity and phenolic profile of some whole-meal African cereals. J Sci Food Agric 93(1):29–36

Ottaway PB (2002) The stability of vitamins during food processing. In: Henry CJK, Chapman C (eds) The nutrition handbook for food processors. Woodhead Publishing, pp 247–264

Piette G, Buteau ML, De Halleux D, Chiu L, Raymond Y, Ramaswamy HS, Dostie M (2004) Ohmic cooking of processed meats and its effects on product quality. J Food Sci 69(2):fep71–fep78

Ramaswamy HS, Marcotte M, Sastry S, Abdelrahim K (eds) (2014) Ohmic heating in food processing. CRC Press, Boca Raton

Ramírez-Jiménez AK, Rangel-Hernández J, Morales-Sánchez E, Loarca-Piña G, Gaytán-Martínez M (2019) Changes on the phytochemicals profile of instant corn flours obtained by traditional nixtamalization and ohmic heating process. Food Chem 276:57–62

Rocha RS, Silva R, Ramos GL, Cabral LA, Pimentel TC, Campelo PH et al (2022) Ohmic heating treatment in high-protein vanilla flavored milk: Quality, processing factors, and biological activity. Food Res Int 161:111827

Sahay KM, Singh KK (2010) Introduction. In: Unit operations of agricultural processing, 2nd edn. Vikas Publishing House, New Delhi, pp 6–21

Sai Prasanna N (2022, April 7) The impact of parboiling on the quality of the millets. Food Infotech. https://www.foodinfotech.com/the-impact-of-parboiling-on-the-quality-of- the-millets/

Sain M, Minz PS, John H, Singh A (2024) Effect of Ohmic Heating on Food Products: An In-Depth Review Approach Associated with Quality Attributes. J Food Process Preserv 2024(1):2025937

Saleh AS, Zhang Q, Chen J, Shen Q (2013) Millet grains: nutritional quality, processing, and potential health benefits. Compr Rev Food Sci Food Safety 12(3):281–295

Sastry SK (2005) Ohmic heating and its applications. J Food Sci 70(1):R64–R66

Sereechantarerk C, Hongsprabhas P, Chanput W, Kamonpatana P (2021) Effects of ohmic heating on structural and physicochemical changes of whey proteins. Agriculture and Natural Resources 55(3):464–472

Sharma R, Sharma S, Dar BN, Singh B (2021) Millets as potential nutri-cereals: a review of nutrient composition, phytochemical profile and techno- functionality. Int J Food Sci Technol 56(8):3703–3718

Shobana S, Malleshi NG (2007) Preparation and functional properties of decorticated finger millet (Eleusine coracana). J Food Eng 79(2):529–538

Shukla K, Srivastava S (2014) Evaluation of finger millet incorporated noodles for nutritive value and glycemic index. J Food Sci Technol 51:527–534

Singh S, Suri S, Singh R (2024) Determination of viscosity of browntop millet grain paste and whole browntop millet flour paste. Curr Nutr Food Sci 20(9):1138–1144

Skudder P, Biss C (1987) Aseptic processing of food products using ohmic heating. Chem Eng (London) 433:26–28

Suma Pushparaj F, Urooj A (2014) Antioxidant activity in two pearl millet (Pennisetum typhoideum) cultivars as influenced by processing. Antioxidants 3(1):55–66

Tadini CC, Gut JA (2022) The importance of heating unit operations in the food industry to obtain safe and high-quality products. Front Nutr 9:853638

Taylor JR, Duodu KG (2017) Sorghum and millets: Grain-quality characteristics and management of quality requirements. In: Cereal grains. Woodhead Publishing, pp 317–351

Tian X, Wu W, Yu Q, Hou M, Jia F, Li X, Dai R (2016) Quality and proteome changes of beef M. longissimus dorsi cooked using a water bath and ohmic heating process. Innov Food Sci Emerg Technol 34:259–266

Veena B, Chimmad BV, Naik RK, Shanthakumar G (2005) Physico-chemical and nutritional studies in Barnyard millet. Karnataka J Agric Sci 18

Wong LL, Xu YW, Lu ZH, Li LT (2011) Correlation of electrical conductivity and thermal properties of native starch during ohmic heating. Int J Food Eng 7(5)

Yogesh K (2016) Pulsed electric field processing of egg products: A review. J Food Sci Technol 53:934–945

Yu S, Ma Y, Sun DW (2009) Impact of amylose content on starch retrogradation and texture of cooked milled rice during storage. J Cereal Sci 50(2):139–144

Zhang H, Mohamed A, Breikin T, Howarth M (2021) Modelling and simulation of an ohmic heating process. Open J Model Simul 9(1):26–42

Infrared Insights: Millets Processing with Heat and Light: Harnessing Infrared for Precision and Energy-Efficient Processing

9

Lalita, Mattaparthi Lakshmi Durga, Aseeya Wahid, Rajendra Hamad, and Puneet Kumar

9.1 Introduction

Millets, tiny powerhouses of nutrition, are gaining popularity as a sustainable and healthy alternative to major staple food crops like rice and wheat. Their resilience, high nutrient content, and gluten-free nature make them ideal for future food security. While they are not consumed at large scale in the developed world, millets are crucial staples in many developing countries due to prominently grown in third world countries and their resilience in dry climates. The worldwide production of millets increased from 27.05 to 30.85 million tons in 2011 to 2021. During the same period, production in India slightly increased from 12.7 to 13.21 million tonnes (FAO 2024). Pearl millet, foxtail millet, proso millet, and finger millet are the most prominent millet varieties. Renowned for their resilience, millets are a staple food in resource-scarce regions, serving as a cornerstone of food security. Millets are enrichment of essential minerals like calcium, iron, magnesium, selenium, and manganese, millets offer a substantial nutritional profile (Fernandez et al. 2003). Their low glycemic index makes them a suitable dietary choice for individuals managing diabetes. While not a primary protein source, millets are notably rich in the

Lalita (✉)
ICAR-Central Institute of Agricultural Engineering, Bhopal, India

Faculty of Agricultural Engineering, Sher-e-Kashmir University of Agricultural Sciences & Technology of Jammu, Jammu, India

M. L. Durga · R. Hamad
ICAR-Central Institute of Agricultural Engineering, Bhopal, India

A. Wahid
ICAR-Central Institute of Agricultural Engineering, Bhopal, India

School of Agricultural Sciences Shri Venkateshwara University, Gajraula, UP, India

P. Kumar
ICAR-Central Institute of Temperate Horticulture, Srinagar, India

S. Jaddu, R. Chandra Pradhan (eds.), *Innovative Millet Processing*,
https://doi.org/10.1007/978-3-031-99864-5_9

amino acid methionine. Traditional processing often relies on conventional heating techniques that can be energy-intensive, leading to inconsistencies in finished product. Additionally, compact equipment allows for precise control over heating (Sakare et al. 2020). IR radiation heats objects directly by transferring energy through their molecules, offering several advantages for millet processing, including precise heating, improved quality, faster processing times, and reduced environmental impact (Puneet et al. 2022). IR technology can be applied to various stages of millet processing, from dehulling and drying to parboiling and roasting. Precise control over heating ensures consistent results, minimizing product variations and ensuring high-quality millet products.

IR radiations are categorized based on its wavelength, encompassing near, mid, and far IR radiations. Interestingly, different components within food absorb IR radiation in unique ways. For instance, proteins preferentially absorb IR within a specific range of wavelengths, between 6 and 11 μm, allowing for easy control and quick adjustments during processing. Unlike traditional methods, IR radiation penetrates directly into the food itself, without heating the surrounding air, resulting in high heat transfer capacity (Riadh et al. 2015). IR heating finds applications in drying, blanching, pasteurizing, thawing food items etc. One challenge associated with IR heating is the limited depth of penetration into food, but selecting the appropriate IR wavelength can significantly improve radiation travel depth within the food Unlike opaque materials like metals, food surfaces exhibit a more complex interaction with IR radiation, as it can interact with the entire bulk of the food, not just the surface. The interaction depends on the food's composition, the chosen IR wavelength, and other factors (Manyatsi et al. 2023). Beyond heating, IR radiation offers other valuable applications. Fourier-transform IR (FTIR) spectroscopy leverages a portion of the IR spectrum for analytical purposes, allowing for the measurement of the chemical composition of a sample. Near-infrared (NIR) spectroscopy, spanning the wavelength range of 380–2500 nm, is of particular interest due to its importance in various applications (Wang et al. 2015). This region of the spectrum is commonly used in multispectral and hyperspectral imaging technologies for assessing fruit quality and other agricultural applications (Wang et al. 2015). Additionally, organic materials with strong absorption in the NIR region are used in areas such as thermal management and energy conversion (Patil and Chaudhary 2020). NIR spectroscopy in association with chemometric methods such as Principal Component Regression and Partial Least-Squares are very good tool for extracting valuable information from raw data. In contrast, mid-infrared (MIR) spectroscopy exploits the principles of fundamental vibration modes which differentiate between functional groups. For analysis point of view, MIR evaluate linear regression analysis at a specific wavenumber can sometimes be effective and sufficient (Qu et al. 2015; Johnson et al. 2023).

9.2 Fundamentals of Infrared Radiation

Infrared radiation, discovered by F. W. Herschel, is of fundamental importance in modern science and technology. The field of IR technology encompasses target-background-atmospheric radiation interactions and electro-optical systems. Understanding these fundamentals is crucial for applying IR technology to systems engineering problems. Despite its significance, the history and scope of IR radiation discovery are not widely known, unlike other scientific breakthroughs such as X-rays. Over the past decades, IR has transitioned from a purely research topic to a field of practical applied science with great potential value. This evolution emphasis the great scope of IR technology in various applications and offers vast opportunities for continued education and research in this area.

IR is an electromagnetic wave generated from heated materials (both naturally and artificially), and its wavelength lies just outside the range of visible light. The wavelength of IR ranges from 0.75 to 1000 μm (Fig. 9.1). However, this entire IR range is further divided into three categories, but the specific ranges for each category are not clear (Huang et al. 2021). This type of radiation encompasses the wavelength range between the visible light and the microwave regions, with NIR, MIR, and far-infrared (FIR) regions being the primary subdivisions. The interaction of IR radiation with materials is a complex phenomenon, involving processes such as reflection, transmission, and absorption. The IR radiation originates from the thermal energy emitted by objects, with the intensity of this emission increasing as the fourth power of the object's absolute temperature, as described by the Stefan-Boltzmann law (Skotnicki and Knap 2018).

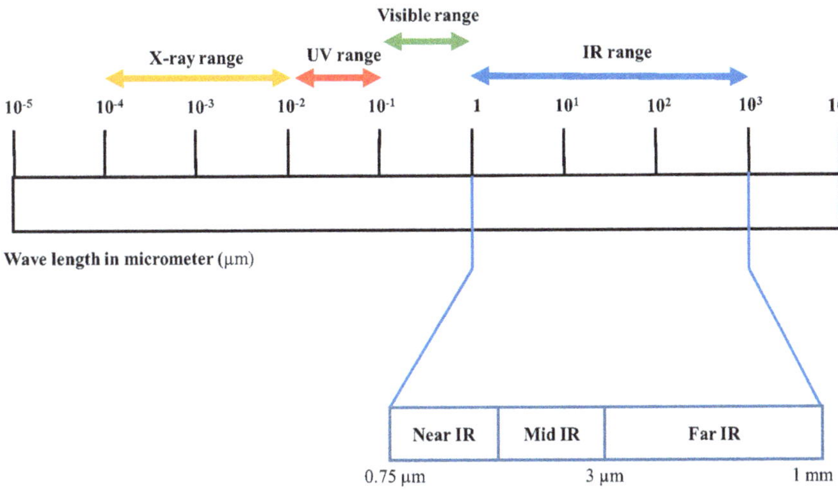

Fig. 9.1 Electro-magnetic spectrum distribution with focus on IR wave range. (Adopted from Puneet et al. (2022), Sakare et al. (2020), Aboud et al. (2019))

9.2.1 Infrared Radiation Heating Mechanism

Electromagnetic (EM) waves can traverse across a vacuum without the need for a medium, IR being part of EM, also don't need to make surface-to-surface contact for heat transmission. The heat source radiates IR radiation, which directly reaches the material's interior layer by penetrating the heated surface. The molecules in the various layers of the material absorb the IR energy, which raises and varies the vibrational energy level of the molecules to produce heat and raise the temperature. One of the prominent benefits of IR heating is its ability to reduce energy losses and significantly preserve the product's original quality. The surface temperature rises swiftly, substantially quicker than that of conduction and convection heat transfer mode. Since IR radiation's energy is immediately absorbed by material's molecules and therefore has a very high energy density. The wavelength of the radiation and the properties of the sample under treatment (such as moisture content, MC) determine how far the radiation can penetrate. The mode of heat transfer of EM radiation, when strikes on food's surface, takes place through change in rotational, vibrational and electronic state of molecules or atoms. The incident IR rays can be reflected, absorbed, or scattered when it comes into contact with food material (Fig. 9.2). Various food components absorb different ranges of IR radiations, so impacts the absorption intensities of IR at different wavelengths. The incident radiation's wavelength dependent energy absorption mechanisms can be divided into three categories: (i) alteration in the electronic state associated with wavelengths between 0.2 and 0.7 μm (visible & ultraviolet rays); (ii) alteration in the vibrational state associated with wavelengths between 2.5 and 1000 μm (FIR); and (iii) alteration in the rotational state associated with wavelengths exceeding 1000 μm (microwave). The efficiency of IR heating is function of absorption behaviours of the material being heated. Materials with a high absorption coefficient in the IR region, such as organic compounds, metals, and certain ceramics, are particularly effective at converting IR radiation into heat (Patil and Chaudhary 2020). This is due to the ability of these materials to interact with the oscillating electric and magnetic fields of the IR radiation, causing the molecules or atoms to vibrate and generate thermal energy.

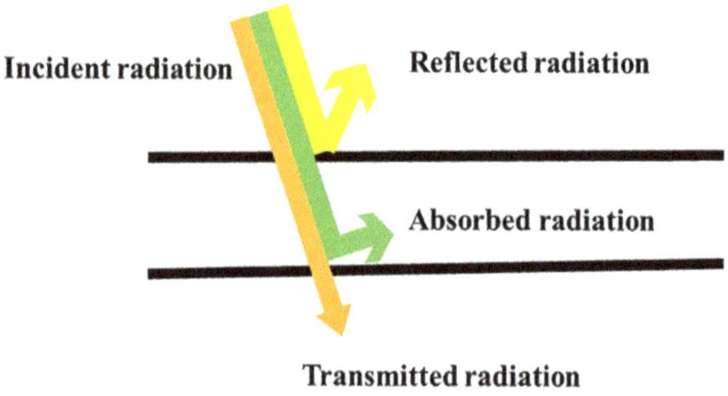

Fig. 9.2 Extinction of total radiation (absorption, transmission and reflection)

9.3 Application of IR Heating on Hulling of Millets

Millets are traditionally processed using methods like wooden pestle and mortar or stone rotary mills in India, while Africa uses abrasive disk dehullers (Murty and Kumar 1995). It involves a two-stage approach: first, decortication to remove bran through moistening, followed by winnowing or sieving to separate bran from grain (Smith 1996). The next is the reducing endosperm to meal or flour. However, mechanical milling now largely supersedes these traditional methods to some extent, improving efficiency but still posing challenges due to the small grain size of millets, which complicates separation of hull from endosperm during dehulling and bran from flour while milling. Significant loss of hull and bran takes otherwise it could be used for extraction of silica, oil and proteins and serve as feed to animals, poultry and fishes. In this case, pre-milling treatment such as IR radiation heating becomes essential for easy and maximize separation of hull and bran from endosperm. The hulling and milling operation greatly depends upon the morphology of millet kernels. Generally, the millet kernels' shape varies from tear-shaped to ovoid, that is similar to pearl millet, sorghum and maize, except finger millet those are spherical in shape (Abdelrahman et al. 1984; McDonough et al. 1986). All millets are also distinct in terms of their colour i.e., creamy white, gray, and purple (Pearl millet), white, red, yellow, brown, or black shades (Foxtail millets) and white to brown (Finger millets). Whereas, the size of mostly all millets lies in range between 1–2 mm in diameter and 1000 kernel weight around 8–15 g.

Millets are two types on the basis of husk, one has outer shiny cover (foxtail, little, kodo, proso, barnyard) called husk and another without husk (finger millet, pearl millet and jowar). In general, millets consist of starchy endosperm, pericarp, aleurone layer, testa and germ and shown in Figs. 9.3 and 9.4.

Pearl millet, predominantly grown in India, has grains with a large germ (up to 21%) (Abdelrahman et al. 1984). Millet grains are primarily composed of the endosperm (about 76%), which is further divided into distinct layers. The outermost layer, the aleurone, surrounds the endosperm. Beneath this, the peripheral endosperm contains dense protein and small starch granules, while the corneous endosperm houses larger, polygonal starch granules embedded in a protein matrix. The innermost part, the floury endosperm, is characterized by loosely packed, large starch granules and a lower protein content.

The remaining portion of the grain (about 37%) is comprised of the bran and husk. Finger millet grains, in particular, have a distinctive five-layered testa that contributes to their color due to the presence of flavonoids and tannins. Millets can be categorized into two main types based on seed structure: utricles and caryopses. Utricles, found in finger millet, proso, and foxtail millets, have a pericarp that loosely surrounds the seed. In contrast, caryopses, like pearl millet, fonio, and teff, have a pericarp tightly fused to the seed. The process of removing the outer layers, known as dehulling, is essential for millet processing. For utricle-type millets, the pericarp can be easily removed by rubbing the grain, either dry or after soaking. However, for caryopses, the fused pericarp and testa form a more robust barrier, requiring more intensive processing methods. Pre-milling treatments can enhance

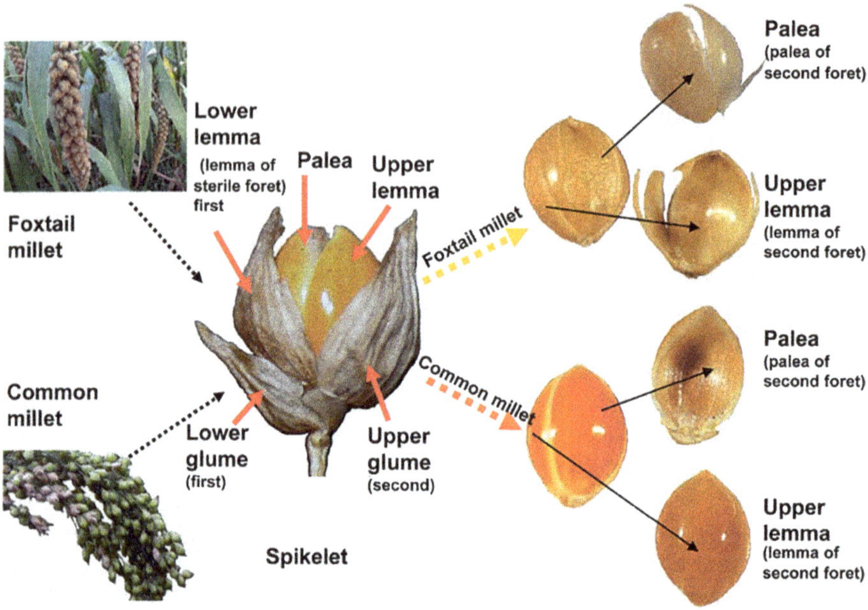

Fig. 9.3 Morphology of millets grain. (Adopted as it from "Phytoliths analysis for the discrimination of foxtail millet (*Setaria italica*) and common millet (*Panicum miliaceum*)" published in open source journal)

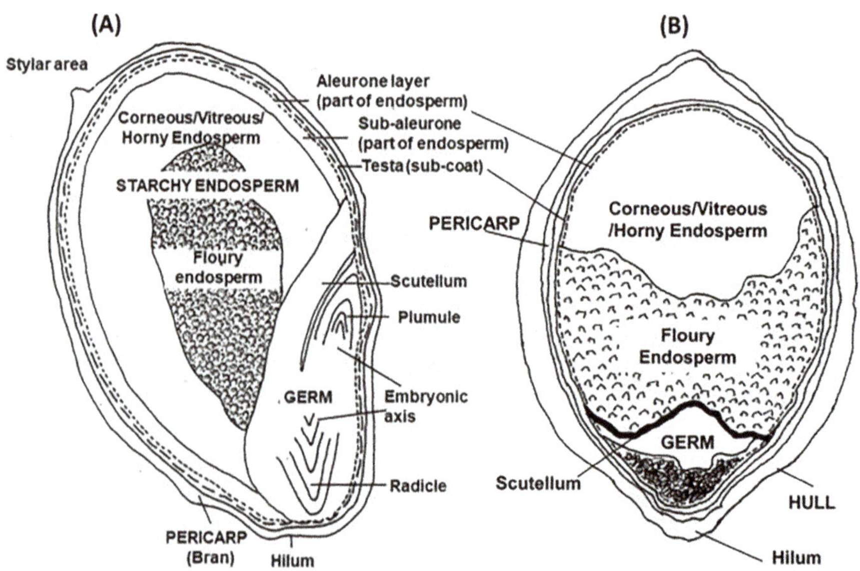

Fig. 9.4 Cross-sectional view of Sorghum and proso millets. (Reprinted with permission from Sorghum and millets. In Ed. ICC handbook of twenty-first century cereal sciences and technology)

the dehulling process and improve the overall quality of millet-based products. When IR radiation hits grain, it causes molecular vibrations, generating intermolecular friction that heats the grain uniformly (Deepa and Hebbar 2016). IR treatment loosens the hull-endosperm attachment by thermally degrading gum and mucilage, making dehulling easier (Kumar et al. 2022). To expose the millets to IR heating it is essential to raise their moisture content, this increased moisture content causes grain expansion in length, width, thickness, and geometric diameter, primarily in the endosperm (Baryeh 2002). Primarily, this expansion occurs in the grain endosperm in comparison to the outer hull. Consequently, when millets are exposed to the IR radiation results in reduction in their moisture content and shrinkage of endosperm and slight shrinkage in hull, creating a gap that eases dehulling. The physical properties of millets are closely linked to their moisture content. Exposure to IR radiation reduces moisture content, altering physical properties and affecting dehulling and milling. Millets with elliptical shapes have lower sphericity compared to cono-spherical shapes (0.783–0.803 vs. 0.905–0.921) (Baryeh 2002). Both sphericity and surface area of millet kernels are directly proportional to their moisture content. Reduced moisture from IR exposure decreases sphericity, leading to uneven surface generation that assist dehulling. The angle of repose of the millets i.e., pearl millet, decreases from 49° at 22.5% to 3.4° at 5% moisture content, this signifies ease of flowability during dehulling in IR pre-milling treated millets. For maximum efficiency in IR heating, food materials should be processed as coarse granules or grains spread out in a thin layer. Continuous agitation of the material is crucial to expose fresh surfaces to the heat source, thereby enhancing heat transfer [Lorentzen 1981]. Hydrothermal processing of grains involves subjecting them to high-pressure steam or hot water to loosen the hull, enhancing milling efficiency (Prashanth et al. 2024). This process includes soaking, steaming, and drying. Post-treatment, the endosperm hardens, reducing grain breakage. Key factors affecting dehulling and nutritional properties include soaking time, steaming duration, drying temperature, and final moisture content. It is found that soaking finger millet at 44 °C for 2.1 hours, followed by steaming at 110 °C at 5 PSI for 5 minutes followed by drying, increased head yield from 70% to 86% and reduced breakage from 15% to 1.3% (Kaur et al. 2020). Rathore and Singh 2017 showed effective decortication of pearl millet after soaking for 8–10 hours, steaming for 30 minutes at 98 °C, and drying, yielding a 65.40% decortication rate. Pre-conditioning followed by heat treatment (i.e., hydrothermal, microwave, gamma irradiation and IR heating) of millets grains result in reduction a* and b*, hue (°h) and colour difference (ΔE) (Mubaiwa et al. 2018; Mwangwela et al. 2007). The IR treated millets milled to flour leads lowest change in lightness, yellowness and redness of moistened millets treated for 5 to 10. minutes. Whereas, application of IR heating can help in higher retaining nutrition and colour pigment in end product. Other methods used for pre-milling treatments poses certain drawbacks, such as in hydrothermal treatment, the nutritional compounds, colour pigments leach out in the soaking solution (Bayram et al. 2004). IR treatment prior to milling reduces the moisture content of millets consequently improves their nutritional composition. The fat content improves with IR treatment, it is might be due to fact that lipids become more extractable because of micro-crack formation was well as pressure difference development. IR treatment may negatively impact the protein solubility or nitrogen solubility index due to protein denaturation and starch aggregation. At higher moisture content, the denatured proteins forms matting like structure around the starch granules and lead to starch aggregation (Ogundele and Emmambux 2018). Whereas, this impact can be reduced by pre-conditioning to moisture content greater than 50% w.b. and exposing to IR over prolong period. In comparison with other dry heat treatments such as microwave & gamma rays, IR have least impact on viscosity and pasting properties of millets flour (Mukwevho and Emmambux 2022). This might be due to microwave and gamma rays denature proteins, restricting their expansion and causing starch-protein aggregation. The denatured protein with exposed hydrophobic sites restricts starch granules access to water for pasting, thus reduction in viscosity. Maltese crosses were visible in untreated samples and evidenced birefringence due to crystalline starch structure. Maltese crosses gradually disappeared with increased treatment time for all samples, suggesting progressive starch gelatinization (Ogundele et al. 2017). Similarly, the water absorption capacity of flour obtained from IR treated millets increases from 1.8 g/g for control to 3.7 g/g for treated millets. This effect might be due to higher heat generation caused by water molecule vibration when exposed to IR heating.

9.4 Application of IR Heating on Milling of Millets

Millet milling transforms raw grains into consumable products. It involves cleaning, removing inedible layers like husk and bran, and polishing. The process differs for shelled and naked millets. Precise milling ensures high-quality, nutrient-rich millet flour or grains. Millet milling machinery is generally less advanced compared to that used for other major cereals like paddy, wheat, and maize, despite the widespread use of traditional milling methods in many countries. Due to the small size of the grains, a major challenge in millet milling is the efficient separation of clean flour from the bran. It is important to employ an efficient grinding technique for convenient incorporation. In last few decades, the consumption of millets has increased due to their numerous benefits, therefore many advancements in milling technologies have been tried in different regions of world. The most remarkable is the fonio dehuller developed was the dehuller had pliable abrasive disks in spiral propeller form with a fixed abrasive tray (Smith 1996) and had capacity of 10 kg per hour. In the 1970s, a horizontal disk-type debranner based on Carter disk designed for sorghum and millet processing. This machine could handle large quantities of grain and achieved a reasonable yield. Another dehulling method involves mimicking the hand-pounding process. The United Milling Systems machine uses a screw to feed pre-moistened grain into a chamber where a rotating rotor with perforated screens removes the husk. This system, combined with a disk mill, produces high flour yields. The SOTRAMIL (Society de Transformation du Mil), developed specifically for pearl millet (Goussault and Adrian 1977; Kent 1983), incorporates a series of steps. After cleaning and washing, the grain is decorticated using a horizontal millstone machine. The resulting product is further cleaned with a brush machine before being ground into flour using an impact grinder. A unique aspect of this process is the use of hot air during grinding, which reduces moisture content, sterilizes the flour, and inactivates enzymes. The final flour is obtained through sieving. The milling of millets yields about 77% and removed husk and bran layer contributes around 23% (Malleshi and Hadimani 1993). Many pre-milling treatments have been employed to improve the dehulling and milled qualities of millets. The dry and wet heat treatments such as microwave, IR, gamma radiation, hot air and hydrothermal treatments extensively evaluated as pre-milling treatments. The milled exposure to temperature above 100 °C is key to improve the dehulling and milling characteristics. In hot air treatment method, the millets are exposed to temperature above 100 °C for 60 to 120 minutes (Kadlag et al. 1995; Nithya et al. 2012), whereas in microwave treatment the millets surfaces may attain 110 °C (Yadav et al. 2011). Under hydrothermal treatment, the millets are first soaked in the grains followed by exposing to high pressure steam i.e., 1.05 kg/ cm^2 (Yadav et al. 2011). Attention must greatly be focused on improvement in size reduction technology for millet grains at commercial scale to ensure the smooth supply of millet flour. Therefore, efforts are initiated to shift millet processing equipment and methods to commercial methods by applying environment-friendly processes in order to ensure the product quality and food sustainability; for instance, commercial methods may result in less viscosity, structural change, and product deterioration under different storage conditions. It is proved that IR heating and milling machine type impact the moisture content and dehulling yield of conditioned legumes. The dehulling yields varied by mill type and temperature. Legumes grains treated at 86 and 123 °C in disc and cracking mills showed yields of 62% and 65%, respectively. In impact mills, yields decreased significantly at higher temperatures. Similarly, the combined effect of microwave and IR on dehulling of various legumes such as black, pinto, red kidney beans have found significant and microwave heating for 4.5 minutes yields satisfactory results. IR heating treated red kidney beans had higher dehulling yields than microwave-treated ones. Overall, IR heating removed more hull than microwave treatment. The mailing pre-treatments enhances the endosperm yields before decortication, consequently raise the nutritional content in millets flour (Mannuramath and Yenagi 2015). The in vitro starch digestibility of millets is about 66%, which can be improved to 74% by hydrolysing millets for 180 minutes causes partial gelatinization of starch, followed by popping them using either microwave, IR and sand or salt heating. (Muralikrishna et al. 1986). Pre-treatment with IR heating further decreases the chances of fat oxidation. Average in vitro protein digestibility of raw and cooked finger millet of 71 and 87%, respectively, which can be further improved with IR pre milling treatment. Millet milling activates lipase enzymes, leading to increased free fatty acid levels and reduced flour shelf life. To address this issue, pre-milling treatments such as infrared radiation, microwave heating, and hydrothermal processing can be employed to inactivate these enzymes and improve flour quality. (Batariuc et al. 2022). These millets contain significant amount of anti-nutritional components i.e., phytic acid, oxalates in pearl millet, foxtail millets and finger millets, and goitrogen only in pearl miller, (Lestienne et al.

2007). Tannins and trypsin inhibitors are also anti-nutrients in millets. These phytic acids decreases the bioavailibity of mineral viz. Ca, Fe and Zn, the goitrogens associated with development of unpleasant odor in milled pearl millet flour (Reddy et al. 1986). The unpleasant odor can be eliminated by dehulling (debranning), soaking, and cooking. The application of thermal treatment prior to milling operation helps in reduction in anti-nutritional compounds in millets (Yadav et al. 2011). Therefore, it is suggested to perform decortication operation, removal of bran layer from millets, significantly reduce flavonoid-type compounds and goitrogen content (Lestienne et al. 2007). The IR heating impacts inhibition of trypsin, denaturation of aflatoxin, and thermal decomposition of haemagglutinins and tannins. Otherwise, without IR pre-milling treatment, the millets take longer time to denature haemagglutinins and trypsin inhibitors. Therefore, IR heating is also assists in reduction in energy required for cooking.

9.5 Application of Infrared Drying on Millet

Infrared drying (ID) takes advantage of the unique qualities of IR radiation to speed up the drying process, with radiation entering the surface and converting into heat within the material. This direct conversion greatly shortens drying time when compared to traditional methods. ID is also highly energy-efficient because the energy is absorbed directly by the material, reducing waste from heating the air. Because of the quick and homogeneous heating, IR drying often improves the quality of dried items by preserving nutritional and sensory qualities like flavour, colour, and texture. Unlike traditional drying methods, which can cause temperature gradients within the material, ID maintains a more consistent temperature, resulting in uniform drying and maintaining product quality (Sakare et al. 2020; EL-Mesery et al. 2022; Nalawade et al. 2018; Jeevarathinam et al. 2021; Nasiroglu and Kocabiyik 2009; Puneet et al. 2022). It requires less airflow around the sample than traditional procedures because radiation, not convection, is the predominant route of heat transfer. The combination of IR with hot air drying enhances drying efficiency by leveraging quick heating and moisture removal capabilities. This hybrid strategy promotes more consistent drying, minimises over-drying or burning, and lowers energy use. The drying time decreases as IR intensity increases because higher intensity causes greater heat absorption, increased sample temperature, and rapid mass transfer, resulting in a faster drying rate and shorter drying period. Higher air temperature while keeping steady IR intensity also shortens the drying time. Due to the supremacy of internal diffusion and the prevalence of bound water during IR drying, the hybrid drying process exhibits no constant rate period but rather a declining rate period. The process starts with removing moisture from larger capillaries and then moves to smaller ones, leading to a slower evaporation rate. (El-Mesery et al. 2022; Jeevarathinam et al. 2021; Nasiroglu and Kocabiyik 2009). The coupling of ID with microwave drying (ID-MD) has advantages as microwaves generate heat by causing water molecules within the material to vibrate, resulting in faster moisture evaporation and substantially shorter drying time. Microwaves, on the other hand, can produce uneven heating and hot spots, putting materials in danger of damage. IR, which heats the surface by direct absorption, provides more even heating. ID-MD uses microwaves to heat the material from within, while IR heats the surface, resulting in uniform heating and reducing the possibility of hot spots and overheating. This combination improves drying efficiency, saves energy, and retains the nutritional and sensory properties of food goods. For example, when drying eggplant, IR-MD shortened drying time while effectively preserving beta-carotene. However, precise regulation of microwave power and IR intensity is required to avoid overheating and provide uniform drying (Aydogdu et al. 2015; Kgonothi et al. 2024). Another approach of hybrid drying of ID with freeze drying (IDFD) combines IR light and standard freeze drying to increase efficiency and save expenses. Freeze drying entails freezing the substance, lowering pressure, and removing ice through sublimation. IDFD increases drying rates and energy efficiency by directly heating the product, reducing energy losses. This approach preserves structural integrity and nutritional quality, making it appropriate for industrial uses in food and pharmaceuticals. For example, IDFD greatly lowers drying time while improving product quality in commodities such as bananas and mushrooms. However, IDFD confronts constraints such as a large initial investment and process complexity, necessitating thorough parameter optimization for each application (Shih et al. 2008; Antal and Kerekes 2016). In summary, ID and IR-assisted hybrid drying systems offer substantial advantages in drying efficiency, energy conservation, and product quality for various food and agricultural products. Each method, whether combined with hot air, microwaves, or freeze drying, presents distinct advantages and factors to consider, indicating the potential for enhanced industrial drying processes.

9.6 Application of IR Heating on Microbial Inactivation

The preservation of food and its derivatives is significantly challenged by the proliferation of mold and other microorganisms. A promising approach to tackle this issue involves the utilization of IR radiation for microbial decontamination. The efficacy of IR radiation in eliminating microorganisms from its differential absorption by various chemical groups present in both food and microorganisms. Exposure to IR radiation induces a heating effect within food, leading to the thermal denaturation of proteins and nucleic acids (Brandl et al. 2008). This thermal stress disrupts the microbial cell wall, shrinks the cytoplasmic membrane, and ultimately results in DNA damage, culminating in the inactivation of microorganisms (Staack et al. 2008). For example, cumin seeds treated with IR radiation exhibited a significant reduction in total mesophilic aerobic bacteria counts (Erdoğdu and Ekiz 2011). Moreover, the combination of IR and UV radiation proved to be even more effective in achieving rapid microbial decontamination (Shavandi et al. 2020). The reduction of B. cereus spores by IR radiation was influenced by factors such as water activity, temperature, and pH. IR treatment, also demonstrated inactivation of Salmonella and Pediococcus spp. and L. monocytogenes. The efficacy of inactivation of microorganism by IR radiation is influenced by factors such as the target microorganism, its growth stage, radiation intensity, exposure time, and moisture content (Gande and Muriana 2003). Its ability to rapidly and effectively inactivate microorganisms, coupled with its potential to preserve food quality, makes it an attractive alternative to conventional methods. However, further research is necessary to optimize IR technology for various food products and to address challenges such as ensuring uniform heat distribution and preventing overheating.

9.7 Introduction to Infrared Spectroscopy

Infrared (IR) spectroscopy is a vital analytical technique used extensively in chemical and food analysis. IR works based on the principle that IR radiation can excite molecular vibrations, which in turn result in characteristic absorption patterns of individual molecular structures and functional groups. IR spectroscopy operates within the IR region, which spans from 0.8 to 1000 micrometers in wavelength, or 12,500 to 10 cm^{-1} in terms of wavenumbers. The underlying working principle of IR spectroscopy is the exciting molecular vibrations. Molecules absorb IR radiation at frequencies that match the vibrational frequencies of their bonds, resulting in characteristic absorption bands that can be recorded as an IR spectrum. These vibrations include stretching, bending and twisting motions of the molecular bonds (Theophanides 2012). Modern IR spectrometers, such as Fourier Transform Infrared (FTIR) spectrometers, have enhanced the resolution and speed of spectral acquisition. FTIR spectrometry, in particular, converts the time-domain data into a spectrum using Fourier transformation, which allows for high-resolution and precise measurements (Ozaki 2021). The basic setup of IR spectroscopy is shown in Fig. 9.5. IR spectroscopy utilized across various fields due to its versatility and non-destructive nature and has been instrumental in characterizing food componets by

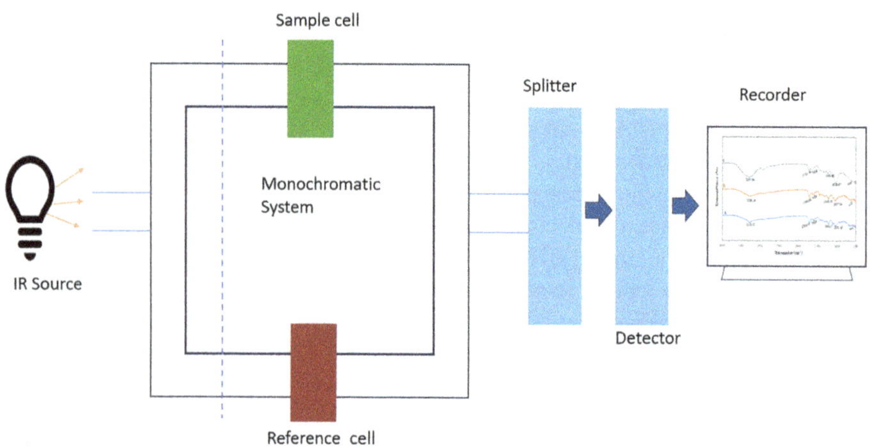

Fig. 9.5 Infrared spectroscopy

identifying specific vibrational fingerprints of different functional groups and fillers within the polymer matrices (El-Azazy et al. 2023). Additionally, it is employed in the analysis of inorganic compounds, food quality assessment, and even in forensic science for substance identification (Ozaki 2021). Therefore, IR spectroscopy becomes an essential and versatile tool in scientific research and industries. Its ability to provide detailed molecular information through non-destructive analysis makes it indispensable in many fields, from chemistry and biochemistry to environmental science, pharmaceuticals, material science, food safety and forensics. In food processing, IR spectroscopy is particularly valuable for assessing the quality and safety of products, ensuring that they meet industry standards and consumer expectations (Qu et al. 2015). The continuous advancements in spectrometric technologies and methods promise to further enhance its applications and efficacy.

9.7.1 Application of Infrared Spectroscopy in Millet Processing

IR spectroscopy has found significant applications in millet processing, serving as a robust analytical tool for quality control, compositional analysis, and process monitoring. One of the primary uses of IR spectroscopy in millet processing is in the assessment of nutritional components such as proteins, fats, carbohydrates, and fibers (Olamiti et al. 2020). By measuring the absorption of IR light at specific wavelengths, IR spectroscopy scan provide detailed information about the molecular composition of millet grains (Wiedemair et al. 2019a, b). This capability is particularly valuable for ensuring the nutritional quality of millet-based products, allowing producers to optimize their processes to retain or enhance these beneficial components (Babu et al. 2019). The various application of IR spectroscopy on millets are sown in Table 9.1.

Quantitative detection of protein, fat, starch, and amino acids in foxtail millet using Fourier Transform Near-Infrared Spectroscopy (NIRS) was investigated by Yang et al. (2013). The FTIR spectra revealed an intense absorbance band in the IR region between 9000 and 4000 cm^{-1}. A prominent band at 8262 cm^{-1} was attributed to the C-H stretching second overtone of $-CH_3$ and $-CH_2-$ groups. Additionally, a band at 5793 cm^{-1} was observed due to the C-H first overtone of $-CH_2-$ groups, and a band at 5180 cm^{-1} is present due to the second overtone of the C=O stretching vibration. A shoulder around 4716 cm^{-1} can be attributed to the C–H and C=O stretching vibrations of -HC=CH- groups. The performance criteria for FT-NIR calibration models are as follows: models with an R^2 value below 0.66 are considered unusable, while models with an R^2 value between 0.66 and 0.83 can be used for sample screening, such as for crude fat and total starch. Models with an R^2 value between 0.83 and 0.92 are suitable for many applications, including the detection of amino acids such as arginine, glycine, histidine, lysine, proline, threonine, tyrosine, and valine. Finally, models with an R^2 value between 0.92 and 0.98 are suitable for most applications and quality control purposes. The study on changes in starch crystallinity of gelatinized foxtail millet using FTIR revealed that crystallization was more likely to occur when the material was dried using

Table 9.1 The various application of IR spectroscopy for millets

S.No.	Millet	Key points	Ref
1.	Proso millet	The developed handheld IR-spectrometer accurately predicted the protein content in proso millet. The multivariate analysis yielded similar results; it showed some drawbacks compared to standard data processing software	Wiedemair et al. (2019a, b)
2.	Fonio, finger and pearl millet	The charterization of polyphenolic fractions of various millets by IR-FTIR spectroscopy exhibits bands at 3286 cm^{-1}, 3274 cm^{-1}, and 3012 cm^{-1}, which were associated with hydroxyl groups (–OH) in all millet types. The broad and distinctive bands in the range 3300–2500 cm^{-1} wavenumber, resulting from –OH stretching. These bands correspond to the stretching vibrations of hydroxyl groups, specifically those attached to phenolic molecules.	Bello et al. (2022)
3.	Foxtail millet	Dehydrated millet starch at high temperatures resulted with V-shaped crystal structure. The V-type structure significantly hinders the starch's ability to absorb water, ultimately impacting its overall hydration properties.	Wang et al. (2017)
4.	Foxtail millet	Developed a cost-effective and rapid method for detecting and quantifying fat in millet. Using a miniature NIR spectrometer connected to a smartphone. A root means square error for prediction of 0.301 g/100 g fat, and a residual predictive deviation (RPD) of 3.225	Wang et al. (2022)
5.	Pearl and finger millet	The differences between the pearl and finger millet flours lie in the 1600–600 cm^{-1} region of the spectrum. A strong peak around 3468 cm^{-1} for finger millet flour, indicating the presence of O-H bonds (likely from hydroxyl groups). Pearl millet flour showed a similar peak but at a slightly lower frequency (3466 cm^{-1}). Both flours displayed a peak around 2925 cm^{-1}, suggesting C-H bonds (from organic molecules).	Gull et al. (2016)
6.	Foxtail millet	A developed calibration models for the prediction of various quality parameters (crude protein content, total starch content, and apparent amylose content) in both whole and dehusked foxtail millet grains. The model designed for crude protein exhibited the strongest prediction, whereas the model for apparent amylose displayed the weakest performance. This discrepancy can be attributed to the presence of waxy millet samples, which were devoid of amylose.	Zhang et al. (2020)

high-temperature methods to reduce the moisture content to moderate or low levels, specifically below 46% (Wang et al. 2017). This high-temperature drying process facilitated the development of a V-type crystalline structure in the millet starch. The formation of this particular crystalline structure had a significant impact on the hydration properties of pregelatinized millet starch. The V-type crystalline structure greatly restricted the ability of the starch to absorb water, which in turn affected its overall hydration properties. As a result, the dried millet products exhibited slower rehydration rates, making it more challenging to quickly rehydrate them. This limitation in rehydration can impact the usability and convenience of the dried millet products, especially in applications where fast rehydration is desirable. Therefore, understanding and controlling the drying process and moisture content is crucial to maintaining the desired hydration properties and ensuring the quality of pregelatinized millet starch products. In addition to nutritional analysis, IR spectroscopy is employed to detect and quantify contaminants and adulterants in millet. It also aids in the evaluation of millet's physical properties, such as moisture content and grain hardness, which are critical parameters in milling and product formulation. The moisture content, for instance, affects the storage stability and milling efficiency of millet. By accurately measuring these physical attributes, IR spectroscopy helps in optimizing the processing conditions, thereby improving the yield and quality of the final products. The technique's sensitivity to molecular vibrations enables the identification of various impurities, including mycotoxins, pesticides, and heavy metals, which can compromise food safety. This rapid and non-destructive method offers a significant advantage over traditional chemical analysis techniques, facilitating real-time monitoring and swift corrective actions during processing. Wang et al. (2022) developed a cost-effective and rapid method for detecting and quantifying fat in millet. Using a miniature NIR spectrometer connected to a smartphone, the collected spectral data were from various origins. The spectral data preprocessed by standard normalization and first derivative techniques. Key wavelengths were identified using several statistical methods, including bootstrapping, iterative shrinkage, and competitive sampling. These selected wavelengths were then used to build a prediction model for fat content using partial least squares regression. The resulting model demonstrated high accuracy in predicting fat content with minimal error and a strong correlation between predicted and actual values. This approach offers a rapid, portable, and cost-effective method for assessing millet quality, making it suitable for on-site inspections in the food industry. Similarly, developed handheld semiautomatic IR spectrometers for the protein content in millets. The semi-automated, consumer-oriented multivariate data analysis based IR spectrometers was compared with sophisticated data evaluation techniques. The performance was assessed using key statistical parameters from established cross- and test set-validated partial least squares regression (PLS-R) models. Both milled and intact samples were analyzed to further understand the significance of sample homogeneity. The findings indicated that this spectrometer could accurately analyze the protein content of millet grains, with root mean square error (RMSEP) values for milled and intact grains being approximately 0.5%. Two PLS-R models developed using handheld instruments also provided good results for milled grains, with RMSEP values around 0.6%. It exhibited some limitations compared to standard data processing software. However, for intact grains, the results were comparable. The charterization of polyphenolic fractions from three millet types (Fonio, Finger millet, and Pearl millet) by FTIR analysis exhibits bands at 3286 cm^{-1}, 3274 cm^{-1}, and 3012 cm^{-1}, which are associated with hydroxyl groups (–OH) in all millet types. The broad and distinctive bands, resulting from –OH stretching (3300–2500 cm^{-1}), suggest the possible presence of polyphenols. These bands are linked to the stretching (υ) vibration of hydroxyl groups and –OH attached to phenolic moieties (Sharaf et al. 2013; Franca et al. 2014). Additionally, in the functional region, Band at 3007 cm^{-1}, 2921 cm^{-1}, 2854 cm^{-1}, and 2851 cm^{-1}, which could correspond to the aliphatic υCH2 symmetric frequency or may be due to ethanol. The fingerprint region, spanning 1600–500 cm^{-1}, includes bands that may be attributed to double bonds (–C=C–), carbonyl bonds (–C=O) stretching vibrations, and aromatic ring deformations associated with polyphenol compounds. These bands could also be related to the C=O stretching vibration observed in many polyphenols and their derivatives, as well as the symmetric bending vibration of N–H, such as in caffeic acid and its derivatives. Bands at 1410 cm^{-1}, 1448 cm^{-1}, 1458 cm^{-1}, and 1462 cm^{-1} observed across all spectra may be linked to CH_2 and CH_3 groups attached to aromatic rings of polyphenols, aromatic stretching vibrations, and C–H bending (δ) vibrations. The C–O group vibrations of polyols, such as hydroxyflavonoids, were indicated by a band at 1264 cm^{-1}. A band identified at 702 cm^{-1} may be related to aromatic ring vibrations or ethanol. FTIR bands at 1022 cm^{-1}, 1045 cm^{-1}, 1058 cm^{-1}, and 1112 cm^{-1} could be associated with sugar moieties attached to polyphenols, aromatic –OH group vibrations, and the symmetrical stretching of –C–C– (Bello et al. 2022). Chen et al. (2013) explored the potential of NIRS, to analyze the nutritional content (protein, total carbohydrates, and crude fat) for foxtail millet. They employed advanced statistical techniques to identify the most informative wavelengths of near IR light for analyzing each component. The results were accurate for protein and total carbohydrates, with the model achieving over 90% accuracy. While the analysis for crude fat was still decent, it showed slightly lower accuracy at around 70%. These findings suggest that NIRS has great potential as a rapid and non-destructive method for assessing the nutritional value of foxtail millet. This application is particularly beneficial in industrial settings where consistency and efficiency are paramount. Another important application is in the authentication and classification of different millet varieties. Each millet type has a unique IR spectral fingerprint, which can be used to distinguish between species and even identify geographic origins. This is crucial for labeling and marketing purposes, as well as for protecting against fraudulent practices (Peng et al. 2018). The ability to rapidly and accurately classify millet varieties ensures that consumers receive products that meet their expectations and comply with regulatory standards. Overall, the integration of IR spectroscopy in millet processing represents a significant advancement in the field, providing a powerful tool for ensuring product quality, safety, and authenticity. Its non-destructive nature, coupled with its ability to deliver rapid and precise results, makes IR spectroscopy an indispensable technique in the modern millet processing industry.

9.8 Conclusions

Infrared radiation has emerged as a pivotal technology in food processing, particularly for delicate grains like millets. Its ability to deliver precise and efficient heat makes it an ideal tool for preserving millet's nutritional value while optimizing processing outcomes. Unlike traditional methods, IR heating allows for targeted energy application, minimizing heat loss and ensuring that the grain's core is heated evenly. The nutritional richness of millets demands careful handling. IR heating excels in this regard by enabling precise control over temperature and moisture levels. By selectively heating specific components within the millet kernel, IR technology helps to preserve essential nutrients and enhance the overall quality of the final product. Beyond its precision benefits, IR heating is synonymous with energy efficiency. The efficiency of IR heating translates to lower operational costs and reduced environmental impact. Another application of IR technology is spectroscopy, plays a vital role in quality control and process monitoring. By analyzing the spectral signature of millet grains, composition, moisture content, and overall quality of millets can be determined. This information is invaluable for ensuring product consistency and identifying potential issues early in the processing chain.

Thus, IR technology offers a comprehensive solution for enhancing millet processing. By combining precise heating, energy efficiency, and advanced analytical capabilities, IR-based methods are revolutionizing the industry. As the global focus on nutrition and sustainability intensifies, it poised to play a pivotal role in producing high-quality millet products that meet the demands of modern consumers.

References

Aboud SA, Altemimi AB, Al-HiIphy RS, Yi-Chen A, Cacciola F (2019) A comprehensive review on infrared heating applications in food processing. Molecules 24(22):4125. https://doi.org/10.3390/molecules24224125

Antal T, Kerekes B (2016) Investigation of hot air- and infrared-assisted freeze-drying of apple. Food Process Preser 40(2):257–269. https://doi.org/10.1111/jfpp.12603

Aydogdu A, Sumnu G, Sahin S (2015) Effects of microwave-infrared combination drying on quality of eggplants. Food Bioprocess Technol 8:1198–1210. https://doi.org/10.1007/s11947-015-1484-1

Babu AS, Mohan RJ, Parimalavalli R (2019) Effect of single and dual-modifications on stability and structural characteristics of foxtail millet starch. Food Chem 15(271):457–465. https://doi.org/10.1016/j.foodchem.2018.07.197

Baryeh EA (2002) Physical properties of millet. J Food Eng 51(1):39–46. https://doi.org/10.1016/S0260-8774(01)00035-8

Batariuc A, Ungureanu-Iuga M, Mironeasa S (2022) Characterization of sorghum processed through dry heat treatment and milling. Appl Sci 12(15):7630. https://doi.org/10.3390/app12157630

Bayram M, Kaya A, Oner MD (2004) Changes in properties of soaking water during production of soy-bulgur. J Food Eng 61(2):221–230. https://doi.org/10.1016/S0260-8774(03)00094-3

Bello OM, Ogbesejana AB, Balkisu A, Osibemhe M, Musa B, Oguntoye SO (2022) Polyphenolic fractions from three millet types (Fonio, Finger millet, and Pearl millet): their characterization and biological importance. J Clin Pharmacol 2(1):100020. https://doi.org/10.1016/j.ccmp.2022.100020

Brandl MT, Pan Z, Huynh S, Zhu YI, McHugh TH (2008) Reduction of salmonella Enteritidis population sizes on almond kernels with infrared heat. J Food Protect 71(5):897–902. https://doi.org/10.4315/0362-028X-71.5.897

Chen J, Ren X, Zhang Q, Diao X, Shen Q (2013) Determination of protein, total carbohydrates and crude fat contents of foxtail millet using effective wavelengths in NIR spectroscopy. J Cereal Sci 58(2):241–247. https://doi.org/10.1016/j.jcs.2013.07.002

Deepa C, Hebbar HU (2016) Effect of high-temperature short-time 'micronization'of grains on product quality and cooking characteristics. Food Eng Rev 8:201–213. https://doi.org/10.1007/s12393-015-9132-0

El-Azazy M, El-Shafie AS, Al-Saad K (2023) Introductory chapter: infrared spectroscopy-principles and applications. In: Infrared spectroscopy-perspectives and applications. IntechOpen. https://doi.org/10.5772/intechopen.109139

El-Mesery HS, Sarpong F, Xu W, Elabd MA (2022) Design of low-energy consumption hybrid dryer: a case study of garlic (Allium sativum) drying process. Case Stud Thermal Eng 33:101929. https://doi.org/10.1016/j.csite.2022.101929

Erdoğdu SB, Ekiz HI (2011) Effect of ultraviolet and far infrared radiation on microbial decontamination and quality of cumin seeds. J Food Sci 76(5):M284–M292. https://doi.org/10.1111/j.1750-3841.2011.02192.x

FAO (2024) Food and Agriculture Organization of the United Nations. Online source culled from https://www.fao.org. Accessed 5 July 2022

Fernandez DR, Vanderjagt DJ, Millson M, Huang YS, Chuang LT, Pastuszyn A, Glew RH (2003) Fatty acid, amino acid and trace mineral composition of Eleusine coracana (Pwana) seeds from northern Nigeria. Plant Foods Hum Nutr 58:1. https://doi.org/10.1023/B:QUAL.0000040323.67339.cb

Franca JR, De Luca MP, Ribeiro TG, Castilho RO, Moreira AN, Santos VR, Faraco AA (2014) Propolis-based chitosan varnish: drug delivery, controlled release and antimicrobial activity against oral pathogen bacteria. BMC Complement Altern Med 14:478. https://doi.org/10.1186/1472-6882-14-478

Gande N, Muriana P (2003) Prepackage surface pasteurization of ready-to-eat meats with a radiant heat oven for reduction of listeria monocytogenes. J Food Proc 66(9):1623–1630. https://doi.org/10.4315/0362-028X-66.9.1623

Goussault B, Adrian J (1977) The milling of Pennisetum millet and the value of the protein in the products. In: Proceedings of a Symposium on Sorghum and Millet for Human Food. Tropical Products Inst., London (p 13)

Gull A, Prasad K, Kumar P (2016) Evaluation of functional, antinutritional, pasting and microstructural properties of Millet flours. J Food Meas Charact 10:96–102. https://doi.org/10.1007/s11694-015-9281-0

Huang D, Yang P, Tang X, Luo L, Sunden B (2021) Application of infrared radiation in the drying of food products. Trends Food Sci Technol 110:765–777. https://doi.org/10.1016/j.tifs.2021.02.039

Jeevarathinam G, Pandiselvam R, Pandiarajan T, Preetha P, Balakrishnan M, Thirupathi V, Kothakota A (2021) Infrared assisted hot air dryer for turmeric slices: effect on drying rate and quality parameters. Lwt 144:111258. https://doi.org/10.1016/j.lwt.2021.111258

Johnson JB, Walsh KB, Naiker M, Ameer K (2023) The use of infrared spectroscopy for the quantification of bioactive compounds in food: a review. Molecules 28(7):3215. https://doi.org/10.3390/molecules28073215

Kadlag RV, Chavan JK, Kachare DP (1995) Effects of seed treatments and storage on the changes in lipids of pearl millet meal. Plant Foods Hum Nutr 47:279–285. https://doi.org/10.1007/BF01088264

Kaur A, Bhise S, Kaur M (2020) Hydrothermal treatments for paddy to improve physicochemical quality of brown rice. J Microbio Biotech Food Sci 9(5):913–926. https://doi.org/10.15414/jmbfs.2020.9.5.913-926

Kent NL (1983) Technology of cereals
Kgonothi D, Mehlomakulu NN, Emmambux MN (2024) Effects of combining microwave with infrared energy on the drying kinetics and Technofunctional properties of Orange-fleshed sweet potato. J Food Process Preserv 2024:1. https://doi.org/10.1155/2024/6336446
Kumar P, Chakraborty SK, Kate A (2022) Influence of infrared (IR) heating parameters upon the hull adherence and cotyledon integrity of whole pigeon pea (Cajanus cajan L.) grain. LWT 154:112792. https://doi.org/10.1016/j.lwt.2021.112792
Lestienne I, Buisson M, Lullien-Pellerin V, Picq C, Trèche S (2007) Losses of nutrients and anti-nutritional factors during abrasive decortication of two pearl millet cultivars (Pennisetum glaucum). Food Chem 100(4):1316–1323. https://doi.org/10.1016/j.foodchem.2005.11.027
Lorentzen J (1981). Freeze drying: the process, equipment and products. Developments in food preservation Vol. 1. London: Applied Science Publishers, pp 153–175
Mannuramath M, Yenagi N (2015) Optimization of hydrothermal treatment for little millet grains (Panicum miliare). J Food Sci Technol 52:7281–7288. https://doi.org/10.1007/s13197-015-1798-z
Manyatsi TS, Al-Hilphy AR, Majzoobi M, Farahnaky A, Gavahian M (2023) Effects of infrared heating as an emerging thermal technology on physicochemical properties of foods. Crit Rev Food Sci Nutr 63(24):6840–6859. https://doi.org/10.1080/10408398.2022
McDonough CM, Rooney LW, Earp CF (1986) Structural characteristics of Eleusine corocana (finger millet) using scanning electron and fluorescence microscopy. Food Struct 5(2):9
Mubaiwa J, Fogliano V, Chidewe C, Linnemann AR (2018) Bambara groundnut (Vigna subterranea (L.) Verdc.) flour: a functional ingredient to favour the use of an unexploited sustainable protein source. PLoS One 13(10):e0205776. https://doi.org/10.1371/journal.pone.0205776
Mukwevho P, Emmambux MN (2022) Effect of infrared and microwave treatments alone and in combination on the functional properties of resulting flours from bambara groundnut seeds. LWT 153:112448. https://doi.org/10.1016/j.lwt.2021.112448
Muralikrishna G, Malleshi NG, Desikachar HS, Tharanathan RN (1986) Effect of popping on the properties of some millet starches. Starch-Stake 38(2):48–51
Murty DS, Kumar KA (1995) Traditional uses of sorghum and millets. Sorghum Millets 221:153–177
Mwangwela AM, Waniska RD, McDonough C, Minnaar A (2007) Cowpea cooking characteristics as affected by micronisation temperature: a study of the physicochemical and functional properties of starch. J Sci Food Agric 87(3):399–410. https://doi.org/10.1002/jsfa.2711
Nalawade SA, Sinha A, Hebbar HU (2018) Infrared based dry blanching and hybrid drying of bitter gourd slices: process efficiency evaluation. J Food Process Eng 41(4):e12672. https://doi.org/10.1111/jfpe.12672
Nasiroglu S, Kocabiyik H (2009) Thin-layer infrared radiation drying of red pepper slices. J Food Process Eng 32(1):1–16. https://doi.org/10.1111/j.1745-4530.2007.00195.x
National Research Council, Policy, Global Affairs, Office of International Affairs, Board on Science, Technology for International Development (1996) Lost crops of Africa: volume I: grains. National Academies Press
Nithya KS, Ramachandramurty B, Krishnamoorthy VV (2012) Effect of processing methods on nutritional and anti-nutritional qualities of hybrid (COHCU-8) and traditional (CO7) pearl millet varieties of India. Yadav DN, Anand T, Kaur J, Singh AK. Improved storage stability of pearl millet flour through microwave treatment. Agric Res 1:399–404. https://doi.org/10.1007/s40003-012-0040-8
Ogundele OM, Emmambux MN (2018) Effect of infrared heating of pre-soaked whole and dehulled bambara groundnut (Vigna subterranea) seeds on their cooking characteristics and microstructure. LWT 97:581–587. https://doi.org/10.1016/j.lwt.2018.07.059
Ogundele OM, Minnaar A, Emmambux MN (2017) Effects of micronisation and dehulling of pre-soaked bambara groundnut seeds on microstructure and functionality of the resulting flours. Food Chem 214:655–663. https://doi.org/10.1016/j.foodchem.2016.07.022
Olamiti G, Takalani TK, Beswa D, Jideani AI (2020) Effect of malting and fermentation on colour, thermal properties, functional groups and crystallinity level of flours from pearl millet

(Pennisetum glaucum) and sorghum (Sorghum bicolor). Heliyon 6(12):e05467. https://doi.org/10.1016/j.heliyon.2020.e05467

Ozaki Y (2021) Infrared spectroscopy—mid-infrared, near-infrared, and far-infrared/terahertz spectroscopy. Anal Sci 37(9):1193–1212. https://doi.org/10.2116/analsci.20R008

Patil SR, Chaudhary AS (2020) Synthesis and optical properties of Near-Infrared (NIR) absorbing azo dyes. In: Chemistry and technology of natural and synthetic dyes and pigments. IntechOpen, London

Peng SH, Wang R, Yang LZ, He L, He X, Liu X (2018) Biosorption of copper, zinc, cadmium and chromium ions from aqueous solution by natural foxtail millet shell. Ecotoxicol Environ Saf 15(165):61–69. https://doi.org/10.1016/j.ecoenv.2018.08.084

Prashanth P, Joshi TJ, Singh SM, Rao PS (2024) Hydrothermal treatment of pearl millet grains: effects on nutritional composition, antinutrients and flour properties. Grain Oil Sci Tech 20:87. https://doi.org/10.1016/j.gaost.2024.04.002

Puneet K, Chakraborty SK, Lalita (2022) Infrared radiation: principles and applications in food processing. Thermal Food Eng Oper 11:349–373. https://doi.org/10.1002/9781119776437.ch12

Qu JH, Liu D, Cheng JH, Sun DW, Ma J, Pu H, Zeng XA (2015) Applications of near-infrared spectroscopy in food safety evaluation and control: a review of recent research advances. Crit Rev Food Sci Nutr 55(13):1939–1954. https://doi.org/10.1080/10408398.2013.871693

Rathore S, Singh K (2017) A critical optimization study on hydrothermal treatment for decortication of pearl millet to improve its consumption efficiency. J Food Measur Char 11(3):1501–1515. https://doi.org/10.1007/s11694-017-9529-y

Reddy VP, Faubion JM, Hoseney RC (1986) Odor generation in ground, stored pearl millet. Cereal Chem 63(5):403–406

Riadh MH, Ahmad SA, Marhaban MH, Soh AC (2015) Infrared heating in food drying: an overview. Dry Technol 33(3):322–335. https://doi.org/10.1080/07373937.2014.951124

Sahay KM, Singh KK (2004) Unit operations of agricultural processing. Vikas Publishing House Pvt Ltd.

Sakare P, Prasad N, Thombare N, Singh R, Sharma SC (2020) Infrared drying of food materials: recent advances. Food Eng Rev 12(3):381–398. https://doi.org/10.1007/s12393-020-09237-w

Sharaf S, Higazy A, Hebeish A (2013) Propolis induced antibacterial activity and other technical properties of cotton textiles. Int J Biol Macromol 59:408–416. https://doi.org/10.1016/j.ijbiomac.2013.04.030

Shavandi M, Taghdir M, Abbaszadeh S, Sepandi M, Parastouei K (2020) Modeling the inactivation of Bacillus cereus by infrared radiation in paprika powder (Capsicum annuum). J Food Safety 40(4):e12797. https://doi.org/10.1111/jfs.12797

Shih C, Pan Z, McHugh T, Wood D, Hirschberg E (2008) Sequential infrared radiation and freeze-drying method for producing crispy strawberries. Trans ASABE 51(1):205–216

Skotnicki T, Knap W (2018, October) Emerging terahertz technologies for security, quality control, vision and medical applications. In: 2018 14th IEEE International Conference on Solid-State and Integrated Circuit Technology (ICSICT). IEEE, pp 1–4

Smith IF (1996) Increasing consumer utilization of minor cereals in West Africa through sustained research and development of appropriate production and processing technologies. In: Dendy DAV (ed) Sorghum and millets: proceedings of the symposia. International Association for Cereal Science and Technology, Vienna, pp 171–197

Staack N, Ahrné L, Borch E, Knorr D (2008) Effect of infrared heating on quality and microbial decontamination in paprika powder. J Food Eng 86(1):17–24. https://doi.org/10.1016/j.jfoodeng.2007.09.004

Taylor JRN (2004) Foods and nonalcoholic beverages. In: Wrigley C, Corke H, Walker CE (eds) Encyclopedia of grain science, vol 1. Elsevier, Amsterdam, pp 380–390

Theophanides T (2012) Introduction to infrared spectroscopy. Infrared Spectroscopy-Materials Science, Engineering and Technology, pp 1–10

Wang H, Peng J, Xie C, Bao Y, He Y (2015) Fruit quality evaluation using spectroscopy technology: a review. Sensors 15(5):11889–11927

Wang R, Chen C, Guo S (2017) Effects of drying methods on starch crystallinity of gelatinized foxtail millet (α-millet) and its eating quality. J Food Eng 207:81–89. https://doi.org/10.1016/j.jfoodeng.2017.03.018

Wang F, Wang C, Song S (2022) Rapid and low-cost detection of millet quality by miniature near-infrared spectroscopy and iteratively retaining informative variables. Foods 11(13):1841. https://doi.org/10.3390/foods11131841

Wiedemair V, Mair D, Held C, Huck CW (2019a) Investigations into the use of handheld near-infrared spectrometer and novel semi-automated data analysis for the determination of protein content in different cultivars of Panicum miliaceum L. Talanta 205:120115. https://doi.org/10.1016/j.talanta.2019.120115

Wiedemair V, Ramoner R, Huck CW (2019b) Investigations into the total antioxidant capacities of cultivars of gluten-free grains using near-infrared spectroscopy. Food Control 1(95):189–195. https://doi.org/10.1016/j.foodcont.2018.07.045

Yadav DN, Balasubramanian S, Kaur J, Anand T, Singh AK (2011) Optimization and shelf-life evaluation of pearl millet based halwa dry mix. J Food Sci Eng 1(4):313

Yang XS, Wang LL, Zhou XR, Shuang SM, Zhu ZH, Li N, Li Y, Liu F, Liu SC, Lu P, Ren GX (2013) Determination of protein, fat, starch, and amino acids in foxtail millet [Setaria italica (L.) Beauv.] by Fourier transform near-infrared reflectance spectroscopy. Food Sci Biotechnol 22:1495–1500. https://doi.org/10.1007/s10068-013-0243-1

Zhang H, Wang X, Wang F, Zhao F, Li X, Fan G, Zhao Z, Guo P (2020) Rapid prediction of apparent amylose, total starch, and crude protein by near-infrared reflectance spectroscopy for foxtail millet (Setaria italica). Cereal Chem 97(3):653–660. https://doi.org/10.1002/cche.10281

10 Pulsed Electric Field and Light-Based Technologies in Millet Processing

Rabin Gyawali, Hema L. Degala, Poliana M. de Souza, and Ajit K. Mahapatra

10.1 Introduction

Millets, cereals belonging to the *Poaceae* family, are among the oldest cultivated crops, known for their ability to thrive in drought-prone areas and endure extreme climatic conditions, such as unpredictable weather and nutrient depletion in soil. Due to their resilience, millets are known as climate-smart crops and a nutritious food that benefits consumers, the planet, and farmers, highlighting their positive impact on the environment, health, and livelihoods. However, cereal grains, such as rice, wheat, and maize, overshadowed their use in many parts of Africa and Asia. Over time, this had led to a decline in investment, research, and support for millet production, relegating their use primarily to be used as fodder for animals (Anitha et al. 2019).

The current challenges regarding sustainable food production, climatic changes, water scarcity, and the growing demand for gluten-free cereals have renewed interest in millets. These challenges present an opportunity for farmers, nutritionists, and food and feed manufacturers to explore the nutritional and functional properties of millet grains (Hassan et al. 2021). As a result, millets, have gained recognition as contributors to food security worldwide, especially in arid and semiarid regions of the world where they can help combat hunger and malnutrition (Saleh et al. 2013).

Globally, millet production is estimated at around 28 million tons, with India being the largest producer (40% of global production), followed by Niger (11% of global production) (FAO 2023). The pearl millet (*Pennisetum glaucum*), finger

R. Gyawali · H. L. Degala · A. K. Mahapatra (✉)
Food Engineering Laboratory, Agricultural Research Station, College of Agriculture, Family Sciences, and Technology, Fort Valley State University, Fort Valley, GA, USA
e-mail: mahapatraa@fvsu.edu

P. M. de Souza
African Sustainable Agriculture Research Institute, University Mohammed VI Polytechnic, Laayoune, Laâyoune-Sakia El Hamra, Morocco

S. Jaddu, R. Chandra Pradhan (eds.), *Innovative Millet Processing*,
https://doi.org/10.1007/978-3-031-99864-5_10

millet (*Eleusine coracana*), and sorghum (*Sorghum bicolor*) contribute a major share of total millet production. Between 2015 and 2020, their productivity reached 9.19, 1.63, and 4.36 million tons, respectively (Sruthi and Rao 2021; Yadav et al. 2024). These grains are major sources of carbohydrates and proteins in many parts of Africa and Asia. Recent research has highlighted potential health benefits, prompting renewed global interest in their use as an alternative to more commonly consumed cereals like wheat and rice. Millets are rich in essential nutrients, including amino acids, minerals, and trace elements (Anitha et al. 2019).

Among the various types of millet, pearl millet and finger millet are particularly known for their nutritional profiles. Pearl millet is an excellent source of energy, providing higher caloric content compared to rice and wheat, and is rich in thiamine, niacin, and riboflavin (Malik 2015). It also contains minerals such as calcium, iron, and phosphorus (Adeola and Orban 1995). Finger millet, on the other hand, has a superior protein profile due to its high content of lysine, threonine, and valine (Ravindran 1991). Beyond their nutritional value, millets offer several nutraceutical benefits, such as improving digestive health, reducing cholesterol levels, preventing heart disease, protecting against diabetes, lowering cancer risks, boosting energy levels, and improving muscular functions (Amadou et al. 2013). Despite these benefits, millet consumption has declined over the past three decades, primarily due to little awareness of its nutritional advantages and lack of advanced processing and product development technologies (ICAR 2023) and challenges, such as grain hardness, poor dehulling efficiency, and susceptibility to rancidity during storage. In response to these challenges, the UN General Assembly declared 2023 as the "International Year of Millet" to promote millet production, enhance public awareness of their health benefits, and address climate change challenges (FAO 2023). Moreover, to enhance millet production for sustainable consumption and use, a variety of millet-based products with enhanced shelf stability through various processing methods and novel technologies is needed. Although an array of processing technologies are available, their specific applications across various millet varieties and the associated nutritional changes warrant further exploration (Mundassery et al. 2024).

10.2 Processing Methods

Millet processing is essential to improve edibility, enhance sensory qualities, and increase nutrient bioavailability by reducing antinutritional factors, such as phytates, tannins, and oxalates, which can inhibit mineral absorption. Millet processing is generally categorized into two main stages: primary and secondary processing, each comprising several specific operations that transform raw millet grains into consumable forms (Mundassery et al. 2024).

Primary processing includes initial steps, such as cleaning, dehulling, polishing, milling, and refining. These operations are designed to remove nonedible components, such as the husk, seed coat, and, in some cases, the germ. Cleaning is the first crucial step, which involves the removal of dust, stones, and other foreign materials to ensure the quality and safety of the final product. Dehulling, also known as decortication, removes the outer husk or hull, which is often indigestible. Polishing may follow after dehulling, which helps to refine the grains by removing the seed coat, enhancing their appearance and reducing cooking time. Milling further breaks down the grains into finer particles, and refining ensures that any remaining impurities or undesired components are removed to produce a uniform product.

Secondary processing involves more advanced methods like extrusion, baking, flaking, and popping, which are used to convert the processed millet into final, ready-to-eat products. Among these, extrusion is a versatile technique that uses heat and pressure to shape millet into various forms, such as pasta or breakfast cereals. And the baking is used to produce bakery products, including bread and biscuits, while flaking and popping are applied to create puffed or flaked products like snacks and ready-to-eat cereals. These secondary processes not only enhance the sensory qualities of millets, such as taste, texture, and appearance, but also potentially improve their nutritional profile by altering the bioavailability of certain nutrients.

10.2.1 Conventional Processing Methods

Conventional primary processing methods for millets encompass both primary and secondary techniques aimed at improving their edibility, nutritional profile, and sensory attributes. Primary processing methods, such as hand pounding, centrifugal dehulling, abrasive, and attrition milling are traditional practices that are widely used because of their simplicity. However, they present several limitations. These methods are often labor-intensive and time-consuming, resulting in a higher loss of bran, endosperm, and overall nutritive value, as well as increased grain breakage. The mechanical forces involved in these methods often strip away outer layers of the grains that are rich in dietary fiber, vitamins, and minerals.

Conventional processing methods include techniques, such as decortication, milling, soaking, germination, malting, fermentation, and puffing (Mahajan et al. 2023; Yousaf et al. 2021), which aim to improve the edibility, nutritional content, and sensory properties of millets while reducing antinutritional factors (Nazni and Shobanadevi 2016; Mahajan et al. 2023). Studies have shown that processing methods can significantly affect the functional, nutritional, and pasting properties of millets (Nazni and Shobanadevi 2016; Yousaf et al. 2021). Germination has been found to reduce antinutritional factors, while roasting increases nutritional compounds (Nazni and Shobanadevi 2016). These processing techniques can enhance nutrient accessibility and bioavailability, making millets more suitable for human consumption (Mahajan et al. 2023; Yousaf et al. 2021). The development of value-added millet products through sustainable processing methods can contribute to food and nutritional security (Mahajan et al. 2023).

Conventional thermal methods used in secondary food processing, such as pan roasting, extrusion cooking, microwave cooking, and ohmic heat that are low cost and energy efficient and enhance digestibility and functional properties. However, nonuniform temperature distribution might affect the quality of the millet grains, such as reduced antioxidants and bioactive compounds (Mundassery et al. 2024). Overall, proper processing of millets is crucial for improving their nutritional value and functional properties.

10.2.2 Modern Processing Methods

Modern processing methods for millets are characterized by advancements in technology that significantly improve efficiency, product quality, and environmental sustainability. Primary processing methods such as the use of double-chamber centrifuges, millet mills, and polishers have revolutionized the dehulling and milling of millets. These technologies enhance dehulling efficiency and milling yield while preserving the nutritional value of the grains. For example, double-chamber centrifuges can rapidly and effectively separate the husk from the grain with minimal loss of edible parts. A double-chamber centrifuge is designed with two separate chambers, allowing for more precise and efficient separation of components. In a conventional single-chamber centrifuge, the entire mixture is processed in one chamber, and separation occurs in one stage. However, in a double-chamber centrifuge, the material is separated in two distinct chambers, which can provide more control over the separation process. In the first chamber, the material might undergo a rougher separation, while in the second chamber, a finer separation can occur, reducing the loss of valuable components, such as edible grain parts. This dual-chamber approach allows for optimized processing, improving the overall efficiency of the machine. The separation can also be adjusted based on the type of material being processed, which results in better precision, minimal waste, and higher-quality output compared to a conventional single-chamber centrifuge. Millet mills and polishers improve the uniformity and quality of the final product. Moreover, these methods are eco-friendly and labor-free, representing a sustainable approach to millet processing. However, the higher initial capital cost for purchasing and installing such modern equipment remains a limiting factor for many small-scale farmers and processors (Mundassery et al. 2024).

Modern secondary processing methods have also emerged as crucial in maintaining and enhancing the quality and safety of millet-based products. These include nonthermal techniques, such as pulsed electric fields (PEF) and light-based technologies (Fig. 10.1), besides the thermal emerging technologies, such as forced convection roasting, infrared hot air drying, and microwave and oven drying (Vieira et al. 2021). Nonthermal methods are particularly advantageous as they preserve the sensory attributes, nutritional content, and bioactive compounds of millets while extending shelf life and enhancing product quality (Mundassery et al. 2024). For instance, PEF and cold plasma are effective in inactivating microbial populations and enhancing shelf life without the detrimental effects of high-temperature treatments, such as nutrient degradation and loss of flavor (Jindal and Nikhanj 2023). These methods are also rapid and energy-efficient, offering substantial environmental benefits. However, their application requires higher investment costs, more

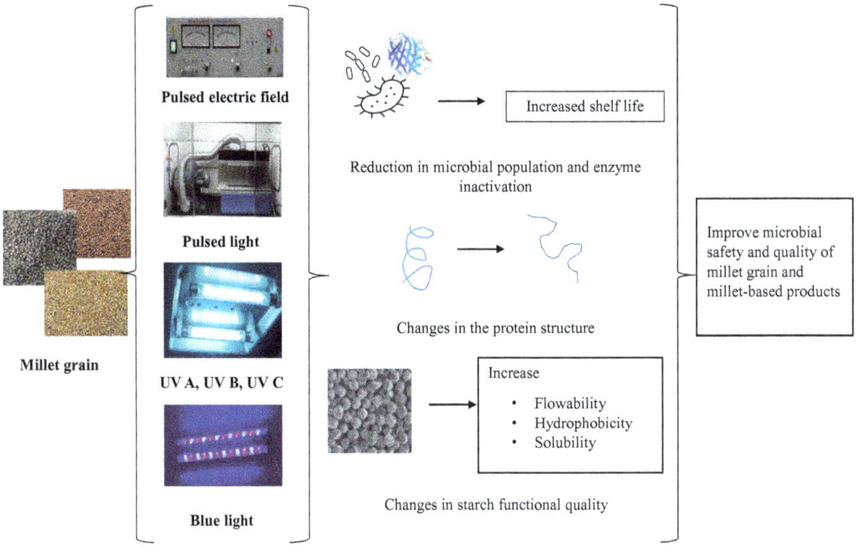

Fig. 10.1 Modern processing methods for millet processing

space, and greater maintenance, which could pose challenges for widespread adoption, especially in resource-limited settings.

The adoption of these nonthermal methods could significantly improve the availability and quality of millet and its products, aligning with the increasing demand for gluten-free alternatives to rice and wheat. As consumer awareness of the health benefits and environmental sustainability of millets grows, modern processing technologies will be crucial in meeting market demands and ensuring product quality (Mundassery et al. 2024).

Nonthermal processing techniques are particularly gaining popularity due to their minimal adverse effects on the nutritional and functional qualities of millets. For example, high-pressure processing and PEF can enhance the textural and organoleptic properties of millet products without compromising their nutritional integrity (Jindal and Nikhanj 2023). Furthermore, these modern methods contribute to the development of value-added millet products that support food and nutritional security, benefiting consumers, farmers, and the environment (Mahajan et al. 2023; Jindal and Nikhanj 2023). Generally, foods are thermally processed to destroy the vegetative microorganisms for preservation. However, only thermal treatment triggers many unwanted biochemical reactions, which leads to undesirable sensorial and nutritional effects. Therefore, several nontraditional preservation techniques are being developed to meet consumers' demand. Ensuring food safety and at the same time meeting such demands for retention of nutrition and quality attributes has resulted in increased interest in emerging preservation techniques (Souza et al. 2019).

10.2.2.1 Pulsed Electric Field

The PEF processing is an innovative nonthermal method that involves the application of brief bursts of high voltage pulses, typically ranging from 20 to 80 kV/cm to a food product held between two electrodes (Singh et al. 2024). This technology is recognized for its sustainability and energy efficiency, and requires no chemicals, and is environmentally friendly. Unlike standard thermal treatment methods, the temperature rise in PEF is much lower, resulting in negligible changes to the nutritional and organoleptic properties of foods (Mollakhalili-Meybodi et al. 2022).

The PEF technology enhances food safety by effectively inactivating microorganisms and enzymes, which makes it a promising alternative to conventional thermal pasteurization methods (Arshad et al. 2020). The process works causing cell membranes to disintegrate, leading to an imbalance in the ion concentration resulting in the cell leakage and cell death (Singh et al. 2024). The primary mechanism behind this microbial inactivation is electroporation, where high-voltage pulses create pores within the cellular membranes, leading to permeabilization and breakdown of cell structures (Yan et al. 2017). It can also cause the electrostatic interactions in protein chains to break resulting in the cleavage or combination of amino acids thus disrupting the mechanisms that maintain their secondary structures. This alteration modifies the composition and functional characteristics of proteins in cereals and grains by disrupting the mechanism that maintain their secondary structures (Giteru et al. 2018), which can improve the composition of proteins in cereals and grains, making them more suitable for food processing.

As a result, PEF is currently being used to improve the applicability of protein in the food applications. The application of high voltage pulses also improves the chances of protein extractability by inducing swelling, pore formation, pore widening, and water inflow (Suri et al. 2024). In addition, PEF can help retain the maximum amount of nutrients and flavor thus improving the nutritional quality of foods. The effectiveness of PEF depends on the pH and conductivity of food samples, electric field power, treatment time, energy generation, pulse width, frequency, and temperature (Arshad et al. 2020). Foods with high electric conductivity reduce the resistance of the chamber and require more energy to achieve a specific electric field (Lohani and Muthukumarappan 2016). The PEF treatment can be applied at high or low intensities. Both methods have been shown to boost protein extraction (Suri et al. 2024). A study on mild intensity PEF in sorghum flour effectively released bound phenolics with minimal damage to other nutrients. On the other hand, higher electric field intensity may cause cell damage resulting in phenolic degradation. The study showed that the mild intensity PEF can be used to produce sorghum flour with better phenolic composition and antioxidant activity. Treatment time influences PEF efficacy, as treatment time increased pulses applied to the sample also increased inducing metabolic response by opening cell pores. However, excessive increase of treatment time could break the cell wall irreversibly thereby destroying the membrane completely resulting in no additional amount of phenolics. Overall, the PEF has several advantages including lower processing time and temperature and lower energy use compared to conventional processing methods (Mollakhalili-Meybodi et al. 2022). The electric field power determines the intensity of the pulses, influencing the degree of disruption in the cell structure. Treatment time plays a critical role in ensuring sufficient exposure to the electric field for effective processing, while energy generation is linked to the efficiency of the pulses delivered. The pulse width and frequency control the timing and frequency of the electric field application, affecting the overall effectiveness of the treatment in achieving desired outcomes. Additionally, temperature can alter the physical properties of food and the conductivity of the samples, further influencing the treatment efficiency.

The PEF supports sustainable food processing by reducing water usage, emissions, and energy consumption while using by-products from waste food. This aligns with sustainable development goals aimed at minimizing food wastage and ensuring nutrient security (Yan et al. 2017). However, despite its potential benefits, the industrial adoption of PEF technology remains limited due to the need for reliable and practical electrical systems and the high cost of implementation installation. Ongoing research and development are required to design cost-effective, application-specific systems suitable for widespread commercial use.

Overall, PEF technology offers a promising alternative to conventional food processing methods by enhancing food safety, nutritional value, and processing efficiency while supporting sustainable practices. As research advances and technological challenges are addressed, PEF will have the potential to play a significant role in the future of the food industry.

10.2.2.2 Light-Based Technology

Light-based technologies are nonthermal alternatives to conventional food preservation and disinfection methods that rely on thermal processing and chemical agents or additives. These technologies have the potential to enhance food safety, prolong shelf life, and preserve the quality characteristics and nutritional composition of foods (Sunita et al. 2022; Souza and Garcia 2012) and act synergically with natural antimicrobials (Possas et al. 2021). Light-based technology involves pulsed light (PL), UV light, blue light emitting diodes, etc. Pulsed light is a promising nonthermal food treatment technology that uses very brief, high-power pulses of a wide spectrum of light (200–1100 nm), comprising infrared, visible, and UV regions (Gyawali et al. 2024). An important advantage of PL and UV-LED technologies is their ability to preserve the nutritional and sensory qualities of foods. Traditional thermal methods often degrade essential nutrients, vitamins, and bioactive compounds, negatively impacting the taste, texture, and appearance of food. In contrast, PL and UV-LED treatments help retain these qualities, making them ideal for applications where maintaining the food's natural properties is crucial.

Pulsed light interacts with several food ingredients to alter the quality and stability of the food during preservation, and the efficacy of PL on food is attributed to photothermal, photochemical, and photophysical effects (Joshi et al. 2023; Bryant et al. 2021).

- Photothermal effects occur when the light pulses cause rapid heating of the food surface, leading to the inactivation of microorganisms without significantly raising the overall temperature of the food.

- Photochemical effects involve the light's ability to break down chemical bonds in microorganisms, enzymes, and other food components, which can enhance food safety and extend shelf life.
- Photophysical effects refer to the physical changes induced by the light pulses, such as the disruption of cell membranes, which can lead to the inactivation of bacteria and yeasts.

The PL technology is particularly effective in inactivating a wide range of microorganisms, including bacteria, spores, viruses, and protozoan oocysts. This microbial inactivation is achieved through a combination of mechanisms including DNA damage, protein denaturation, and disruption of cellular structures that disrupt microbial DNA, proteins, and cellular structures, leading to their death or inactivation (Jindal and Nikhanj 2023). The UV-LED technology is effective in microbial and enzyme inactivation, significantly enhancing the safety and stability of various food products. Unlike conventional methods, these light-based technologies do not require chemicals or high temperatures, making them a cleaner alternative for ensuring food safety.

Pulsed light has several potential benefits, such as a very short processing time, no toxic residues (Souza et al. 2013), easy adaptability, high flexibility, and low energy costs (Mollakhalili-Meybodi et al. 2022). Pulsed light can interact with different food ingredients, thus modifying the food's quality and stability during storage. One such example is lipase inactivation caused by the photothermal and photochemical effects of PL, with pulse fluence playing a major role (Jeon et al. 2019).

The PL can therefore be effectively used to improve the stability of millet flour during storage as it can cause the desirable changes to the factors that cause rancidity and lipid oxidation (Sruthi and Rao 2021). Although PL is the most widely used light-based nonthermal food preservation technique and has the potential to be used in millet processing, there has been limited research on its application. However, its flexibility allows it to be effectively combined with other nonthermal methods, such as PEF and thermosonication, improving decontamination efficiency while maintaining or enhancing food quality (Mahajan et al. 2023; Jindal and Nikhanj 2023). For instance, coupling PL with PEF can provide synergistic effects, ensuring more comprehensive microbial inactivation while preserving sensory and nutritional properties (Andaluz-Mejía et al. 2022).

Beyond food safety and quality, light-based technologies are environmentally friendly as they do not require chemical additives or generate excessive heat, making them suitable for sustainable industrial applications. This aligns with the growing consumer and regulatory demand for greener food processing methods. The adoption of PL and UV-LED technologies can help reduce the carbon footprint of food processing operations while maintaining high standards of food quality and safety. As the demand for high-quality, safe, and sustainable food products increases, the use of light-based technologies is likely to expand, presenting new opportunities for innovation in the food industry.

10.3 PEF and Light-Based Technologies for Millet Quality

Novel nonthermal technologies such as PEF and other light-based technologies are being increasingly explored to improve food safety and quality (Mahapatra et al. 2024; Souza et al. 2025). These techniques could also offer potential application in improving microbial safety and quality of millet grains. Table 10.1 shows the effect of PEF and different light-based technologies that can be used to improve safety and quality of millets.

10.3.1 Pulsed Electric Field

The effects of the PEF on the microbial safety and quality on millets have not been extensively investigated (Sruthi and Rao 2021). More investigations are necessary to understand how PEF can be used in millet and millet-based product processing.

Table 10.1 Effect of pulsed electric field (PEF) and light-based technologies on millet quality

Types	Nonthermal methods	Processing conditions	Findings	References
Foxtail millet	PEF	3 kV/cm	Increased hydrophobicity and solubility Led to an enlargement of protein particles	Zhang et al. (2023)
Sorghum flour	PEF	45% (w/v) flour treated at 2 kV/cm for 875 μs	Increased total phenolic content and antioxidant activity	Lohani and Muthukumarappan (2016)
Foxtail millet	UV-B	Low intensity, 43.9 μW/cm^2	Decrease in total carotenoid content	Huo et al. (2024)
Glutinous millet seeds	PL	Fluence, 54.43 J/cm^2	Reduction of 0.66 log/g microbial count No noticeable changes in water activity	Hwang et al. (2018)
Foxtail millet	Different lights	Dark (control), white (380 nm), red (650 nm), and blue (450 nm) light	Blue light lowered MDA and fatty acids, volatile compounds, lipoxygenase and lipopolysaccharide activity	Liang et al. (2021)
Ogi slurry from millet	Sunlight	Slurry was placed in oven at 105 °C for 60 min and then sun dried for 90 min	AFB1 and total aflatoxins content decreased	Adefunke et al. (2017)
Foxtail millet flour	UV-A	Autoclaved flour (110 °C, 48 kPa, 1 h) treated with 906,120 μW.s.cm^{-2} UV dose	Showed improved flowability	Harshitha et al. (2024)

PL Pulsed light, *MDA* Malondialdehyde, *AFB1* Aflatoxin B1

Pulsed electric field can be used to modify proteins to improve the functionality of foods (Ran et al. 2019). Zhang et al. (2023) studied the effects of moderate-intensity PEF (3 kV/cm) treatment on the structure and physicochemical properties of foxtail millet. The PEF increased millet's hydrophobicity and solubility in alcohol solution without changing the distribution of subunit composition. It also aids in the formation of soluble aggregates leading to an enlargement of protein particles thereby making protein structure more organized after PEF treatment. This functional changes in protein structure make it a promising ingredient for food applications. In another study, mild PEF was used to treat sorghum flour (45%, w/v) for 500, 875, and 1250 μs at 1, 2, and 3 kV/cm to improve the phenolic antioxidants. Total phenolic content and antioxidant activity were 24.8% and 33.9% higher, respectively, than the control flour at 2 kV/cm when treated for 875 μs. This suggests that the phenolic antioxidants amount in PEF treated flour can be increased with PEF (Lohani and Muthukumarappan 2016). Studies have reported that the presence of higher phenolics in finger millet grain was shown to have a positive effect on malt quality by reducing the fungal load on the germinating grain (Siwela et al. 2010). The PEF treatment can also disrupt electrostatic interaction within protein chains resulting in the cleavage or combination of amino acid chains thereby contributing to the modification and structural and functional properties of proteins (Giteru et al. 2018). This changes in protein structure can have several functional applications in the development of millet-based products in food applications.

10.3.2 Light-Based Technologies

Light-based preservation methods can increase food safety and shelf life with minimal changes in nutritional and functional characteristics. Enzyme activity plays a role in reducing the shelf life of millet flour during storage. The PL has been shown to decontaminate food surfaces through microbiological and enzymatic inactivation thereby extending the shelf life (Joshi et al. 2023). The photothermal and photochemical effects of PL have been shown to inactivate lipase activity in foods (Jeon et al. 2019). Hwang et al. (2018) investigated the antimicrobial effects of PL on the surfaces of glutinous millet seeds and found that at fluence of 54.43 J/cm^2, there was a reduction of 0.66 log/g in microbial count. Furthermore, there were no significant changes in the water activity of the samples up to this level of exposure.

Foxtail millet grain is usually yellow in color primarily due to the presence of carotenoid that plays an important role in its quality. Huo et al. (2024) examined the effect of UV-B (43.9 μW/cm^2) on foxtail millet cultivars in terms of carotenoid content and color. Authors observed a decrease in the b* values, indicating reduced yellowness, when the millet plants were exposed to lower UV-B intensities. Furthermore, they noted a decrease in total carotenoid content under reduced UV-B exposure, suggesting that an optimal intensity is needed for carotenoid synthesis. The total carotenoid of both millet varieties exposed to reduced UV-B was lower than that of control samples. These results showed the important role of UV-B intensity for promoting pigment accumulation in foxtail millets influencing their color and carotenoid content. This study thus provides valuable information for the development of new millet varieties with enhanced carotenoid content.

It is reported that various light qualities exert distinct effects on the chemical composition of foods (Deng et al. 2017). Liang et al. (2021) investigated the effects of dark (control), white (380 nm), red (650 nm), and blue (450 nm) light on the quality of foxtail millet stored for 90 days at 35 ± 1 °C to see if lights can delay oxidative rancidity. Blue light showed a significantly slower increase in malondialdehyde (MDA) content and fatty acid value than that stored under white or red light. Furthermore, blue light also lowered lipoxygenase and lipopolysaccharide activity responsible for quality degradation. The millet that was kept under blue light showed high levels of both catalase and superoxide dismutase activities. Since these enzymes are efficient scavengers of peroxides and free radicals, blue light-treated millets can have better storage stability and a longer shelf life. Volatile compounds and fatty acids are the indicator of changes in the quality of millets. Under the blue light, volatile compounds and the fatty acid values increased more slowly than the other light conditions. This could be possibly due to the increased antioxidant capacity caused by the blue light (Ávila-Sosa et al. 2017). To limit the adverse effects of these volatile components on millet quality, authors recommended blue light as a clean and inexpensive technology to extend the shelf life and quality. Adefunke et al. (2017) evaluated various processing techniques to reduce the aflatoxin level of ogi slurry made from millet. Samples treated with sunlight resulted in a significant decrease in both aflatoxin B1 (AFB1) and total aflatoxin content. The degree of reduction was related to the duration of sunlight exposure and how evenly millet particles were sized, as this directly affected the penetration of solar irradiation and thus the level of aflatoxins.

Harshitha et al. (2024) evaluated the effects of UV-A on the properties of foxtail millet flour. The UV-A (906,120 μW.s.cm^{-2}) exposed flour from autoclaved (110 °C, 48 kPa, 1 h) treated millet showed improved flowability while UV-A exposed flour from untreated millet demonstrated a higher water swelling capacity. With the UV-A exposure, color changes were not significant. These changes in quality properties of millet flour can contribute to the industrial product development and culinary applications. Light-based technology such as pulsed UV light can lead to chain oxidation, structural fragmentation, and the formation of cross-links and aggregates of protein functional groups. This process can thus enhance the functional property of millet grains such as improvement of solubility and foaming ability (Mollakhalili-Meybodi et al. 2022).

10.4 Environmental Sustainability

The use of nonthermal technologies such as PEF and various light-based technologies (UV-A, B, C) in millet processing offers innovative solutions for improving environmental sustainability. These technologies can increase energy efficiency and decrease waste generation while greatly enhancing the quality and safety of millet grains and millet-based products. For example, PEF can effectively inactivate microbiological contaminants resulting in a higher nutritional value and safer product without requiring energy-intensive high temperature thermal treatments (Zhang et al. 2019). Similarly, light-based technologies have germicidal properties and can be used to disinfect millet grains and increase their shelf life thereby reducing the use of water and generation of chemicals and waste that are usually associated with other conventional methods (Zhang et al. 2019). Since these methods use less energy and have a smaller carbon footprint, they fall under eco-friendly processing practices and thus are sustainable options. One potential use of sustainable wastewater management plan is the reusability of contaminated water. Surendra et al. (2020) assessed the use of $ZnFe_2O_4$ nanoparticles in photocatalytic degradation of pollutants exposed to UV light and sunlight and used the treated water to grow finger millet. Results showed decrease in wastewater's pollution level under UV and sunlight, and the growth of millet was comparable to the control plants that were treated with DI water. Thus, the use of UV light treatment for the wastewater reuse could be a sustainable solution for the growth of millet crops. These environmentally friendly processing methods can reduce the amount of organic solvent usage and wastewater generation and thereby contribute to the reduction of negative environmental effect. The PEF and light-based technologies offer significant advantages for sustainable millet processing. They improve energy efficiency, reduce resource usage, and maintain the nutritional and sensory quality of foods. Additionally, these technologies enhance food safety and allow for the use of by-products, making them environmentally friendly alternatives to traditional millet processing methods.

10.5 Conclusion

The integration of PEF and light-based technologies represents a significant advancement in the processing of millet, offering promising benefits in terms of food safety, nutritional quality, and environmental sustainability. The PEF enhances protein extractability and functionality while maintaining flavor and nutrients, although its industrial application is hindered by high costs and the need for practical electrical systems. Light-based technologies, such as PL and UV-LEDs, provide nonthermal alternatives that effectively inactivate microorganisms and enzymes, thereby prolonging shelf life and preserving the nutritional and sensory attributes of millet and millet products. Despite these advancements, research gaps remain, particularly in understanding the full extent of PEF's effect on millet's microbial safety and protein functionality and optimizing light-based technologies for various millet varieties and processing conditions. Future research should focus on developing cost-effective and scalable PEF systems and exploring the synergistic effects of combining light-based technologies with other nonthermal methods. Additionally, further studies are needed to evaluate the long-term effects of these technologies on millet quality and their potential for industrial-scale applications.

References

Adefunke JAY, Joseph OO, Kayode OA et al (2017) Influence of different processing methods on aflatoxin level in Ogi slurry produced from millet (*Pennisetum typhoideum*). Am J Biomed Eng 5(4):83–87

Adeola O, Orban JI (1995) Chemical composition and nutrient digestibilityof pearl millet (*Pennisetum glaucum*) fed to growing pigs. J Cereal Sci 22:177–184

Amadou I, Gounga ME, Guo-Wei L (2013) Millets: nutritional composition, some health benefits and processing—a review. Emir J Food Agr 25(7):501–508

Andaluz-Mejía L, Ruiz-De Anda D, Ozuna C (2022) Non-thermal technologies combined with antimicrobial peptides as methods for microbial inactivation: a review. PRO 10:995. https://doi.org/10.3390/pr10050995

Anitha S, Govindaraj M, Kane-Potaka J (2019) Balanced amino acid and highermicronutrients in millets complements legumes for improved humandietary nutrition. Cereal Chem 97:74–84

Arshad RN, Abdul-Malek Z, Munir A et al (2020) Electrical systems for pulsed electric field applications in the food industry: an engineering perspective. Trends Food Sci Technol 104:1–13

Ávila-Sosa R, Ávila-Crisóstomo E, Reyes-Arcos EA et al (2017) Effect of blue and UV-C irradiation on antioxidant compounds during storage of Hawthorn (*Crataegus mexicana*). Sci Hortic 217:102–106

Bryant MT, Degala HL, Mahapatra AK et al (2021) Inactivation of *Escherichia coli* K12 by pulsed UV light on goat meat and beef: microbial responses and modelling. Int JFood Sci Technol 56(2):563–572

Deng M, Qian H, Chen L et al (2017) Influence of pre-harvest red light irradiation on main phytochemicals and antioxidant activity of Chinese kale sprouts. Food Chem 222:1–5

Food and Agricultural Organization (2023) Millet year. https://www.fao.org/millets-2023/en Accessed 27 Aug 2024

Giteru SG, Oey I, Ali MA (2018) Feasibility of using pulsed electric fields to modify biomacromolecules: a review. Trends Food Sci Technol 72:91–113

Gyawali R, Mahapatra AK, Bardsley CA (2024) Nonthermal techniques, antimicrobial agents, and packaging to improve the microbial safety of nuts. Trends Food Sci Technol 104363

Harshitha T, Ramesh J, Gawali PP et al (2024) Impact of ultraviolet radiation coupled with thermal treatment on the foxtail millet flour properties. J Food Meas Charact 18(8):1–15

Hassan ZM, Sebola NA, Mabelebele M (2021) The nutritional use of millet grain for food and feed: a review. Agric Food Secur 10:16. https://doi.org/10.1186/s40066-020-00282-6

Huo Y, Pang X, Feng H et al (2024) Effect of ultraviolet B radiation on the biosynthesis of carotenoids in foxtail millet grains. Agric 14(2):289

Hwang HJ, Cheigh CI, Chung MS (2018) Comparison of bactericidal effects of two types of pilot-scale intense-pulsed-light devices on cassia seeds and glutinous millet. Innov Food Sci Emerg Technol 49:170–175

Indian Council of Agricultural Research (2023) Millet. https://icar.org.in/crop-science/millets. Accessed 27 Aug 2024

Jeon MS, Park KM, Yu H et al (2019) Effect of intense pulsed light on the deactivation of lipase: enzyme-deactivation kinetics and tertiary structural changes by fragmentation. Enzym Microb Technol 124:63–69

Jindal P, Nikhanj P (2023) A review on processing technologies for value added millet products. J Food Process Eng 46(10):e14419

Joshi TJ, Singh SM, Rao PS (2023) Novel thermal and non-thermal millet processing technologies: advances and research trends. Eur Food Res Technol 249(5):1149–1160

Liang K, Liu Y, Liang S (2021) Analysis of the characteristics of foxtail millet during storage under different light environments. J Cereal Sci 101:103302

Lohani UC, Muthukumarappan K (2016) Application of the pulsed electric field to release bound phenolics in sorghum flour and apple pomace. Innov Food Sci Emerg Technol 35:29–35

Mahajan M, Singla P, Sharma S (2023) Sustainable postharvest processing methods for millets: a review on its value-added products. J Food Process Eng 47(1):e14313

Mahapatra AK, Degala HL, de Souza PM (2024) Processing of food using light sources. In: Future crops and processing technologies for sustainability and nutritional security. CRC Press, Florida, pp 64–85

Malik S (2015) Pearl millet nutritional value and medicinal uses. IJARIIE-ISSN (O) 1(3):2395–4396

Mollakhalili-Meybodi N, Nejati R, Sayadi M (2022) Novel nonthermal food processing practices: their influences on nutritional and technological characteristics of cereal proteins. Food Sci Nutr 10(6):1725–1744

Mundassery A, Ramaswamy J, Natarajan T et al (2024) Modern and conventional processing technologies and their impact on the quality of different millets. Food Sci Biotech 33:2441–2460

Nazni P, Shobana Devi R (2016) Effects of processing on the characteristics changes in barnyard and foxtail millet. J Food Process Technol 7(3):1–8

Possas A, Valero A, Garcìa-Gimeno RM, Souza PM (2021) Combining UV-C technology and caffeine application to inactivate *Escherichia coli* on chicken breast fillets. Food Control 129:108206

Ran XL, Zhang M, Wang Y et al (2019) Novel technologies applied for recovery and value addition of high value compounds from plant byproducts: a review. Crit Rev Food Sci Nutr 59(3):450–461

Ravindran G (1991) Studies on millets: proximate composition, mineral composition, and phytate and oxalate contents. Food Chem 39(1):99–107

Saleh ASM, Zhang Q, Chen J et al (2013) Millet grains: nutritional quality, processing, and potential health benefits. Compr Rev Food Sci Food Saf 12:281–295

Singh SM, Joshi J, Rao PS (2024) Technological advancements in millet dehulling and polishing process: an insight into pretreatment methods, machineries and impact on nutritional quality. Grain Oil Sci Technol 7(3):186–195

Siwela M, Taylor JRN, de Milliano WAJ et al (2010) Influence of phenolics in finger millet on grain and malt fungal load, and malt quality. Food Chem 121(2):443–449

Souza PM, Garcia A (2012) Study of short-wave ultraviolet treatments (UV-C) as a non-thermal preservation process for liquid egg products. Doctoral Thesis, Instituto de Agroquímica y Tecnología de Alimentos, Valencia

Souza PM, Briviba K, Müller A et al (2013) Cyto-genotoxic and oxidative effects of a continuous UV-C treatment of liquid egg products. Food Chem 138(2–3):1682–1688

Souza PM, Melo R, Santos MAA et al (2019) Risk management of egg and egg products: advanced methods applied T.E. Coldea (Ed.). In: Food engineering. Intech Open, London, pp 1–23

Souza PM, Mahapatra AK, Silva EP (2025) Emerging technologies for water decontamination: ensuring safe and clean water. Emerg Trends Technol Water Manag Conserv 52. https://doi.org/10.4018/979-8-3693-6920-3.ch002

Sruthi NU, Rao PS (2021) Effect of processing on storage stability of millet flour: a review. Trends Food Sci Technol 112:58–74

Sunita T, Sharma AM, Kaukab S et al (2022) Light-based processing technologies for food. In: Current developments in biotechnology and bioengineering. Elsevier, Amsterdam, pp 183–218

Surendra BS, Shekhar TS, Veerabhadraswamy M et al (2020) Probe sonication synthesis of ZnFe2O4 NPs for the photocatalytic degradation of dyes and effect of treated wastewater on growth of plants. Chem Phys Lett 745:137286

Suri S, Balasaheb KS, Yadav DK, Malakar S, Choudhary P, Mohapatra A, Dhurve P (2024) Overview of millet proteins: quality characteristics, effect of thermal/non-thermal processing and applications. Food Biosci 57:103434

Vieira KH, Lima FR, de Melo R, Pereira KC, Oliveira CD, Mendes CF, Pinto NAVD, de Souza PM (2021) Caracterização da farinha de semente de abóboraobtidaporsecagemem micro-ondas e estufa/characterization of pumpkin seed flour obtained by drying in microwaves and oven. Braz J Dev 7(3):22267–22283. https://doi.org/10.34117/bjdv7n3-100

Yadav OP, Singh DV, Kumari V, Prasad M, Seni S, Singh RK, Mohapatra T (2024) Production and cultivation dynamics of millets in India. Crop Sci 64(5):2459–2484

Yan LG, He L, Xi J (2017) High intensity pulsed electric field as an innovative technique for extraction of bioactive compounds—a review. Cri Rev Food Sci Nutr 57(13):2877–2888

Yousaf L, Hou D, Liagat H et al (2021) Millet: a review of its nutritional and functional changes during processing. Food Res Int 142:110197

Zhang ZH, Wang LH, Zeng XA et al (2019) Non-thermal technologies and its current and future application in the food industry: a review. Int J of Food Sci Technol 54(1):1–13

Zhang Y, Jing X, Chen Z et al (2023) Effects of moderate intensity pulsed electric field on the structure and physicochemical properties of foxtail millet (*Setaria italica*) prolamin. Cereal Chem 100(2):360–370

11 Extrusion Technology: A New Approach for the Utilization of Millets

Zia Parveen, Sunita Mishra, and Hema Deupa

11.1 Introduction

The current demand for novel and nutritious foods, combined with increasing lifestyle diseases, has formed a substantial new market for millets, pseudocereals, and legumes (Patil and Kaur 2018). Harper first introduced extrusion processing to food material. Extrusion combines a number of unit operations such as mixing, cooking, shearing, puffing, final shaping, and drying in one energy-efficient rapid continuous process. Extrusion technology is highly beneficial from the component of nutritional value as there is a lower nutrient loss, lower operating cost, and higher productivity along with presenting versality and energy efficiency (Thilagavathi et al. 2015; Kaur et al. 2020) in comparison to the thermal processing technique. Extrusion is a low-cost, nutritious production technique which employs a high-temperature, short-time cooking process in which grains are fed into an extruder with the appropriate feed rate, temperature, residence time, mechanical shear, and pressure settings maintained in the equipment. Various varieties and types of millets (Shree Anna) are grown in India. Millets are nutritionally equivalent to cereals and provide a great source of protein, minerals, and phytochemicals. Soaking, germinating, malting, decortication, and cooking each possess an effect on antioxidant activity (Shakuntala Manay and Shadaksharaswamy 2020). Various characteristics of millet-based extruded products can be produced with varying grain combination, feed rate, temperature, pressure, and residence time during the cooking process. The millet-based extruded products such as pasta, noodles, snacks, variety of dough, etc. get popular day by day. Extruded snacks were prepared by mixing flour of foxtail millet (60%), rice (5%), cowpea (5%), and Bengal gram (20). The expansion ratio tends to be less in millet-based extruded products, especially when the foxtail millet flour content was high. Further, noodles being one of the popular products among

Z. Parveen (✉) · S. Mishra · H. Deupa
Department of Food and Nutrition, Babasaheb Bhimrao Ambedkar University (A Central University), Lucknow, Uttar Pradesh, India

S. Jaddu, R. Chandra Pradhan (eds.), *Innovative Millet Processing*,
https://doi.org/10.1007/978-3-031-99864-5_11

all age groups and having higher shelf life is most frequently developed using cold extrusion technique in combination with different millet flour in varying composition. The glycemic index value of such products is extremely lower (Kate and Singh 2021).

These extrusion techniques have two primary benefits: they can create complex cross-sections and work with brittle materials that are exposed to compressive and shear stress. It has become an important food processing technique in the production of ready-to-eat cereals, pasta, snacks, pet foods, and textured vegetable protein. Extrusion is a process where food ingredients are driven through a die to produce a desired shape (Kharat et al. 2011). Extruded millet products are not only convenient and easy to consume, but they are also easy to store, transport, and prepare.

11.2 Types of Food Extruder in Millet Processing

Food extruded products can be obtained using three common machineries as discussed below (Fig. 11.1):

11.2.1 Single-Screw Extruder

This is the most common type of extruder, which is available in a variety of patterns and have a consistent pitch. Here, one screw pushes the ingredients through die. The container is located in the feed region which is used to supply the raw materials, and the material is then transferred to the progressive segment by a pivoting screw. A rise in the material temperature is caused by the shallower screw channel and compressed material. Often, gelatinization of starch results in increase material strength. By the metering region and forced through the die opening, it is transported further (Thejaswini et al. 2015).

11.2.2 Twin-Screw Extruder

These extruder uses two screws for extruding which allows for a more complex forms and higher production rates. Twin-screw extruders use high moisture extrusion and more complicated material that contain higher capacities of fiber, lipid, and other ingredients or components. Four different setups are possible depending on the location of the screws and the direction of their pivot: coaxial screws (I), coaxial non-intermeshing screws (II), counter-axial intermeshing screws (III), and counter-axial non-intermeshing screws (Maurya and Said 2014).

Fig. 11.1 Demonstrate three types of food extruders: (**a**) single-screw extruder, (**b**) twin-screw extruder, and (**c**) cold extruder

11.2.3 Cold Extruder

These types of extruders are suitable for a temperature-sensitive features employed to effectively blend and shape batter, without direct warming or cooking inside the extruder. In food handling, it is utilized generally for producing pasta.

11.3 Parameters of Millet Extrusion Technology

The process and its parameters for the millet extrusion technology are moisture percentage and blend ratio of the melt, temperature, screw speed, die shape and size, cutting speed, extruder type, barrel temperature, pressure, and residence time. Hence, the value of processing parameters, encompassing their codes, is reported below (Chakraborty et al. 2009).

Moisture content: 15–25%, depending on the millet type and the desired outcome. Moisture is a normal component of raw materials. However, it can be added directly into the grains in the form of water or steam in the preconditioner or drum. Moisture acts as a plasticizer and participates in a number of reactions. So, moisture softens the mass obtained and gives or modifies its rheology and also promotes starch gelatinization and protein denaturation. Further, moisture increases the possibility of a reaction between protein molecules, creating novel links and improving cross-linking (Riaz 2011). The moisture level to be handled during extrusion changes according to the type of extruder used, whether single or twin screw.

Temperature: Critical process parameters in millet processing, influencing the final products, include quality, texture, and nutritional content. Optimal temperature ranges for millet processing:

(a) *Extrusion*: 100–150 °C (212–302 °F)
(b) *Drying*: 50–100 °C (122–212 °F)
(c) *Roasting*: 150–200 °C (302–392 °F)
(d) *Puffing*: 200–250 °C (392–482 °F)

Screw speed: The extruder type and product define the speed of rotation, which ranges from 100 to 500 rpm. Screw speed impacts shear action, mechanical energy dissipation, melt viscosity, pressure, and residence time, as it has been reported by Dong et al. (2022). Finally, the increase in screw speed led to change from a long-branched chain of starch to short-branched chains.

Feed rate: 10–50 kg/h (22–110 lbs./h). It is one of the important parameters. A constant and properly working feeder is essential in order to offer a continuous feed rate. As the bulk of raw materials fed into the extruder are in particle form, understanding of the multiple particulate-scale properties and flow parameters is necessary. The particle forms, fine or coarse structure, determine the energy requirements, with the coarser particles demanding higher energy flows (Mcguire et al. 2022).

Die shape and size: Normally, the extruder has a series of shaped holes from which the extrudate emerges. These die holes may have a number of shapes to form the food material into rods, spheres, donuts, tubes, strips, etc. Expansion occurs as the product leaves the die because of the rapid lowering of pressure from the end of the extruder to ambient conditions.

Cutting speed: Cutting speed is a critical parameter in millet process technology, particularly in extrusion cutting process. Speed ranges for millet processing:

(a) *Low cutting speed*: (50–100 rpm) appropriate for delicate products, such as puffed snacks.
(b) *Medium cutting speed*: (100–200 rpm) appropriate for most millet products, such as flakes and pellets.
(c) *High cutting speed*: (200–300 rpm) appropriate for high-volume production, such as cereals and crackers.

Drying temperature: The drying temperature has an impact on the quality, texture, and health benefits of the final product. Optimal temperature ranges for drying millet:

(a) *Low-temperature drying*: (50–70 °C/122–158 °F) preserves nutrient and color, suitable for sensitive millets.
(b) *Medium-temperature drying*: (70–90 °C/158–194 °F) balances nutrient retention and energy efficiency, suitable for most millets.
(c) *High-temperature drying*: (90–110 °C/194–230 °F) fastest drying method but may compromise nutrient content and color.

Drying time: Drying time ranges for millet process technology:

(a) *Short drying time*: (30 min–1 h) suitable for heat-sensitive millets.
(b) *Medium drying time*: (1–2 h) suitable for most millets.
(c) *Long drying time*: (2–4 h) suitable for high-moisture millets.

Extruder type: In millet extrusion technology, there are three types of food extruders that are used: single screw extruders, twin-screw extruders, and cold extruders. Extruder selection is determined by a variety of parameters, including millet type and content, desired product texture and form, nutrient retention needs, product capacity and scale, and energy efficiency and cost issues.

Barrel temperature: (Optimum 50–150 °C). The extruder barrel temperature has the greatest impact on the material present inside. Variations in temperature in the barrel lower viscosity, mechanical energy, pressure, and torque values (Guyony et al. 2022). This enhances extrudate characteristics, particularly starch gelatinization and expansion. At the same time, the temperature increase may also promote protein denaturation and texturization, leading to an increase in viscosity (Lin et al. 2002).

Pressure: 1–10 bar (14.5–145 psi). The level of pressure in the extruder barrel is determined by the barrel section and screw design, but other parameters additionally play a crucial part. Severe processing conditions, such as high rotational speed and temperature, lead to substantially reduced pressure during high moisture extrusion (Pietsch et al. 2019).

Residence time: 30 s to several minutes depending on the extruder and product. The definition of the residence time of feed substance particles in an extruder is essential as it impact several parameters and properties of the continuous reactor. Further, residence time could vary depending on the extruder type used, single or twin screw. A comparison of the residence time distribution in counterrotation and corotation twin-screw extruders shows that a sharper residence time distribution is obtained in the counterrotating twin-screw extruder (Shah and Gupta 2004).

Raw material quality: The quality of millets, starches, and other raw ingredients throughout the extrusion process can have a big impact on the end product. The components should have the required rheological characteristics, particle size, and moisture content (Yadav et al. 2023).

Additives: Millet products have additives, which enhance their quality, texture, and nutritional value. Common additives used in millet processing are vitamins and minerals, amino acids and protein powders, emulsifiers and stabilizers, flavorings and colorants, antioxidants and antimicrobials, texturizers and fillers, dough conditions, and strengtheners.

11.4 Millet Extrusion Technology

Millet production technology encompasses the following practices as illustrated in Fig. 11.2.

Preprocessing

Grinding or milling grains usually occurs to separate the endosperm, bran, and germ, reduce particle size, and facilitate the production of refined flour. The chemical composition of millet can alter considerably during the milling process. During milling, the bran, which contains greater quantities of phytic acid and polyphenols than whole millet grains, is removed. This reduces the levels of these compounds and improves protein and starch digestibility in final processed goods like chapati and bread (Rathore 2016).

Sieving is an essential process in food processing and other industries that allows for efficient particle size separation and refining. It gives exact control over particle size distribution, which enhances the quality and consistency of the final products.

Mixing

Mix millet flour with water and optional ingredients (starches, proteins, fibers) to create a uniform blend. Mixing is a critical step in millet extrusion technology, ensuring uniform distribution of ingredients, additives, and moisture. Various

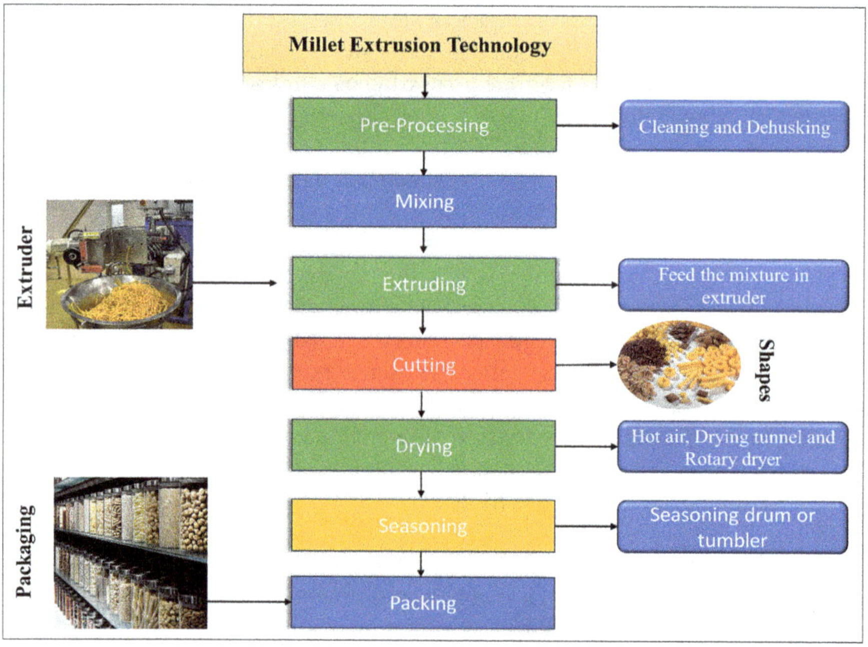

Fig. 11.2 Shows processing of millets by extrusion technology

mixing methods in millet extrusion include batch mixing, continuous mixing, paddle mixing, ribbon mixing, and conical mixing.

Extrusion

Feed the mixture into the extruder, and utilize heat, moisture, and pressure to break down the starches and reform them into a new shape. Extrusion is a method for manufacturing food with a single screw at temperatures below 100 °C. The raw material is heated to a steady temperature as it is hydrated, mixed, and formed (Shelar and Gaikwad 2019). Popular cold-extruded products include pasta, vermicelli, noodles, flakes, and extruded rice, among others. Proso millet pasta, with an amylose content of approximately 20%, offers potential for gluten-free pasta manufacturing because of its gelling ability and high cold-paste consistency. Vermicelli made from barnyard millet exhibits significant levels of beta carotene (1039 μg/100 g) and iron (3.81 mg/100 g) (Goel et al. 2021). Millet-based cold-extruded products generally exhibit a superior nutritional profile compared to conventional cold-extruded products (Sarojani et al. 2021).

Cutting

Cutting the extruded product into acceptable shapes (puffs, flakes, pellets) with a revolving blade or cutting die modifies the final product's texture, shape, and quality. Changing the cutting angle and orientation affects the product's shape and texture.

Drying

Dry the extruded product by removing extra moisture by the use of hot air, drying tunnel, or rotary drier. Uniform airflow allows equal drying and prevents hot spots. Different types of drying processes are employed in millet extrusion technology, including hot air drying, infrared drying, microwave drying, and fluidized bed drying.

Seasoning

Use a seasoning drum or tumbler to add flavorings, vitamins, or minerals as required. Seasonings are selected to produce specific flavor and aroma profiles, and precise seasoning dosage handling assures consistent taste intensity.

Packaging

Package the final product in bags, boxes, or other containers; store in a dry and cool place. Packaging is a final step in millet extrusion technology, protecting the final products' quality, texture, and freshness. Types of packaging in millet extrusion are:

(a) *Bags and pouches*: Flexible packaging for snacks, cereals, or flour.
(b) *Boxes and cartons*: Rigid packaging for high-moisture products or long-shelf-life requirements.
(c) *Cans and jars*: Airtight packaging for high-moisture products or long-shelf-life requirements.
(d) *Wrappers and films*: Flexible packaging for individual servings or multipacks.

11.5 Structural and Nutritional Value Changes During Millet Extrusion Technology

Extrusion cooking has distinct impacts in different food systems because of the various amounts of starch, proteins, moisture, and other components that influence the structure and texture of the end results (Fellows 2012). Exposure to the combined impacts of high pressure, high temperature, and shear creates various responses and variances in the nutritional profile of the food components, resulting in changes in their functional characteristics (Brennan and Grandison 2012). Gelatinization, solubilization and dextrinization or complex starch formation, protein denaturation, polymerization or cross-linking and texturization, partial or complete deactivation of enzymes, browning reactions, vitamin denaturation, and inactivation of antinutritional factors are some of the changes taking place during processing (Riaz 2010; Steel et al. 2012).

There are various experimental studies that have been reported by different authors based on the structural and nutritional changes occurring during processing of extruded millets products.

According to Allai et al. (2023) and Tej Bhan Singh et al. (2024), the crystalline structure aligns via the relationship between amylose and lipid helix, which occurs due to extrusion process variations in the relative starch crystallinity and can be attributed to the main causes: (1) the size of crystals, (2) the number of crystalline areas (affected by amylopectin chain length and its content), (3) alignment of double helices within crystalline regions, and (4) the level of interaction among double helices. During the extrusion process, shear and impact stresses produce changes in native starch structure These findings highlighted millet blends' potential for manufacturing nutrient-dense extruded snack products.

According to Raju et al. (2025), the development of foxtail millet-based extruded products provides a more nutritious alternative to wheat pasta, focusing enhanced dietary benefits. Millet integration, particularly at 50% substitution, increased fiber, fat, and mineral content (iron, copper, and zinc) while maintaining appropriate protein levels. Although sensory characteristics such as appearance and texture reduced substantially with increasing millet levels, and the nutritional superiority of millet products was clear, they produced greater importance to minerals and fiber than control samples, as well as a quicker cooking time. In addition, foxtail millet pasta demonstrated improved shelf stability and reduced bacterial and yeast counts over 90 days. This made foxtail millet an excellent component to produce health-conscious, fiber-rich extruded products with a high consumer acceptance and shelf life.

According to (Kishore et al. 2024), the study revealed that die temperature and material moisture content showed a major effect on the dependent parameters. The best procedure parameters of 28%wb moisture content, 12 rpm screw speed, 7 rpm feeder speed, and 80 °C die temperature led to 5.67% CL, 942.82 s CT, 2.155% WSI, 7.093 g/g WAI, and 6.815 WAR, with a desirability value of 0.858. Millet analogue rice shows significant variance ($p < 0.05$) in textural characteristics, proximate analysis, and physical attributes compared with raw rice. Millet analogue rice had a poorer crystallinity index than raw rice, although presenting similar FTIR spectral patterns, and SEM examination revealed surface fissures and internal spaces.

Practical Applications

The global prevalence of diabetes continues to rise as a result from consuming food with a high glycemic index (GI), which improves blood sugar levels. Rice flour-based products like white rice are not a healthy food choice for those with diabetes due to their high GI value. Therefore, millet analogue rice was developed from pearl millet, sorghum, and parboiled rice as a substitute for rice using the hot extrusion method. This produced millet analogue rice can be an alternative food for raw rice and has a scope for commercialization in the market.

Millet extrusion technique causes several structural and nutritional changes:

Effect of Structural Changes

(a) Starch gelatinization: Starches are broken down, becoming more accessible to enzymes.
(b) Protein denaturation: Protein unwind, making them more digestible.
(c) Fiber modification: Fiber structure is altered, potentially increasing solubility.
(d) Moisture redistribution: Moisture is redistributed, affecting texture and shelf life.
(e) Particle size reduction: Particles are broken down, increasing surface area.

Effect on Nutritional Changes

(a) Increased bioavailability: Nutrients become more accessible to the body.
(b) Improved protein digestibility: Protein becomes more easily absorbed.
(c) Enhanced starch digestibility: Starches are more easily broken down.
(d) Retention of vitamins and minerals: Extrusion helps retain vitamins and minerals.
(f) Reduced antinutritional factors: Phytates, tannins, and other inhibitors are reduced.

Mode of Action, Mechanism, and Potential Changes in Millet Extrusion Technology (Fig. 11.3 and Table 11.1)

11.6 Future Perspective

The global extruded millet product market is expected to grow quickly as new products, companies, and markets emerge. Extruded millet products seem to have an exciting future. In the sphere of product innovation, new product shapes, flavors, and textures will emerge, such as puffed snacks, millet-based cereals, and extruded

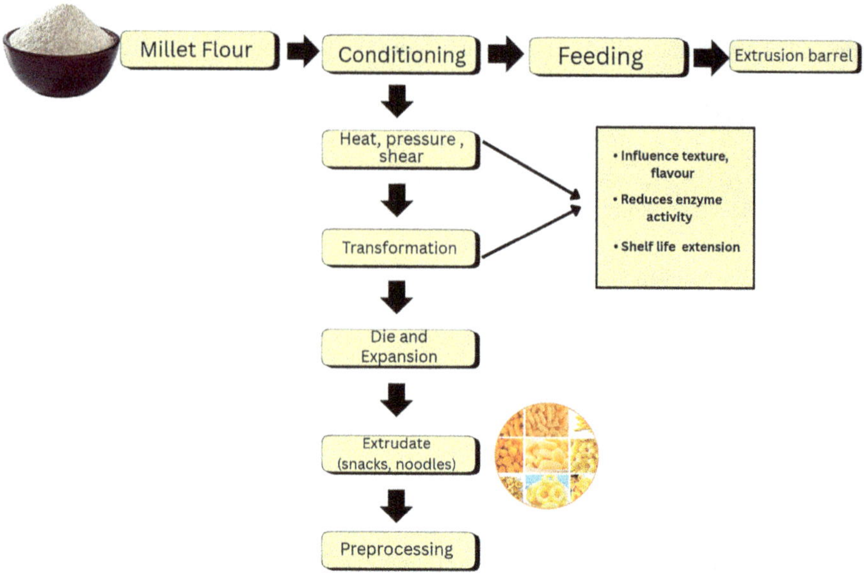

Fig. 11.3 The mode of action, mechanism, and potential changes of millet extrusion

millet flour blends. Nutritional enhancement will be formulated with the value-added nutrients, vitamins, and mineral to enhance their nutritional profile. Sustainable production, i.e., extrusion technology, is focusing on energy efficiency, reduced waste, and eco-friendly packaging to minimize environmental impact. Continuous research and development will improve extrusion technology, processing conditions, and ingredient combinations to enhance product quality and functionality. Innovative millet processing methods that preserve nutritional value while minimizing environmental impact and decreasing waste, such as high-pressure processing, improve millets' sustainability, earning potential, and broad public acceptance of millet-based extruded products such as snacks, pasta, noodles, and furthermore.

11.7 Summary

Extrusion technology is an effective way to enhance the nutritional value of food through the use of various basic raw materials with different colors, shapes, textures, flavors, and improved nutritional qualities. This process has transformative implications for millets, providing benefits such as improved nutritional content, increased texture, and new uses in food production. Extrusion can produce a wide range of processed items from all species of millets, including sorghum millet, finger millet, barnyard millet, foxtail, tiny millet, and pearl millet. The adoption of appropriate type of processing technology with suitable type of equipment surely decreases the antinutritional constituents, off-taste, and flavors of the millet, which

Table 11.1 The health benefits obtained from the millet extruded products

S. no.	Millets	Common name	Extruded products	Health benefits	References
1.	Pearl millet *(Pennisetum glaucum)*	Bajra	Extruded flakes	Improve respiration, gallstones, inhibit the growth of the phytopathogenic fungi	-
2.	Finger millet *(Eleusine coracana)*	Ragi	Pasta Vermicelli	Increases hemoglobin and bone mass, inhibit cataract eye lens	-
3.	Foxtail millet *(Setaria italica)*	Kangni/ kakum	Noodles Pasta Flakes	Slow glucose release, heart-friendly food, colon cancer management	Shan et al. (2015)
4.	Kodo millet *(Paspalum scrobiculatum)*	Kodo	Noodles Vermicelli	Reduce knee and joint pain, aids in weight loss	https://www.netmeds.com/health-library/post/kodo-millet-nutrition-health-benefits-ayurvedic-uses-and-recipes
5.	Sorghum *(Sorghum bicolor)*	Jowar	Puffs Flakes Pasta	Aids weight loss, reduces cholesterol, antimutagenic properties.	Grimmer et al. (1992)
6.	Little millet *(Panicum sumatrense)*	Kutki	Pasta Puffs Flakes	Nutrient dense, aids weight loss	-
7.	Barnyard millet *(Echinochloa Frumentacea)*	Sanva	Pasta Soup Sticks Noodles Flakes	Helps lower blood glucose, cholesterol and triglyceride, improved the level of HDL	Chaurasia and Anichari (2023)

will increase the commercial level production of various value-added millets products. The advancement in the processing technology of millets opens a new horizon, and it will help to raise millets at competitive level of staple food. By carefully controlling process parameters and considering packaging, manufactures can create high-quality products that meet consumer demands and contribute to as sustainable food system.

References

Allai FM, Junaid PM, Azad ZRAA, Gul K, Dar BN, Siddiqui SA, Loenzo JM (2023) Impact of moisture content on microstructural, thermal, and technofunctional characteristics of extruded whole-grain-based breakfast cereal enriched with Indian horse chestnut flour. Food Chemistry X, 20:100959

Brennan JG, Grandison AS (2012) Food processing handbook. Wiley

Chakraborty SK, Singh DS, Kumbhar BK, Singh D (2009) Process parameter optimization for textural properties of ready-to-eat extruded snack food from millet and legume pieces blends. J Texture Stud 40(6):710–726

Chaurasia RK, Anichari N (2023) Nutritional and health benefits of millets: a review. Pharma Innov 12:3360–3363

Dong Y, Wang B, Fang Y, Xu X, Yan S, Cui B, Abd El-Aty AM (2022) Effect of different rotational speeds of the extruder on the structure of corn starch. Starch – Stärke 74(3–4):2100234

Fellows P (2012) Extrusion of foods. Practical Action Technology Challenging Poverty

Goel K, Goomer S, Aggarwal D (2021) Formulation and optimization of value-added barnyard millet vermicelli using response surface methodology. Asian J Dairy Food Res 40(1):55–61

Grimmer HR, Parbhoo V, McGrath RM (1992) Antimutagenicity of polyphenol-rich fractions from sorghum bicolor grain. J Sci Food Agric 59(2):251–256

Guyony V, Fayolle F, Jury V (2022) High moisture extrusion of vegetable proteins for making fibrous meat analogs: a review. Food Rev Intl 38:1–26

Kharat SA, Navale SA, Ingale DV (2011) Extrusion Processing: Novel Technology for the Utilization of Millets. Small, 491.

Kate A, Singh A (2021) Processing technology for value addition in millets. In: Millets and millet technology, pp 239–254

Kaur J, Singh B, Singh A, Sharma S (2020) Characterization of extruded products from millets-legumes in combinations. Pharma Innov J 9(4):369–374

Kishore SG, Dwivedi M, Dalbhagat CG, Thota N, Singh CN, Saravanan S, Vignesh V (2024) Development of extruded millet analogue Rice: a comparison of rheological, structural, physiochemical, and cooking properties with conventional Rice. ACS Food Sci Technol 4(10):2504–2516

Lin S, Huff HE, Hsieh F (2002) Extrusion process parameters, sensory characteristics, and structural properties of a high moisture soy protein meat analog. J Food Sci 67(3):1066

Maurya AK, Said PP (2014) Extrusion processing on physical and chemical properties of protein rich products. J Bioresources Eng Technol 2:61–67

Mcguire C, Siliveru K, Ambrose K, Alavi S (2022) Food powder flow in extrusion: role of particle size and composition. Processes 10(1):178

Patil SS, Kaur C (2018) Current trends in extrusion: development of functional foods and novel ingredients. Food Sci Technol Res 24(1):23–34

Pietsch VL, Bühler JM, Karbstein HP, Emin MA (2019) High moisture extrusion of soy protein concentrate: influence of thermomechanical treatment on protein-protein interactions and rheological properties. J Food Eng 251:11–18

Raju C, Begum SS, Nagraja T, Shobha D (2025) Development and quality evaluation of foxtail millet (Setaria italica) based extruded product. Mysore J Agric Sci 59(1):278–291

Rathore S (2016) Millet grain processing, utilization and its role in health promotion: a review. Int J Nutr Food Sci 5(5):318

Riaz N (2010) Extrusion of cereals. New Food Magazine Issue 4, Kent

Riaz MN (2011) texturized vegetable proteins. In: Phillips GO, Williams PA (eds) Handbook of food proteins. Woodhead Publishing, Sawston/Cambridge, pp 395–418

Sarojani JK, Suvarna CH, Hilli JS, Sneha S (2021) Development of ready to cook Kodo millet pasta. Pharma Innov J 10(6):910–915

Shah A, Gupta M (2004) Comparison of the flow in co-rotating and counter-rotating twin screw extruders. Antec 50:443–447

Shakuntala Manay N, Shadaksharaswamy M (2020) Foods facts and principles. New Age International Publishers. ISBN-13: 978-9389802405 ISBN-10: 9389802407.

Shan S, Shi J, Li Z, Gao H, Shi T, Li Z et al (2015) Targeted anti-colon cancer activities of a millet bran-derived peroxidase were mediated by elevated ROS generation. Food Funct 6(7):2331–2338

Shelar GA, Gaikwad ST (2019) Extrusion in food processing: an overview. Pharma Innov J 8(2):564–568

Singh TB, Kaushik R, Sasikumar R, Kambhampati V, Singha P, Singh SK (2024) Development and characterization of multi-millet extruded food snack. Cereal Res Commun:1–15

Steel C, Leoro M, Schmiele M, Ferreira R, Chang Y (2012) Thermoplastic extrusion in food processing. In: Adel E-S (ed) Thermoplastic elastomers. InTech, Croatia

Thilagavathi T, Banumathi P, Kanchana S, Ilamaran M (2015) Effect of heat moisture treatment on functional and phytochemical properties of native and modified millet flours.

Thejaswini PH, Shetty HS, Mahesha HB (2015) Elicitation of induced resistance in pearl millet against Sclerospora graminicola through seed treatment with medicinal plant extracts. Int J Pure App Biosci 3(2):54–62

Yadav KC, Karhale HB, Singh M (2023) Food extrusion: a highly popular technology to process. Indian J Food Eng (IJFE) 2(3):1–8

Applications of Modified Millet Starches Using Emerging Technologies

12

Vishal Sharma, Samuel Jaddu, Venkatraman Bansode, Dhananjay Kumar Singh, and Rama Chandra Pradhan

12.1 Introduction

12.1.1 Overview of Millet Starches

Millets, commonly cultivated in dry and hot climatic regions, are cereal crops with small seeds belonging to the *Poaceae* family. These crops are highly resilient to adverse climatic stresses such as poor soil fertility and drought, which has led to their use since ancient Asian and African civilizations, making them an important food source in many developing countries (Annor et al. 2017). Major millet species include pearl millet (*Pennisetum glaucum*), foxtail millet (*Setaria italica*), finger millet (*Eleusine coracana*), proso millet (*Panicum miliaceum*), and sorghum (*Sorghum bicolor*), alongside minor millets such as barnyard millet (*Echinochloa* spp.), kodo millet (*Paspalum scrobiculatum*), little millet (*Panicum sumatrense*), teff (*Eragrostis tef*), browntop millet (*Brachiaria ramosa*), and fonio (*Digitaria exilis*) (Punia et al. 2021). Millet grains are naturally gluten-free and rich in proteins, B vitamins, dietary fibers, and minerals including magnesium, phosphorus, and iron

V. Sharma · V. Bansode · D. K. Singh
Department of Food Process Engineering, National Institute of Technology, Rourkela, Odisha, India

S. Jaddu
Department of Food Process Engineering, National Institute of Technology, Rourkela, Odisha, India

Department of Food Technology, School of Basic & Applied Sciences, Hindustan Institute of Technology and Science, Chennai, India

R. C. Pradhan (✉)
Professor & Associate Dean (Research), Department of Food Process Engineering, National Institute of Technology, Rourkela, Odisha, India
e-mail: pradhanrc@nitrkl.ac.in

S. Jaddu, R. Chandra Pradhan (eds.), *Innovative Millet Processing*,
https://doi.org/10.1007/978-3-031-99864-5_12

(Sharma et al. 2024). The starch content in millet grains constitutes 60–75% of their dry weight, composed primarily of amylose and amylopectin, similar to other cereal starches. However, millets are distinct due to their higher amylose content (20–40%) and smaller starch granules, which directly influence their behavior during digestion, cooking, and film formation (Zhu 2014). The higher resistant starch percentage in millet starch contributes to improved gut health and better blood sugar management, due to its slower digestibility compared to other cereal starches (Annor et al. 2017), leading to the increased demand for millet-based starch products. The variation in crystallinity also affects the swelling capacity, water solubility, and pasting properties of millet starches, making them valuable ingredients in the food sector (Q. Sun et al. 2014).

12.1.2 Importance and Benefits of Millet Starches

Millet starches, which constitute about 60–75% of the millet grains, are nutrient-rich, providing significant amounts of dietary fiber, B complex vitamins, and essential minerals like iron, magnesium, and phosphorus (Sonkar et al. 2023). Variations in starch content depend on the genetic, environmental, and postharvest factors, as well as the maturity stages of the grains. These characteristics, including highly resistant starch content, low glycemic index, and high nutritional content, make millet starches particularly valuable for individuals with dietary restrictions, such as those on gluten-free diets (Kam et al. 2016). Millet starches contain antioxidants such as catechins and ferulic acids, which protect cells from oxidative damage by scavenging the free radicals and inhibiting enzymes that produce them. Ferulic acid, being lipid-soluble, integrates into cell membranes and protects them from peroxidation (Chandrasekara and Shahidi 2010). The functional properties of millet starches, including their significant amylose content, influence their gelatinization and retrogradation properties, tending to form crystalline structure by reassociating and making them suitable for forming firmer gels in dairy desserts like puddings (Preetham Kumar et al. 2016). The swelling power and solubility of millet starches also impact the texture, stability, and consistency of food products by altering the water absorption and starch dispersibility, making them ideal for creating smooth sauces and gravies, as well as for thickening stews and soups due to their higher peak viscosity (Annor et al. 2014; Wu et al. 2014).

12.1.3 Introduction to Modification Techniques and Emerging Technologies

Modifying native starches is essential to overcome their limited pasting properties, enhance functionality, improve texture, and reduce retrogradation, thereby expanding their applications, such as smoother sauces, baked goods, and versatile food products in the food industry. Traditional methods of modification include enzymatic, physical, and chemical techniques, which have disadvantages including variable activity of enzymes, requirement of high energy input, and leftover chemical residues (Ashogbon and Akintayo 2014). Emerging technologies such as high-pressure processing (HPP), microwave, cold plasma, and ultrasound provide new, sustainable approaches to modifying the functionality of millet starches by altering their properties and structures (Gao et al. 2019; Zhu 2014).

These advanced modification techniques lead to improved product texture, shelf life, and nutritional value, along with enhancing functional properties of millet starches, such as relatively higher freeze-thaw stability, increased resistant starch content, extraction, characterization (Padhi et al. 2024), and modified rheological properties (Punia et al. 2021). As research continues, these technologies are expected to open new areas for utilization of millet starches, making them competitive alternatives to traditional starches from corn, wheat, and potatoes in different industrial applications (Table 12.1) (Annor et al. 2017).

12.2 Properties of Millet Starches

12.2.1 Chemical and Physical Properties

Millet starches, like other cereal starches, are primarily composed of two glucose polymers: amylose and amylopectin, whose balanced percentage composition affects the millet starches' digestibility, stability, and texture, along with other functional properties. The amylose content in millet starches generally varies between 11.01% and 16.61%, with some varieties reaching as high as 33.9%, depending on the millet cultivar, owing to the different maturity indices, cultivars, and environmental conditions (Zhu 2014). The amylopectin content, typically 70–80%, consists of a highly branched structure that allows for tight molecular packing, leading to ordered crystalline structures within the granule (Annor et al. 2017). This organization typically results in an 'A-type' X-ray diffraction pattern, which is common for cereal starches and indicative of a densely packed monoclinic crystal lattice. The primary physical properties relate to the granule's structure and morphology. Millet starch granules are characterized by their relatively small and polyhedral shape, typically measuring between 4 and 12 μm in diameter (Annor et al. 2014). Under magnification, the surfaces of these granules are generally smooth, lacking the pores or fissures seen in some other starch types. Structurally, the granules are composed of alternating semi-crystalline and amorphous regions. The crystalline regions consist of highly organized amylopectin chains, while the amorphous regions contain both amylose and the branching points of amylopectin (Magallanes-Cruz et al. 2017) (Table 12.2).

12.2.2 Functional Properties

Millet starches, contributing to 60–70% by weight of the grain, have benefits that translate into several applications within food products. Key functional properties

Table 12.1 Impact of modification and emerging technologies on millet starch functionality and applications

Modification/ technology	Mechanism of action	Key changes in properties	Food applications	Non-food applications	References
Heat-moisture treatment (HMT)	Heating at 90–140 °C, limited moisture (10–35%), below gelatinization, 1–16 h. Rearranges amylose/amylopectin; alters crystalline and amorphous regions; increases chain mobility. Granules remain intact.	– Amylose: ↓ in pearl millet; ↑ in proso millet (16.3%). – Swelling power: ↓ (sweet potato, banana); solubility ↑ (except pearl millet). – WAC: ↑ (e.g. proso millet 172.7%). – Gelatinization temp: ↑ (pearl millet To 59.9→78.1 °C). – ΔH: ↓ (pearl millet 7.9–10.8 J/g). – Pasting viscosities: ↓ (pearl millet PV 2825.7→1671 mPa·s). – RS & SDS: ↑. – Relative crystallinity: ↓ by ~20% (proso millet). – Morphology: cracks, holes.	– Noodles, pasta, candy, fillings. – Bread: ↑ volume, ↓ crumb stiffness (with wheat). – Soups, gravies, frozen foods. – Functional foods (↑ SDS, RS). – Gluten-free products.	– Biodegradable films (↑ strength). – Nanocrystals/nanoparticles (↑ thermal stability).	Zheng et al. (2020), Sandhu et al. (2020), Fonseca et al. (2021), Ashogbon (2021), Punia et al. (2021), Punia Bangar et al. (2021), Siwatch et al. (2022), and Amarnath et al. (2024)

Annealing	Hydrothermal treatment with excess water (>60%), below gelatinization, above glass transition, prolonged. Promotes chain interaction, melts imperfect crystallites, repairs granule defects.	– Amylose: ↑ with dual modification (foxtail 27.9%; finger millet 31–33%). – Swelling power: ↓ (foxtail 6.3 g/g at 90 °C). – Solubility & WAC: ↓ (e.g. WFM 2.13 mL/g). – Gelatinization temp: ↑ (WFM To 52.8 °C, Tp 56.8 °C, Tc 60.5 °C). – ΔH: ↓ (WFM AS 1.70 vs. NS 2.27 mJ/g). – Retrogradation: ↑ RS3 (foxtail). – RS: ↑ to 42.3% (foxtail). – Relative crystallinity: ↑ to 58.8% (WFM). – Freeze-thaw stability: ↓ (↑ syneresis). – Texture: ↑ hardness, gumminess.	– Noodles, pasta, fillings. – Bread, baked goods. – Functional foods (↑ RS, SDS; diabetes-friendly).	– Thermoplastics, resins, films. – Nanoparticles.	Ai and Jane (2015), Babu et al. (2019), Fonseca et al. (2021), and Amarnath et al. (2023)
Cross-linking	Formation of di-esters and inter/intramolecular bonds, strengthens starch, restricts swelling.	– Swelling & solubility: ↓. – Gelatinization temp: ↑. – Pasting viscosities: ↓ (PV, BV, SV, PT). – Amylose: ↑ (barnyard millet alkali-treated 19.9%). – Moisture: ↓ up to 20%. – Morphology: little change. – Tablet properties: ↑ crushing/compacting, ↓ friability/disintegration. – Freeze-thaw stability: limited (needs dual modification).	– Thickening agent. – Emulsifiers, binders.	– Colon-targeted drug delivery. – Excipient for tablets. – Heavy metal adsorbent.	Ai and Jane (2015), Ashogbon (2021), Punia et al. (2021), Raghunathan et al. (2021), Bhatt et al. (2023), and Siwatch et al. (2022)

(continued)

Table 12.1 (continued)

Modification/ technology	Mechanism of action	Key changes in properties	Food applications	Non-food applications	References
Oxidation	Introduces carbonyl/ carboxyl groups; causes depolymerization, pore expansion, altered microstructure.	– Viscosity & molecular size: ↓. – Stability: ↑. – Clarity & film properties: ↑. – ΔH: ↓ (pearl millet 9.9→2.5 J/g). – Morphology: surface pores enlarged, granules intact. – Cold storage: improved (↓ hardness, chewiness). – Gelatinization temp & ΔH: ↓ at high levels.	– Jellies, fruit pastes. – General food thickener.	– Tablet/capsule fillers (high density, flowability).	Zhu (2014), Ai and Jane (2015), Punia et al. (2021)
Acetylation	Introduction of acetyl groups into starch chains.	– Retrogradation: ↓ markedly. – ΔH: ↓. – Cold storage stability: ↑ (good for frozen foods). – Swelling & solubility: ↑ with temperature (finger millet). – Tablet properties: ↑ crushing strength, ↓ disintegration. – Gelatinization temp & ΔH: ↑.	– Frozen foods. – Baked products.	– Edible films. – Pharmaceutical excipient (fast disintegration, high strength).	Afolabi et al. (2012), Magallanes-Cruz et al. (2017), Shaikh et al. (2017), Punia et al. (2021)
Enzymatic hydrolysis	Debranching enzymes (pullulanase, isoamylase) hydrolyse α-1,6 linkages; produces linear chains; retrogradation ↑ RS/SDS.	– Paste clarity: ↑. – Syneresis: ↓ (↑ freeze-thaw stability, pearl millet). – RS & SDS: ↑ (short time → SDS; longer time → RS). – Swelling & solubility: ↓ for pearl millet; solubility ↑ for foxtail (combined). – Pasting: ↓ PV, BD, SV, FV; PT ↑. – ΔH: ↓ (proso millet).	– Jellies, fruit pastes. – Functional foods (↑ SDS, RS). – Thickener, stabilizer, fat replacer.	– Enzyme production.	Magallanes-Cruz et al. (2017), Pratiwi et al. (2018), Punia et al. (2021), and Amarnath et al. (2024)

Transglycosylation	Dual enzymatic (β-amylase + transglucosidase). Increases short chains, α-1,6 linkages → slower digestion.	– Digestion: slowed. – SDS: ↑.	– Functional foods (slow digestion, ↑ SDS).	–	Magallanes-Cruz et al. (2017), Hu et al. (2019)
Ultrasonication	Ultrasound waves in aqueous starch; cavitation, shear, chain fragmentation, pore formation; protein removal. Can create annealing effect.	– Viscosity: ↓ (general); millet shows slight change. – Pasting: ↓ PT (some cases), ↑ BD; WFM PT ↑ (73.4 vs. 72.8 °C). – ΔH: ↓ (WFM US 0.81 vs. NS 2.27 mJ/g). – Granule size: ↓, more homogeneous. – SP, solubility, WAC: ↑. (WFM SP 2.3 g/g at 70 °C; WAC 3.16 mL/g). – Relative crystallinity: ↓ (WFM, potato). – Amylose: ↑ (foxtail UAS 27.9%). – Protein: ↓ (foxtail). – Gelatinization: often ↓; WFM To/Tp/Tc ↑. – RS: ↑ (e.g. foxtail dual US + annealing 45.6%). – Freeze-thaw: ↑ stability (dual). – Antinutrients: ↓ tannins (62.8%), phytates (67%).	– Stabilizer, filler. – Functional foods (↑ RS, ↓ GI). – Soups, ketchup, fried snacks. – Pasta, noodles (↑ crispiness).	– Biodegradable packaging. – Bio-nanocomposites (e.g. pearl millet).	Hu et al. (2019), Babu et al. (2019), Punia Bangar et al. (2021), Amarnath et al. (2024), and Kumari et al. (2024)

(continued)

Table 12.1 (continued)

Modification/ technology	Mechanism of action	Key changes in properties	Food applications	Non-food applications	References
Microwave treatment	Non-ionizing electromagnetic radiation; dielectric heating; compresses double helices; disrupts glycosidic bonds; induces reassembly.	– Amylose: ↓ (pearl millet), ↑ (proso millet 22.1%, taro ~20%). – RS: ↑ (proso 18.7%). – Digestion: slowed (↑ SDS, ↓ hydrolysis rate). – WAC & OAC: ↑ (little millet). – Foaming capacity: ↓ (little millet). – Crystalline type: B→A (proso). – Gelatinization: ↑ To (proso). – ΔH: ↓ (proso). – Pasting: ↓ PV (pearl millet 995→630 mPa·s), PT ↑ (72.8→77 °C). – MW: ↓ (proso 3.4 × 10^4 g/mol). – SP: ↓ (pearl millet). – Film solubility: ↑ (34–38%).	– Thickener, stabilizer, fat replacer. – Extruded and baked products. – Bread with reduced staling. – Functional foods (high RS).	– Films (↑ solubility, opacity).	Zheng et al. (2020), Punia Bangar et al. (2021), Wang et al. (2022), and Amarnath et al. (2024)
High-pressure processing (HPP)	Non-thermal, eco-friendly. Pressure modifies non-covalent bonds; preserves quality. Causes swelling, solubility ↑, gelatinization, granule damage.	– SP & solubility: ↑. – Gelatinization: ↓ onset temp; birefringence loss. – Pasting: altered (pearl millet PT ↑, BD ↓). – RS: altered (dual mod ↑ RS; retrogradation ↓ hydrolysis 10–15%). – Retrogradation: inhibited. – Polymorphism: A→B transition. – Morphology: eruptions, cracks (600 MPa). – Relative crystallinity: ↓ (WFM at 200–600 MPa). – Hydration/gelation: ↑ (WFM). – Paste clarity: ↓ (WFM).	– Tailored functional foods. – Frozen, canned foods. – Bakery goods (↓ onset temp, energy-saving).	–	Colussi et al. (2018), Dominguez-Ayala et al. (2022), and Amarnath et al. (2024)

Cold plasma	Non-thermal plasma discharge; introduces carbonyl/carboxyl groups; cross-linking at low voltage, depolymerization at high voltage; alters microstructure.	– WAC: ↑ (1.43→1.55 g/g). – OAC: ↑ (1.01→1.34 g/g). – Amylose: ↓ (31.1→25.9%). – pH: ↓ (7.5→6.9). – Whiteness index: ↑ (82.7→84.5). – Relative crystallinity: ↓ (23.1→21.6%). – Morphology: cracks, scratches. – Paste clarity: ↑ (stable clear paste, pearl millet).	– Foods with improved quality (diabetes-friendly, low GI). – Canned foods, dressings. – Baking (softening cakes). – Frozen battered dishes (↑ adherence). – Beverages, pasta, puddings, jellies.	–	Sonkar et al. (2023), Amarnath et al. (2024)

To onset temperature of gelatinization, *Tp* peak temperature of gelatinization, *Tc* conclusion temperature of gelatinization, *ΔH* gelatinization enthalpy, *PV* peak viscosity, *BV* breakdown viscosity, *SV* setback viscosity, *FV* final viscosity, *PT* pasting temperature, *BD* breakdown, *RS* resistant starch, *SDS* slowly digestible starch, *RDS* rapidly digestible starch, *WAC* water absorption capacity, *OAC* oil absorption capacity, *SP* swelling power, *NS* native starch, *AS* annealed starch, *UAS* ultrasonicated + annealed starch, *US* ultrasonicated starch, *WFM* white finger millet, *LCAM* long-chain amylose, *SCAM* short-chain amylose

Table 12.2 Physicochemical and functional properties of native millet starches compared with other cereal starches

Millet type	Starch content (% dry basis)	Amylose (%)	Granule size (µm)	Gelatinization temperature (°C) (To–Tp–Tc)	Resistant starch (%)	Distinctive functional property	Comparison with rice/wheat/corn/potato	References
Pearl millet	60.5–70% (up to 70% of grain)	3.96–32.5%	Small to large; spherical and polygonal. Peak ~7.6 µm. Large granules: 10.91–20.45 × 9.09–17.72 µm. Pores and indentations present.	60.9–78.0 (e.g. To: 63.4–67.7, Tp: 69.3–71.6, Tc: 74.5–76.3). Pasting temp: 70–90.	1.4–16.5%. Reported values include 9.7–16.5 g/100 g. RDS: 46.3–51.6%; SDS: 37.2–38.7%.	Rich in amylopectin; XRD peak at 23.1° (2θ). Shear thinning and more elastic than viscous. Porous surface aids modification and digestibility.	Superior physicochemical properties vs. other cereals. Thermal profile close to corn starch. Higher gelatinization temp than potato starch. Lower GI (62.58) than finger millet starch (74.17).	Chandra et al. (2016), Punia et al. (2021), Sandhu and Siroha (2017), and Zhu (2014)
Foxtail millet	52.0–69.1%	11.2–32.0%. Native starch: 13.22%.	Largest granules among millets; avg. 6.8–11.8 (range 3–15).	69.65–77.68 (e.g. To: 69.65, Tp: 73.47, Tc: 77.68).	18.2%. RDS: 36.9–50.7%; SDS: 38.3–45.0%.	Highest peak viscosity (2203 cP). Higher gelatinization temp than pearl/finger millet. High iodine binding (K/S values).	More perfect crystalline organization vs. pearl and finger millets. Swelling/solubility lower than potato.	Annor et al. (2014), Chandra et al. (2016), Zhu (2014)
Finger millet	53.4–70%	16.23–38.6%. Native WFM: 16.23%.	Small granules; peak ~5.7 (range 3–10).	63.41–74.00 (e.g. To: 63.41, Tp: 68.64, Tc: 74.00).	High RS. TS: 39.7–50.1%; RDS: 8.3–11.1%.	High LCAM (LCAM:SCAM ratio 1.27). Lowest iodine binding. Highest solubility among some millet starches.	Less susceptible to α-amylase than rice starch. WAC and OAC lower than corn and potato starch.	Chandra et al. (2016), Lawal (2009), Amarnath et al. (2023), Zhu (2014), Annor et al. (2014), and Annor et al. (2017)

Proso millet	56.1–93.7%	1.5–33.9%. Waxy type: 0.38 g/100 g.	Small granules; peak ~5.7 (range 3–10). Waxy/non-waxy: 9.91–12.92.	62.8–82.43 (e.g. To: 68.65, Tp: 71.37, Tc: 80.4). Non-waxy: higher pasting temp.	11.5%.	Highest gelatinization enthalpy (5.16 J/g). Good paste clarity and freeze–thaw stability. High purity (93.7% yield).	Susceptibility to α-amylase similar to maize. Amylose (33.9%) higher than corn (20–30%).	Chandra et al. (2016), Wu et al. (2014), Zhu (2014), Annor et al. (2014), Sun et al. (2014), Han et al. (2021), and Zheng et al. (2020)
Barnyard millet	60.3%	17.6–19.96%. Alkali-treated: 19.96%.	Range 3–10. Similar morphology to alkali-isolated starch.	64.66–74.71 (e.g. To: 64.66, Tp: 69.65, Tc: 74.71).	Not specified for native.	Lowest sediment volume. Higher final viscosity among studied millets. Swelling power of alkali-extracted starch 194.3%.	Swelling/solubility lower than potato.	Chandra et al. (2016), Wu et al. (2014), Annor et al. (2017), Bhatt et al. (2023), and (Zhu 2014)
Little millet	60.9%	Up to 34%	Very small; 0.8–10	~60–80	Not specified for native.	–	Swelling/solubility lower than potato.	Chandra et al. (2016), Annor et al. (2017), Annor et al. (2014), and Zhu (2014)
Sorghum	69.0–73.8%	–	Granules with pores.	65.0–75.0	Not specified for native.	Pores on surface. Protein 0.15% (pharmaceutical starch).	Gelatinization range comparable to corn (68.5–76.0) but wider than millet (55.5–65.0).	Eze and Alozie (2015), Sandhu and Siroha (2017), and Wu et al. (2014)

(continued)

Table 12.2 (continued)

Millet type	Starch content (% dry basis)	Amylose (%)	Granule size (μm)	Gelatinization temperature (°C) (To–Tp–Tc)	Resistant starch (%)	Distinctive functional property	Comparison with rice/wheat/corn/potato	References
Other cereals (rice, wheat, corn, potato)	Rice: 77.2%; Wheat: 64%; Corn: 62.3%; Potato: N/A	Rice: 6.59–24.38%. Wheat: 3.6–27.3% (reported values up to 25.6%). Corn: 0–80% (normal 20–30%; high 50–80%; waxy 0–8%). Potato: 18.8–22.9%.	Rice/Wheat/Corn: N/A. Potato: large (47.39–50.25), smooth, non-porous.	Corn: 68.5–76.0; Potato: 66.47–75.83; Rice/Wheat: N/A	Wheat: 0.39%	Corn and sorghum have pores; potato starch granules are smooth, large, and non-porous.	Millet starches lower swelling/solubilization than potato. Native millet has higher gelatinization temp than potato. Kodo millet has higher RS and lower SDS than rice/wheat.	Almeida et al. (2023), Chandra et al. (2016), Abdel-Aal et al. (2002), Heydari et al. (2021), Kang et al. (2016), Kringel et al. (2020), Han et al. (2021), and Zhu (2014)

To onset temperature of gelatinization, *Tp* peak temperature of gelatinization, *Tc* conclusion temperature of gelatinization, *ΔH* gelatinization enthalpy, *RS* resistant starch, *SDS* slowly digestible starch, *RDS* rapidly digestible starch, *TS* total starch, *GI* glycemic index, *WAC* water absorption capacity, *OAC* oil absorption capacity, *XRD* X-ray diffraction, *LCAM* long-chain amylose, *SCAM* short-chain amylose, *PV* peak viscosity, *BV* breakdown viscosity, *SV* setback viscosity include water solubility index, swelling capacity, degree of gelatinization, rheology, digestibility, and structural properties, which impact the consistency, texture, and stability of foods. Besides acting as a binding agent to provide structural integrity in products like bakery and sausages, by forming a cohesive network of ingredients through the matrix, starch helps maintain the suspensions and emulsions by acting as a stabilizer. Their water and oil absorption capacities are also valuable, helping to improve yield and moisture retention in products like meat fillings and bakery goods. Several key physicochemical properties further enhance the millet starch functionality, for instance, solubility and good swelling power, which are crucial for gelling and thickening abilities. Upon cooling after gelatinization, they form firm, opaque gels, a property that is essential for creating the desired structure in foods such as puddings and custards. Millet starches exhibit unique functional properties that make them valuable in various food applications. These properties are influenced by factors such as the botanical source, processing methods, and chemical composition.

Millets are known for their slower starch and protein digestibility, unique grain structure, high fiber, and phytic acid content compared to other cereals (Annor et al. 2017). This characteristic is attributed to the complex structure of millet starches, which includes higher amylose content and compact granules, contributing to their resistance to enzymatic hydrolysis, by forming tightly packed, more crystalline structures, therefore limiting the enzyme access (Dhital et al. 2017). The functional properties of millet starches, such as gelatinization, pasting behavior, and retrogradation, are crucial for their application in food products. Gelatinization involves the disruption of starch granules in the presence of heat and water, leading to increased viscosity; however, the gelatinization temperature and viscosity of millet starches can vary depending on the millet type and its processing (S. Wang et al. 2015). Pasting behavior, which describes the changes in viscosity during the heating and cooling of starch, is another important property of millet starches, which typically exhibit high peak viscosity and breakdown values, indicating their potential for use in products requiring thickening and gelling properties (Ai and Jane 2015). Additionally, retrogradation, the process of recrystallization of gelatinized starch upon cooling, affects the texture and shelf life of starch-based products. Millet starches have shown lower retrogradation tendencies, making them suitable for applications in refrigerated and frozen foods (Chi et al. 2023). Chemical modifications such as acetylation, cross-linking, and oxidation can further enhance the functional properties of millet starches. These modifications can improve the water absorption capacity, thermal stability, and resistance to shear, which are desirable traits for various industrial applications (Ashogbon and Akintayo 2014; Sandhu and Siroha 2017). Overall, the functional properties of millet starches, combined with their nutritional benefits, make them an important and versatile ingredient in the development of functional foods and industrial products (Babu et al. 2019; Hu et al. 2019).

12.2.3 Comparison with Other Cereal Starches

Millet starches exhibit unique structural and physical properties that distinguish them from other cereal starches, such as rice, wheat, corn, and sorghum. Millet starch granules are predominantly polygonal with some spherical shapes and exhibit deep indentations, which is similar to corn and sorghum starches but different from the smooth surfaces of potato starch granules (Wu et al. 2014). Millet starches exhibit A-type crystallinity (23.69% to 30.89%) and a monoclinic structure, along with rapid digestibility, comparable to wheat and rice starches, but distinct from the B-type crystallinity of potato starch, having comparably slower digestibility and a hexagonal crystalline structure. Additionally, the chemical composition of millet starches varies significantly in amylose content, from 0.38 to 28.4 g/100 g (dry weight basis), which is on par with other cereal starches but includes notable waxy varieties with very low amylose content (Wu et al. 2014).

The physical properties of millet starches, such as water solubility index (WSI), swelling power (SP), paste clarity, sediment volume, and freeze-thaw stability, also demonstrate significant variation. Finger millet starch has high solubility due to its amylose content, similar to high-amylose cornstarch, while waxy proso millet starch shows exceptional paste clarity and freeze-thaw stability, outperforming non-waxy rice and wheat starches (Abdel-Aal et al. 2002). The gelatinization and retrogradation behaviors of millet starches are influenced by amylose content, with waxy varieties exhibiting higher gelatinization temperatures and enthalpies, akin to waxy maize starch but differing from rice starch. The pasting properties of millet starches, measured using a Rapid Visco Analyzer, show variations affected by amylose and amylopectin content, aligning with the characteristics of other cereal starches like wheat and rice. Overall, millet starches' distinctive properties make them suitable for various food and industrial applications, highlighting their potential as a valuable starch source (Wu et al. 2014).

12.3 Modification Techniques

12.3.1 Physical Methods

Physical modification of starch avoids the use of chemicals and is an environmentally friendly approach for applications in food industries to maintain product consistency and to improve the quality of low-fat and gluten-free products. These modification methods typically involve altering the physical structure of starch granules through annealing and heat-moisture treatment (HMT), which can significantly impact the functional properties of starch.

12.3.1.1 Heat-Moisture Treatment

Heat-moisture treatment (HMT) is a hydrothermal method that maintains the granular structure of the starch and involves treating it with a limited or insufficient amount of water (10–35% moisture content) at elevated temperatures (80–120 °C) for a specific period of exposure ranging up to 16 h. This process induces changes in the crystalline and amorphous regions of the starch granules, leading to modifications in their physicochemical properties. HMT alters the thermal stability and pasting properties of starch, making it more resistant to thermal and mechanical stress. For example, studies have demonstrated that HMT can increase the gelatinization temperature owing to the increased crystalline order, requiring more energy to disrupt the structured regions. HMT also decreases solubility and swelling power of starch (Sandhu and Siroha 2017). Additionally, HMT can improve the resistant starch content, which is beneficial for creating functional foods aimed at improving gut health and slowing the release of glucose into the bloodstream. The HMT process also results in reduced retrogradation, which is the tendency of gelatinized starch to recrystallize over time, thus improving the textural properties and shelf life of starch-based products. The extent of these modifications depends on the specific conditions applied during HMT, including moisture content, temperature, and treatment duration (Siwatch et al. 2022).

12.3.1.2 Annealing

Annealing is another hydrothermal technique used to modify starch without disrupting the granular structure, which involves incubating starch in excess water (typically over 40%) at temperatures below its gelatinization point for extended periods. This method promotes molecular rearrangement within the starch granules, enhancing their crystallinity and thermal stability, and also increasing pore formation, which in turn influences solubility, swelling, and digestibility. Annealing increases the gelatinization temperature and decreases the enthalpy change associated with gelatinization, indicating a more stable granular structure (Pratiwi et al. 2018). The annealing process results in starch with improved functional properties, such as increased resistance to enzymatic hydrolysis and reduced swelling and solubility. These changes are attributed to the realignment of starch molecules, which creates a more ordered and compact structure. Annealed starches are particularly useful in food applications where controlled viscosity and stability are desired, such as in sauces, soups, and dairy products (Fonseca et al. 2021). HMT and annealing are effective physical methods for modifying starch, offering improvements in thermal stability, resistant starch content, and overall functional properties without chemical agents. These methods are particularly valuable in the development of starches for food and industrial applications where environmental sustainability and consumer safety are important.

12.3.2 Chemical Methods

Chemical modifications of starch allow for a broad range of functionalization, which introduces chemical groups for chemical and physical properties, extending its applications across various food industries. These methods alter starch at the molecular level, introducing functional groups such as acetyl, carboxyl, and ethyl groups, thereby enhancing its properties like enzymatic hydrolysis resistance, gelatinization temperature, and water retention. The efficiency of these millet starch modifications depends on the starch source, amylose-to-amylopectin ratio, granule morphology, reaction conditions, and the nature and amount of the modifying reagent (Apriyanto et al. 2022).

12.3.2.1 Cross-Linking

Cross-linking leads to the formation of inter- and intramolecular bridges between the millet starch molecules using specific multifunctional reagents like epichlorohydrin (EPI), phosphoryl chloride, adipic acid mixed with anhydride, and sodium tripolyphosphate and improves the stability of starch granules against swelling, high temperature, shear forces, and acidic conditions. The resulting cross-linked starches exhibit decreased retrogradation, increased viscosity, and improved paste clarity. Also, it affects the millet starch morphology, impacting larger starch granules with significant damage as compared to the smaller starch molecules (Sandhu et al. 2021b). Cross-linking is widely used in the food industry for stabilizing dairy products, in pharmaceuticals for transporting molecules and excipients, in wastewater treatment for pollutant chelation, and packaging for bioplastics (Blennow et al. 2000; Carpenter et al. 2015; Mahlow et al. 2016). Other influencing factors for the efficiency of cross-linking include reagent concentration, starch parameters, pH, reaction time, and temperature.

12.3.2.2 Oxidation

Oxidation, a chemical modification technique, involves modifying the hydroxyl groups of starch into carbonyl or carboxyl groups using oxidizing agents like potassium permanganate, sodium hypochlorite, hydrogen peroxide, or ozone. It introduces mainly carbonyl and carboxyl groups into the molecular structure of starch through oxidative reactions, leading to altered film-forming capabilities and solubility, which disrupts the molecular structure, along with reducing retrogradation, enhancing hydrophilicity, and decreasing gelatinization enthalpy. Oxidized starches exhibit lower viscosity, higher clarity, and better low-temperature stability, making them ideal for use in batters, film formers, and food coatings in confectionery and dairy products. Additionally, oxidized starch is used in the paper and textile industries to improve strength and printability, along with better desizing of fabrics and cost-effectiveness (Lang et al. 2025; He et al. 2023). The utilization of ozone for oxidation has gained popularity due to its ability to react with starch without catalysts or special conditions, eliminating the complex reaction conditions or the requirement of additional chemicals, affecting mainly the amorphous regions and preserving the granule morphology, making it suitable for food, textile, and paper processing (He et al. 2023).

12.3.2.3 Acetylation

Acetylation involves introducing (-$COCH_3$) acetyl groups into the starch molecule, altering its functional properties. This procedure involves the addition of vinyl acetate or acetic anhydride to the millet starch, leading to the -OH substitution, along with decreasing the gelatinization temperature and retrogradation while increasing the swelling capacity, viscosity, and water retention properties. Acetylated starches are used as emulsion stabilizers in refrigerated and frozen foods that require stability during freeze-thawing, as film-forming polymers in textiles, and in paper production for packaging applications. The degree of substitution (DS), which refers to the average number of hydroxyl groups being replaced, determines the extent of modification and influences the retrogradation behavior because of the slower starch's recrystallization (Afolabi et al. 2012; Kringel et al. 2020). Studies have shown that acetylated cornstarch exhibits improved thermal stability and improved resistance to retrogradation, making it suitable for various food applications in terms of improved stability and shelf life of starch-based products (Ayucitra 2012). The process of acetylation is influenced by factors such as the molar ratio of reagents, as their higher number leads to an increased degree of substitution. Also, factors like temperature and reaction duration determine the effectiveness and degree of modification (Chen et al. 2017; Pratiwi et al. 2018).

12.3.3 Enzymatic Methods

Enzymatic modifications of starch are crucial for adapting its properties to meet specific industrial requirements, particularly in food, either for hydrolysis, cross-linking, phosphorylation, oxidation, or transglycosylation (Amarnath et al. 2024). These methods utilize the precision and efficiency of enzymes, along with their excellent specificity, to alter the molecular structure compared to the chemical modification techniques for starch. Owing to the risk of high-temperature denaturation and pH sensitivity, the enzymatic modifications are operated under mild conditions minimizing environmental impact, thereby preserving the starch's structural integrity.

12.3.3.1 Hydrolysis

Hydrolysis is one of the most common enzymatic modification processes for the modification of millet starch by breaking down the starch molecules into simpler sugars, including dextrins, oligosaccharides, and monosaccharides. Hydrolysis can involve different enzymes, each with distinct action mechanisms, with one of the key enzymes in hydrolysis being α-amylase (AMY; EC 3.2.1.1), which catalyzes the endohydrolysis of α-(1,4) glycosidic linkages in polysaccharides with three or more α-(1,4) linked D-glucose units, along with its utilization in the liquefaction of starch during the production of sweeteners, like glucose syrups and maltodextrins, and is also used in brewing and baking industries to enhance product quality (Hii et al. 2012, 2002). Another similar enzyme, glucoamylase (AMG; EC 3.2.1.3), hydrolyzes both α-(1,6) and α-(1,4) glycosidic bonds, thereby producing glucose from starchy substrates, having a vital role in the production of bioethanol and high-glucose syrups (van der Maarel et al. 2002).

β-amylase (BAM; EC 3.2.1.2) is another important enzyme found in plants, microorganisms, and animals that cleaves the nonreducing α-(1,4) glycosidic bonds of glucan chains, producing maltose, and is, therefore, particularly useful in the production of high-maltose syrups, essential in brewing and confectionery (Wang et al. 2018). Also, a debranching enzyme, pullulanase (PULA; EC 3.2.1.41), hydrolyzes the α-(1,6) glycosidic bonds in pullulan, which facilitates the production of high-glucose and high-maltose syrups, and is also used to improve the texture and shelf life of bakery products and as an additive in detergents (Hii et al. 2012).

The hydrolysis process significantly impacts the food industry, involving the hydrolyzed starches to produce sweeteners, stabilizers, and thickeners, enhancing the texture and shelf life of processed foods and beverages. Also, in biofuel production, enzymatic hydrolysis is utilized to convert starchy biomass into fermentable sugars, later to be fermented to produce ethanol (Pejin et al. 2012). In pharmaceuticals, drug formulations utilize enzymatically modified starches for controlled release and improved solubility of APIs.

12.3.3.2 Transglycosylation

Transglycosylation involves transferring glycosyl units to acceptor molecules from the donor molecules, forming new glycosidic bonds, and producing modified starches with unique properties. The enzymes amylosucrase (AMYS) cyclodextrin glycosyltransferase (CDGT) facilitate the transglycosylation. Cyclodextrins are cyclic oligosaccharides, formed by the cyclization of parts of an α-(1,4)-D-glucan chain through the action of cyclodextrin glycosyltransferase (CDGT; EC 2.4.1.19). Amylosucrase (AMYS; EC 2.4.1.4), the enzyme belonging to the glycoside hydrolase family, transfers glucose units from sucrose to starch chains, creating modified starches with altered branching patterns, mostly the alpha-glucans. These modifications result in starches that are more resistant to retrogradation and have improved freeze-thaw stability, making them valuable in various food applications, along with the low glycemic and fiber-rich food products, including bakery products and dairy desserts (Seo et al. 2020). Amylomaltase enzyme (EC 2.4.1.25), belonging to the family of glycoside hydrolase which is involved in transglycosylation, forms cyclic or branched structure by transferring segments of a (1,4)-α-D-glucan to new positions in an acceptor molecule, which may be glucose or α-(1,4)-α-D-glucan, thereby catalyzing the formation of the cyclic oligosaccharides. This modification enhances the creaminess of products like yogurt and mayonnaise by rearranging the glucose units in the starch molecules, thereby enhancing the water binding capacity, and is also used to produce plant-based alternatives to gelatin and to prevent staling in bakery products by reducing the amylose retrogradation (Mun et al. 2009).

Transglycosylation is widely utilized in the food industry, to produce modified starches that with resistant retrogradative properties, thereby meeting the enhanced texture, digestibility, stability, and shelf life of food products, along with the utilization of cyclodextrins for encapsulating flavors and enhancing the solubility of food additives. Also, the cyclodextrins made from transglycosylation are used as carriers for sparingly soluble drugs in pharmaceuticals, thereby enhancing their bioavailability and stability (Seo et al. 2020). For utilization in skin and hair care products, modified starches from transglycosylation processes are used to improve texture and stability by providing a nongreasy, smooth feel, which makes the lotions more spreadable and for enhanced conditioning effects.

12.4 Emerging Technologies in Starch Modification

12.4.1 Ultrasonication

Ultrasonication, an emerging nonthermal processing technique, alters the starch physicochemical properties by applying high-frequency sound waves, thereby inducing chemical and physical modifications, including those derived from millet. The ultrasonication mechanism involves the generation of cavitation bubbles in a liquid medium, which collapse violently, creating localized high temperatures and pressures that can significantly alter the molecular and crystalline structures of starch granules (Estivi et al. 2022). Studies on foxtail millet starch have shown that ultrasonication can increase the amylose content significantly, with an amylose content of 21.05% compared to 13.22% in native starch, attributed to the partial depolymerization of amylopectin, which leads to a higher proportion of linear amylose chains (Babu et al. 2019). This modification enhances the functional properties of the starch, making it more suitable for specific applications that require high amylose content. The purity of millet starch is also enhanced by reducing undesirable components such as crude protein and fat, owing to the ultrasonic waves' mechanical force generation, which disrupts the protein-starch interactions, thereby liberating starch molecules from non-starch components (Su et al. 2024; Yadav et al. 2021). In an another study, the ultrasonic treatment of white finger millet starch resulted in increased amylose content and decreased protein and fat content (Amarnath et al. 2023). The modifications induced by ultrasonication further extend to the functional properties of millet starch by enhancing the solubility and swelling power, primarily due to the disruption of the crystalline molecular arrangement, which, in turn, allows water molecules to interact more readily with the hydroxyl groups of amylopectin and amylose. This effect is helpful for applications requiring high water retention and solubility, such as in soups and sauces (Huang et al. 2020). The cavitation effect involved in ultrasonication also helps remove the protein layer from starch granules, further enhancing their swelling power (Karwasra et al. 2020) (Nigar et al. 2024).

Moreover, ultrasonication has been shown to improve the freeze-thaw stability of millet starch, a critical factor for products that undergo freezing before storage or transportation, and the treatment disrupts the amorphous regions of the starch chain, leading to a significant reordering that limits reassociation during freezing, thereby reducing syneresis. This enhanced stability is especially noted in foxtail and finger millet starches (Amarnath et al. 2023; Dey and Sit 2017). Ultrasonication also affects the pasting properties of millet starch, with some studies report the increased peak and final viscosities after low to moderate ultrasonication treatments, attributed to the enhanced swelling power of the starch granules (Bento et al. 2021). However, other studies have noted a decrease in viscosity, possibly due to partial degradation of glycosidic bonds and a weakened polymer network (Chi et al. 2023; Kang et al. 2016). The thermal properties, such as gelatinization temperatures and enthalpy, are also impacted by ultrasonication, as the treatment facilitates easier penetration of water molecules into the starch granules, thereby lowering gelatinization temperatures. While Jambrak et al. (2010) did not observe a statistically significant increase in gelatinization temperature, slight variations at higher ultrasound intensities suggest that surface structural changes might influence water penetration. In some reports, such effects have been attributed to partial gelatinization on the granule surface, which can hinder hydration. Ultrasonication can increase the resistant starch (RS) content in millet starch, which is beneficial for developing foods with a low glycemic index, and the process promotes the fragmentation of longer starch chains, facilitating the formation of RS, which is advantageous for health, as RS functions similarly to dietary fiber and aids in controlling blood sugar levels (Ding et al. 2020). Additionally, increased RS content can improve the nutritional profile of millet-based foods, making them more appealing to health-conscious consumers (Wang et al. 2022). Ultrasonication significantly alters starch morphology by creating surface pores and grooves that enhance water permeability. This is accompanied by a reduction in molecular weight, due to chain scission, and a general decrease in the degree of crystallinity (Falsafi et al. 2019; Wei et al. 2023). Recent advancements have also explored the potential of combining ultrasonication with other treatments, such as acid hydrolysis, to produce nano-starch particles with unique properties, and this dual treatment can significantly reduce the particle size and modify the crystallinity and thermal stability of the starch, depending on its amylose content (Bajer 2023). Research shows that ultrasonication can physically modify starches like cassava and waxy maize, creating surface defects, reducing particle size into the nano-scale, and altering properties such as solubility and crystallinity (Airlangga et al. 2021; Kim et al. 2013). The degree of modification achieved through ultrasonication depends on various factors, including the ultrasound parameters and the botanical origin of the starch. This variability offers the potential for customized applications in the food industry, where specific functional properties are desired. Future research should focus on optimizing these parameters to maximize the benefits of ultrasonication and explore new applications for ultrasonicated millet starch in both food and non-food sectors.

12.4.2 Microwave Treatment

Microwave treatment (MW) is a nonconventional physical method widely utilized to modify starches, including millet starch, due to its efficiency in altering structural and functional properties, by utilizing electromagnetic waves to induce dielectric heating, which causes rapid internal heating and therefore physical damage to starch granules (Saini et al. 2021; Wang et al. 2017). The modification enhances the gel hydration properties of millet starch, including water solubility index (WSI), water absorption index (WAI), and swelling power (SP), which are crucial for the starch's application in food products requiring specific textural characteristics. The WAI and WSI are improved due to the disruption of hydrogen bonds and exposure of hydroxyl groups, facilitating better interaction with water. However, SP may reduce due to the damage of crystalline regions, which reduces the swelling ability of starch granules (Ananthu et al. 2023; Ashogbon 2021). Also, it modifies the pasting properties of millet starch, like peak and final viscosities, and the reduction in peak viscosity, attributed to the breakdown of starch granules and the weakening of the polymer network (Sun et al. 2014), which therefore results in a lower pasting temperature, improving the starch's suitability for applications requiring lower gelatinization temperatures (Guo et al. 2017; Yashini et al. 2024). The thermal properties of millet starch are affected by the microwave treatment, which increases the peak, onset, and conclusion gelatinization temperatures, indicating improved thermal stability, mainly attributed to the starch molecules' reorganization and the formation of a more stable gel network. The enthalpy of gelatinization also decreases, implying the crystalline regions are disrupted, facilitating easier gelatinization. The microwave process can also induce morphological changes, including surface deformation and the creation of pores or cracks, which affects properties like water solubility and permeability (Thangaraju et al. 2023). This structural alteration is often reflected in a change in relative crystallinity, confirming a shift from an ordered to a more amorphous state (Ashogbon 2021). In a broader context, thermal modifications are used to improve the nutritional profile of millet starch by increasing its resistant starch (RS) content. The formation of RS, which is beneficial for controlling the glycemic index, has been demonstrated in proso millet using methods like heat-moisture treatment (Han et al. 2021). The microwave treatment shows the potential to be a promising method for modifying millet starch, by enhancing its functional properties (Singh et al. 2024), making it more suitable for various food applications. However, the degree of modification depends on the specific microwave parameters, providing opportunities for tailored applications in the food industry (Han et al. 2021).

12.4.3 High Pressure Processing (HPP)

High pressure processing (HPP) is an innovative nonthermal processing technique utilized to modify the properties of starches, including millet starch, by applying extremely high pressures ranging from 200 to 600 MPa at various temperatures (25–50 °C), therefore, inducing significant changes in the functional and structural properties of the starch. By altering the starch properties without utilizing any chemical reagents, it provides an environmentally friendly option to the conventional chemical modification methods (de Oliveira et al. 2024).

The structural integrity of starch granules is altered with significant disruption at increasing pressure levels, leading to the loss of the characteristic Maltese cross pattern visualized under polarized light microscopy. This pattern indicates the granular crystalline structure; thus, its loss suggests a decrease in structural order and crystallinity (Colussi et al. 2018). Additionally, the swelling of granules leads to an increased particle size, as confirmed by size distribution analyses, along with a marked reduction in crystallinity, evidenced by the attenuation of the XRD peaks. A general decrease in the enthalpy of gelatinization of modified millet starches indicates a disruption of the crystalline structure and the formation of a more amorphous state (Dominguez-Ayala et al. 2022; Liu et al. 2016). Fourier-transform infrared (FTIR) spectroscopy is utilized for probing the short-range molecular order within starch granules. Structural modifications are typically reflected by changes in two key areas: the O-H stretching band (around 3000–3600 cm^{-1}), which indicates alterations in the hydrogen bonding network, and the 'fingerprint' region (900–1200 cm^{-1}). Peaks within this latter region are sensitive to the short-range order of the double helices. Therefore, a reduction in intensity or a shift in these characteristic peaks after treatment can signify a disruption of the ordered structures and a transition to a more amorphous state (Pozo et al. 2018). The transformation from an "A"-type to a "B"-type crystalline pattern, especially prominent at pressures above 400 MPa, shows the significant impact of HPP on the structural reorganization of starch granules as mentioned by Almeida et al. (2023) for the HPP treatment on rice starch. High-pressure processing (HPP) can enhance the techno-functional properties of rice starch, particularly its water absorption capacity. This improvement is attributed to HPP-induced structural modifications, including granule surface disruption and a decrease in crystallinity, which create a more amorphous structure that facilitates greater water uptake (Almeida et al. 2023).

Conversely, HPP results in softer gel formation by decrease in gel strength, which is beneficial where a less rigid texture is preferred, such as in certain desserts and puddings (Colussi et al. 2018). The reduced gel strength correlates with a lower setback viscosity, indicating reduced retrogradation and better paste stability, useful for products requiring stable viscosity under varying thermal conditions, making HPP-treated starches ideal for gravies, soups, and other processed foods subjected to cooling and heating cycles. High-pressure processing (HPP) can significantly alter the digestibility profile of starches, but the effect is highly dependent on the starch's botanical origin and amylose content. For example, HPP treatment of high-amylose sorghum starch led to a desirable decrease in the rapidly digestible starch (RDS) fraction and an increase in slowly digestible (SDS) and resistant starch (RS), making it suitable for developing low-glycemic-index foods (Liu et al. 2016). In contrast, the same process applied to amylose-free waxy rice starch resulted in the opposite effect, increasing the RDS content while decreasing SDS and RS (Zeng et al. 2018). This demonstrates that HPP is a versatile non-thermal option, but its parameters must be adjusted to the specific type of starch to achieve desired functional and nutritional properties for different food applications.

As a versatile tool, HPP can modify cereal starches for diverse applications, from enhancing dough consistency in gluten-free breads to creating hydrogels for creams and ointments (Raghunathan et al. 2021). Functionally, HPP can also be used to control texture. Treatment of litchi kernel starch, for example, was found to lower its setback viscosity, indicating a reduced tendency for retrogradation, which is useful for products requiring stable viscosity (Sandhu et al. 2021a). Future research should focus on exploring these applications and further optimizing the HPP parameters to maximize the benefits and potential uses of modified millet starch. This technology not only improves the functional properties of starch, such as gel strength, water absorption, and digestibility, but also eliminates the need for chemical reagents to meet the environmental sustainability goals. As research progresses, the full potential of HPP-modified millet starch in various sectors, including food technology and biodegradable materials, will likely offer new opportunities for product development and innovation.

12.4.4 Cold Plasma

Despite the advantages of being sourced from various millet species such as kodo, pearl, and proso millets, along with the nutritional benefits, native millet starches often lack the necessary functional properties for specific industrial applications, which necessitates the modifications to enhance characteristics like stability, solubility, and viscosity. Cold plasma technology, a nonthermal and environmentally friendly technique, has emerged as a promising method for modifying millet starch, with demonstrated improvements in structural alterations and hydration properties. For instance, the water absorption capacity (WAC) of kodo millet starch increased from 1.43 to 1.55 g/g, while the oil absorption capacity (OAC) rose from 1.01 to 1.34 g/g after treatment, with the increment mainly attributed to plasma-induced surface etching and the introduction of polar functional groups, which increase the starch's interaction with water and oil molecules (Sonkar et al. 2023). Furthermore, cold plasma treatment has been shown to significantly reduce the turbidity of pearl millet starch, leading to clearer pastes that exhibit greater stability over a five-day storage period (Kaur and Annapure 2023). The formation of new functional groups post cold plasma treatment, including carboxyl and carbonyl groups, indicates oxidation and introduction of reactive sites on the starch molecules, cold plasma (CP) treatment led to the formation of new carbonyl and carboxyl functional groups, indicating oxidative modification of starch molecules. In proso millet starch, CP—especially when combined with ultra-high pressure (e.g., UHP at 600 MPa)—partially disrupted crystalline regions and reduced short-range order, which in turn enhances solubility and reactivity in various industrial applications (Sun et al. 2022). The underlying mechanisms by which cold plasma modifies starch include depolymerization (Gupta and Routray 2025), cross-linking, and surface etching, where the depolymerization refers to the breakdown of starch molecules into smaller fragments, thereby increasing their solubility, which reduces the viscosity of starch pastes. Cross-linking involves the formation of covalent bonds between starch molecules, enhancing the stability and mechanical strength of starch gels, whereas the plasma etching, a physical process, alters the surface morphology of starch granules, making them more porous and increasing their surface area. This increased surface area facilitates greater interaction with other molecules, such as water or enzymes, improving the starch's functional properties (Subrahmanyam et al. 2024). Enhanced water and oil absorption capacities make these starches ideal for use as thickeners, emulsifiers, and stabilizers in a range of food products, from bakery items to dressings and sauces. Additionally, the eco-friendly nature of cold plasma technology aligns with the growing consumer demand for sustainable and clean-label ingredients (Sahoo et al. 2023). Apart from applications in food, cold plasma-modified millet starch shows promise in pharmaceutical formulations as a drug carrier, in cosmetics for improved texture, and even in biodegradable materials for packaging (Subrahmanyam et al. 2023). As research into cold plasma technology advances, further optimization of treatment parameters—such as gas composition, exposure time, and power—will be crucial. Moreover, exploring the synergistic effects of cold plasma with other modification techniques, such as enzymatic treatment or chemical modification, could yield even better versatile and functional starch derivatives (Heydari et al. 2021).

12.5 Applications of Modified Millet Starches

Modified millet starches have gained considerable attention across multiple industries due to their enhanced physicochemical and functional properties. This section explores these applications, particularly in the food and nonfood sectors.

12.5.1 Food Industry

The food industry extensively utilizes modified millet starches for their unique textural and stabilizing properties, making them ideal for various applications.

12.5.1.1 Texturizers

Modified millet starches can act as an efficient texturizer and contribute to better sensory experience, enhancing the mouthfeel and texture of food products, contributing to a superior sensory experience. For instance, hydroxypropylated and succinylated millet starches have been found to inhibit the starch's inherent ability to retrogradation to prevent the product from being excessively firm or sticky in nature, thereby improving the texture and stability of white sauces by increasing cold storage stability and where water separates from the food matrix (Shaikh et al. 2017). In gluten-free baked goods, the texturizers can assist in achieving the desired crumb structure and softness involving the fluffier nature with suitable mouthfeel, critical for marketability and consumer satisfaction (Nagarajaiah et al. 2019). Additionally, waxy proso millet starch (PMS) is known for its high amylopectin content (95–98%), along with high stability and low retrogradation, making it widely utilized in convenient frozen foods and as a raw material in producing beverages, offering improved shelf life and consistency (Bangar et al. 2021). Ultrasonic modification of the millet starch alters the amylose content (AC), which plays an important role in the textural modification of ready-to-eat or ready-to-cook snacks, because of the shear forces generated due to the cavitation bubbles, leading to the disintegrated structure, which finally corroborates to the homogeneous and smoother gel formation. Also, this increased amylose content correlates negatively with cohesiveness and positively to hardness, making it appropriate for extruded products like noodles and pasta, along with enhancing the crispiness of extruded and fried products (Kumari et al. 2024b).

12.5.1.2 Stabilizers

Modified millet starches effectively stabilize various food products by forming a protective layer around oil droplets in emulsions, preventing the droplets' coalescence, thereby maintaining consistency and extending shelf life. The ability of millet starches to form stable emulsions and gels makes them ideal for sauces, custards, and dairy products. For instance, Shaikh et al. (2017) reported that chemically modified pearl millet starches enhanced the stability of custards by reducing syneresis during storage, because the cross-linked pearl millet starch molecules create a robust gel network which prevents its contraction, ultimately releasing less water. Also, in combination with carrageenan, these pearl millet starches improve the tensile strength, especially for moisture-sensitive foods, and reduce the water vapor permeability of edible films, enhancing their utilization in packaging applications (Sandhu et al. 2020). The pasting properties of non-waxy PMS, characterized by high shear resistance and pasting temperature, suggest potential applications in medical formulations, such as drug delivery systems, specifically controlled drug release, tablet binders, and injectable hydrogels to name a few (Bangar et al. 2021). Notably, ultrasonicated millet starch also demonstrates improved solubility and swelling power, which further enhances its functionality as a stabilizer and thickener in products like ketchup and soups (Kumari et al. 2024b).

12.5.1.3 Thickeners

Modified millet starches, like those from barnyard and pearl millet, are widely utilized as effective thickeners in gravies, soups, and pie fillings, not only because of their desirable texture and viscosity inside the product, but also because they help preventing ingredient separation. For instance, and acid-modified and pregelatinized millet starches enhance the thickening power of pie fillings, ensuring a smooth and consistent texture and preventing the filling from becoming runny and its excessive breakdown during high-temperature baking (Mahajan et al. 2021). The low breakdown viscosity and high pasting temperature of non-waxy pearl millet starch (PMS) indicate a strong resistance to mechanical shear during processing, making it suitable for food processing applications such as in the production of sauces, bakery products, and soups, where prolonged stirring or high shear conditions are common, thereby preventing thinning out too much (Bangar et al. 2021).

Additionally, barnyard millet starch exhibits smaller granular size, with high amylose content, and enhanced gel-forming ability, making it a promising stabilizer and thickener in various food products, including dairy alternatives, soups, sauces, and bakery items. Its high amylose content suits the products requiring strong gel structures, while its small starch granule size contributes to smooth textures. Moreover, barnyard millet starch possessing the capability to swell, provide viscosity, and form gels gives a creamy mouthfeel without the addition of fat and is used widely in low-calorie foods as a fat replacer (Siroha and Bangar 2024). Hydroxypropylated and succinylated starches, which are chemically modified starches, improve the natural thickeners. For instance, custards including hydroxypropylated starch show exceptional cold storage stability, with negligible syneresis observed even after several days, making them ideal for dairy desserts that require to maintain a creamy and consistent texture over time (Shaikh et al. 2017). Such modifications, involving oxidation, acetylation, and cross-linking significantly improves the swelling capacity and stability and texture of these starches, enhancing their utilization in the food industry.

12.5.1.4 Fat Replacers

Owing to the growing demand for healthier low-fat foods, modified millet starches are widely utilized as fat replacers. For instance, octenyl succinylated pearl millet starch has been included in reduced-fat ice cream, providing a similar mouthfeel and creaminess to that of fats, and is helpful in producing and developing dietetic and low-calorie products (Sharma et al. 2017). Notably, it is documented that ultrasonicated high-amylose millet starches not only improve texture but also help lower the glycemic index, making them ideal for health-oriented food products, along with having relatively higher resistant starch (RS) content, which are utilized for developing fiber-rich food products (Kumari et al. 2024b).

12.5.2 Non-food Applications

Modified millet starches are also utilized widely in nonfood sectors due to their functional and biodegradable properties.

12.5.2.1 Biodegradable Packaging Materials

The wide environmental advantages of biodegradable packaging are favoring the utilization of modified millet starches. Edible films prepared from millet starch, when formulated with different biopolymers like carrageenan or curcumin additives, provide excellent barrier properties against oxygen and moisture, protecting packaged foods from oxidative damage (Baek and Song 2019). These films offer barrier properties by forming flexible and cohesive layers to decrease the permeability and are particularly promising for cheese and other dairy products (Yang et al. 2018). Additionally, Al-Hashimi et al. (2020) reported the inclusion of clove essential oils to pearl millet starch films with improved mechanical and barrier properties, along with antimicrobial protection and extended shelf life. Other applications of biodegradable packaging involve the preparation of films from pearl millet starch mixed with borage seed oil, which is gamma-linolenic acid (GLA)-rich and provides an effective barrier against moisture and light by acting as an antioxidant, thus extending the shelf life of packaged foods (Cao et al. 2017). Also, the millet starches with high amylose content treated and modified by ultrasonication are utilized in the preparation of biofilms with higher tensile strength, superior mechanical properties, and a higher modulus of elasticity, compared to the conventional biofilms. These listed properties are important for developing biodegradable packaging materials that are environmentally friendly and robust in nature (Kumari et al. 2024b).

12.5.2.2 Pharmaceuticals

Modified millet starches are widely utilized as excipients in tablet formulations, and also as binders, disintegrants, and fillers, for enhanced stability and superior release profile of active pharmaceutical ingredients (APIs). Odeku and Alabi (2007) documented that pre-gelatinized, acid-modified variants and acetylated forms of modified millet starches show effective disintegrant properties. For example, modified starches utilized in chloroquine phosphate tablet formulations have shown modified dissolution and disintegration, with faster drug release and quick onset of action, compared to conventional cornstarch BP. A comparative study on the disintegrant prepared from pearl millet starch (PMS) utilized in paracetamol tablets demonstrated its comparable efficacy to potato and maize starches, suggesting that PMS can serve as an economical alternative to conventional starches in millet-rich regions (Osman et al. 2021). Modified millet starches also work as tablet binders in tablet pharmaceutical formulations, along with the acid-modified millet starch mucilage, which has been documented for its binding properties in paracetamol tablets, exhibiting comparable friability and hardness to those made with maize starch BP (Meko et al. 2018a, b). Additionally, the dissolution profile was enhanced by acid-modified starch for tablets, with all the batches achieving more than 70% drug release within 30 min, and thereby making it a favorable binder for different applications in pharmaceutical industry (Meko et al. 2018a, b). The acid modification significantly enhanced the millet starch functionality as a super-disintegrant, as demonstrated in a study on diclofenac sodium tablets, which outperformed native starch in terms of dissolution rate and disintegration time, with more than 90% drug release within 30 min, unlike sodium starch glycolate, a commonly used super-disintegrant (Eraga et al. 2017).

Additionally, Eze and Alozie (2015) conducted an assessment of pharmaceutical grade starch (PGS) produced from corn, millet, and sorghum, nothing that millet produced a pharmaceutical quality starch despite having a lower gelatinization temperature range (55.50–65.00 °C) compared to corn and sorghum. The study demonstrated unique properties for millet starch, including the highest amount of iron (9.00 ppm), the lowest protein content, and a substantial loss on drying (13.00%), making it a viable alternative to more commonly used starches (Eze and Alozie 2015). Odeku and Alabi (2007) further emphasize that tropically grown starches like from millets have shown potential as valuable substitute for commercially available chemically modified starches, adding worth to underutilized crops and offering specialized properties for specific pharmaceutical applications. In addition, Bhatt et al. (2023) examined the tableting and techno-functional properties of starch extracted from barnyard millet (Echinochloa esculenta) in its chemically modified (phosphorylated) and native (alkali-treated) states, finding improved flow properties in the phosphorylated barnyard millet starch, as shown by a Hausner ratio of 1.25, along with enhanced tableting characteristics such as lower friability and improved compressibility relative to its native form. Further, the swelling power of alkali-treated starch was 194.3%, while phosphorylated starch showed reduced swelling due to a compact molecular structure, specifying its utilization in controlled-release processes providing sustainable and efficient solutions for drug delivery systems (Malik et al. 2023).

12.5.2.3 Miscellaneous Applications of Millet Starch

The cosmetic industry also utilizes modified millet starches as stabilizers, thickeners, and in lotions, creams, and other skin care products. Being hypoallergenic and nontoxic, and possessing an antioxidant nature, when combined with other active ingredients, millet starches provide protective benefits against oxidative stress, valuable in antiaging and skin-protective products (Ju and Song 2019). The modified millet starch, for instance, from pearl millet, has shown favorable impact in biofuel production, and also can be processed to make enzymes like α-amylase, crucial in the starch liquefaction process, playing a vital role in breaking the starch molecules into fermentable sugars during the production of bioethanol. The utilization of millet starch as a low-cost feedstock for enzyme production provides a cost-effective and sustainable alternative to traditional sources (Ikram-ul-Haq et al. 2005). Besides applications in biofuel, production of industrial enzymes such as α-amylase has also been explored, which are widely used across various industries, including textiles, food, fermentation, and paper manufacturing. The multifaceted utilization of millet starch both in enzyme production and industrial applications highlights its economic importance and versatility.

12.6 Conclusion

12.6.1 Key Points

The exploration of modified millet starches has shown substantial enhancements in their functional properties, positioning them as versatile ingredients in various applications. Chemical modifications such as octenyl succinylation and succinylation significantly alter paste clarity, swelling power, and paste viscosity of pearl millet starches. These modifications create new opportunities to be used as effective thickening agents in culinary products like sauces, gravies, and ketchups. Structural analyses via scanning electron microscopy (SEM) indicate that the chemical alterations do not drastically alter the granular morphology of the starch, except in certain cases like succinylated starch. Additionally, acetylation has been shown to eliminate retrogradation peaks and lower the gelatinization enthalpy, making acetylated starches highly suitable for frozen food applications due to their low temperature stability. The unique physicochemical properties of various millet starches, including those from pearl and proso millet, emphasize their potential not only in food but also non-food industrial applications.

12.6.2 Importance of Continued Research and Development

The scope of research on millet starches, particularly their application and modification, is still emerging, and given the contrast in their physical and chemical properties, continuous research is vital to understand these differences and optimize their functionalities for specific uses. For instance, the solubility, swelling power, and pasting properties of these starches can be customized through combined or dual modifications involving physical, chemical, and enzymatic treatments. Also, such developments are important for improving the starch' resistance to digestive enzymes, thereby enhancing their suitability for developing functional foods with specific health benefits. The lack of comprehensive data on lesser-known millet varieties and the need for standardization in experimental methods present opportunities for future research. This research can help in creating a standardized framework for evaluating and utilizing millet starches, thereby expanding their industrial and commercial applications.

12.6.3 Future of Millet Starch Applications

Millet starches, with their adaptability to harsh growing conditions and their potential for diverse functional modifications, present an untapped resource for the food and industrial sectors. As research continues to exhibit their full capabilities, millet starches are poised to become competitive alternatives to traditional starches derived from wheat, corn, and potatoes. Their potential uses extend beyond food thickeners and binders to include applications in pharmaceuticals, biodegradable packaging, and health-oriented food products. The development of new modification techniques and the exploration of underutilized millet varieties will further enhance their functionality and marketability. The future of millet starch applications is promising, with the potential to contribute significantly to food security, economic growth, and sustainable agricultural practices, and, therefore, will require a collaborative effort from researchers, industry stakeholders, and policymakers to fully utilize the benefits of millet starches and integrate them into global food and industrial systems.

Acknowledgments The authors are thankful to the National Institute of Technology for providing research and literature collection facilities.

Data Availability Statement The data that support the findings of this study are available from the corresponding author upon reasonable request.

Conflict of Interest Statement The authors report no conflict of interest.

References

Abdel-Aal E-S, Hucl P, Chibbar R, Han H, Demeke T (2002) Physicochemical and structural characteristics of flours and starches from waxy and nonwaxy wheats 1. Cereal Chem 79:458. https://doi.org/10.1094/CCHEM.2002.79.3.458

Adebowale KO, Afolabi TA, Olu-Owolabi BI (2005) Hydrothermal treatments of finger millet (Eleusine coracana) starch. Food Hydrocoll 19(6):974–983. https://doi.org/10.1016/j.foodhyd.2004.12.007

Afolabi TA, Olu-Owolabi BI, Adebowale KO, Lawal OS, Akintayo CO (2012) Functional and tableting properties of acetylated and oxidised finger millet (Eleusine coracana) starch. Starch – Stärke 64(4):326–337. https://doi.org/10.1002/star.201100151

Ai Y, Jane J (2015) Gelatinization and rheological properties of starch. Starch – Stärke 67(3–4):213–224. https://doi.org/10.1002/star.201400201

Airlangga B, Sugianto AM, Parahita G, Puspasari F, Mayangsari NE, Trisanti PN, Sutikno JP, Sumarno S (2021) Study of cassava starch degradation using sonication process in aqueous sodium chloride. J Sci Food Agric 101(6):2406–2413. https://doi.org/10.1002/jsfa.10864

Al-Hashimi AG, Ammar ABGL, Cacciola F, Lakhssassi N (2020) Development of a millet starch edible film containing clove essential oil. Foods 9(2):Article 2. https://doi.org/10.3390/foods9020184

Almeida RLJ, Santos NC, Feitoza JVF, dos Santos Pereira T, da Silva Eduardo R, de Almeida Mota MM, de Almeida Silva R, de Alcântara Silva VM, de Alcântara Ribeiro VH, Muniz CES, Ribeiro CAC, da Costa GA, de Figueiredo MJ (2023) Evaluation of the technological properties of rice starch modified by high hydrostatic pressure (HHP). Innovative Food Sci Emerg Technol 83:103241. https://doi.org/10.1016/j.ifset.2022.103241

Amarnath MS, Muhammed A, Antony AK, Malini B, Sunil CK (2023) White finger millet starch: physical modification (annealing and ultrasound), and its impact on physicochemical, functional, thermal and structural properties. Food Humanit 1:599–606. https://doi.org/10.1016/j.foohum.2023.07.010

Amarnath MS, Muhammed A, Antony AK, Yashini M, Sherin AJ, Sunil CK (2024) Millet starch modification: effect of thermal and non-thermal methods on its properties and their application – a review. Food Humanit 3:100356. https://doi.org/10.1016/j.foohum.2024.100356

Ananthu R, Buvaneswaran M, Meena L, Sunil CK (2023) Microwave treatment effect on functional and pasting properties, and storage stability of white finger millet. J Food Process Eng 46(7):e14341. https://doi.org/10.1111/jfpe.14341

Annor GA, Marcone M, Bertoft E, Seetharaman K (2014) Physical and molecular characterization of millet starches. Cereal Chem 91(3):286–292. https://doi.org/10.1094/CCHEM-08-13-0155-R

Annor GA, Tyl C, Marcone M, Ragaee S, Marti A (2017) Why do millets have slower starch and protein digestibility than other cereals? Trends Food Sci Technol 66:73–83. https://doi.org/10.1016/j.tifs.2017.05.012

Apriyanto A, Compart J, Fettke J (2022) A review of starch, a unique biopolymer – structure, metabolism and in planta modifications. Plant Sci 318:111223. https://doi.org/10.1016/j.plantsci.2022.111223

Ashogbon AO (2021) Dual modification of various starches: synthesis, properties and applications. Food Chem 342:128325. https://doi.org/10.1016/j.foodchem.2020.128325

Ashogbon AO, Akintayo ET (2014) Recent trend in the physical and chemical modification of starches from different botanical sources: a review. Starch – Stärke 66(1–2):41–57. https://doi.org/10.1002/star.201300106

Ayucitra A (2012) Preparation and characterisation of acetylated corn starches. Int J Chem Eng Appl 3(3):156

Babu AS, Mohan RJ, Parimalavalli R (2019) Effect of single and dual-modifications on stability and structural characteristics of foxtail millet starch. Food Chem 271:457–465. https://doi.org/10.1016/j.foodchem.2018.07.197

Baek S-K, Song KB (2019) Characterization of active biodegradable films based on Proso millet starch and curcumin. Starch – Stärke 71(3–4):1800174. https://doi.org/10.1002/star.201800174

Bajer D (2023) Nano-starch for food applications obtained by hydrolysis and ultrasonication methods. Food Chem 402:134489. https://doi.org/10.1016/j.foodchem.2022.134489

Bangar SP, Ashogbon AO, Dhull SB, Thirumdas R, Kumar M, Hasan M, Chaudhary V, Pathem S (2021) Proso-millet starch: properties, functionality, and applications. Int J Biol Macromol 190:960–968. https://doi.org/10.1016/j.ijbiomac.2021.09.064

Bento JAC, Ferreira KC, Fidelis MC, de Souza Neto MA, de Paixão e Silva GL, Bataus LAM, Caliari M, Soares Júnior MS (2021) Ultrasound modification of white Garland-lily starch:

functional, thermal, and pasting properties. Starch – Stärke 73(3–4):2000129. https://doi.org/10.1002/star.202000129

Bhatt P, Kumar V, Rastogi H, Malik MK, Dixit R, Garg S, Kapoor G, Singh S (2023) Functional and tableting properties of alkali-isolated and phosphorylated barnyard millet (Echinochloa esculenta) starch. ACS Omega 8(33):30294–30305. https://doi.org/10.1021/acsomega.3c03158

Cao TL, Yang S-Y, Song KB (2017) Characterization of barnyard millet starch films containing borage seed oil. Coatings 7(11):Article 11. https://doi.org/10.3390/coatings7110183

Carpenter MA, Joyce NI, Genet RA, Cooper RD, Murray SR, Noble AD, Butler RC, Timmerman-Vaughan GM (2015) Starch phosphorylation in potato tubers is influenced by allelic variation in the genes encoding glucan water dikinase, starch branching enzymes I and II, and starch synthase III. Front Plant Sci 6. https://doi.org/10.3389/fpls.2015.00143

Chandra D, Chandra S, Pallavi, Sharma AK (2016) Review of finger millet (Eleusine coracana (L.) Gaertn): A power house of health benefiting nutrients. Food Sci Human Wellness 5(3):149–155. https://doi.org/10.1016/j.fshw.2016.05.004

Chandrasekara A, Shahidi F (2010) Content of insoluble bound Phenolics in millets and their contribution to antioxidant capacity. J Agric Food Chem 58(11):6706–6714. https://doi.org/10.1021/jf100868b

Chen Y, Sun X, Zhou X, Hebelstrup KH, Blennow A, Bao J (2017) Highly phosphorylated functionalized rice starch produced by transgenic rice expressing the potato GWD1 gene. Sci Rep 7(1):3339. https://doi.org/10.1038/s41598-017-03637-5

Chi C, Xu K, Wang H, Zhao L, Zhang Y, Chen B, Wang M (2023) Deciphering multi-scale structures and pasting properties of wheat starch in frozen dough following different freezing rates. Food Chem 405:134836. https://doi.org/10.1016/j.foodchem.2022.134836

Colussi R, Kaur L, da Zavareze ER, Dias ARG, Stewart RB, Singh J (2018) High pressure processing and retrogradation of potato starch: influence on functional properties and gastro-small intestinal digestion in vitro. Food Hydrocoll 75:131–137. https://doi.org/10.1016/j.foodhyd.2017.09.004

de Oliveira LC, Gouseti O, Macnaughtan B, Clerici MTPS, Sampaio U, Bakalis S, Muttakin S, Cristianini M (2024) Application of thermally assisted high hydrostatic pressure to modify sorghum starch: multi-scale structure, techno-functional properties and digestibility. J Food Sci Technol 62:109. https://doi.org/10.1007/s13197-024-06014-z

Dey A, Sit N (2017) Modification of foxtail millet starch by combining physical, chemical and enzymatic methods. Int J Biol Macromol 95:314–320. https://doi.org/10.1016/j.ijbiomac.2016.11.067

Dhital S, Warren FJ, Butterworth PJ, Ellis PR, Gidley MJ (2017) Mechanisms of starch digestion by α-amylase—structural basis for kinetic properties. Crit Rev Food Sci Nutr 57(5):875–892. https://doi.org/10.1080/10408398.2014.922043

Ding Y, Liang Y, Luo F, Ouyang Q, Lin Q (2020) Understanding the mechanism of ultrasonication regulated the digestibility properties of retrograded starch following vacuum freeze drying. Carbohydr Polym 228:115350. https://doi.org/10.1016/j.carbpol.2019.115350

Dominguez-Ayala JE, Soler A, Mendez-Montealvo G, Velazquez G (2022) Supramolecular structure and technofunctional properties of starch modified by high hydrostatic pressure (HHP): a review. Carbohydr Polym 291:119609. https://doi.org/10.1016/j.carbpol.2022.119609

Eraga S, Nwajuobi V, Iwuagwu M (2017) Super-disintegrant activity of acid-modified millet starch in diclofenac tablet formulations

Estivi L, Brandolini A, Condezo-Hoyos L, Hidalgo A (2022) Impact of low-frequency ultrasound technology on physical, chemical and technological properties of cereals and pseudocereals. Ultrason Sonochem 86:106044. https://doi.org/10.1016/j.ultsonch.2022.106044

Eze SOE, Alozie KN (2015) Assessment of pharmaceutical grade starch produced from some local grains-CORN, sorghum and millet. J Chem Soc Nigeria 40(1):Article 1. https://journals.chemsociety.org.ng/index.php/jcsn/article/view/18

Falsafi SR, Maghsoudlou Y, Rostamabadi H, Rostamabadi MM, Hamedi H, Hosseini SMH (2019) Preparation of physically modified oat starch with different sonication treatments. Food Hydrocoll 89:311–320. https://doi.org/10.1016/j.foodhyd.2018.10.046

Fonseca LM, Halal SLME, Dias ARG, da Zavareze ER (2021) Physical modification of starch by heat-moisture treatment and annealing and their applications: a review. Carbohydr Polym 274:118665. https://doi.org/10.1016/j.carbpol.2021.118665

Gao X-W, Liu H-Y, Lv J, Cui M (2019) A novel element differential method for solid mechanical problems using isoparametric triangular and tetrahedral elements. Comput Math Appl 78(11):3563–3585. https://doi.org/10.1016/j.camwa.2019.05.026

Guo Q, Sun D-W, Cheng J-H, Han Z (2017) Microwave processing techniques and their recent applications in the food industry. Trends Food Sci Technol 67:236–247. https://doi.org/10.1016/j.tifs.2017.07.007

Gupta KK, Routray W (2025) Cold plasma: A nonthermal pretreatment, extraction, and solvent activation technique for obtaining bioactive compounds from agro-food industrial biomass. Food Chem 472:142960. https://doi.org/10.1016/j.foodchem.2025.142960

Han M, Dang K, Wang J, Gao L, Wang H, Ivanistau A, Yang Q, Feng B (2021) New type of food processing material: the crystal structure and functional properties of waxy and non-waxy Proso millet resistant starches. Molecules 26(14):Article 14. https://doi.org/10.3390/molecules26144283

He X-H, Xia W, Chen R-Y, Dai T-T, Luo S-J, Chen J, Liu C-M (2020) A new pre-gelatinized starch preparing by gelatinization and spray drying of rice starch with hydrocolloids. Carbohydr Polym 229:115485. https://doi.org/10.1016/j.carbpol.2019.115485

He, Y., Ye, F., Tao, J., Zhang, Z., & Zhao, G. (2023). Ozone exposure tunes the physicochemical properties of sweet potato starch by modifying its molecular structure. Int J Bio Macro 236:124002 https://doi.org/10.1016/j.ijbiomac.2023.124002

Heydari A, Razavi SMA, Hesarinejad MA, Farahnaky A (2021) New insights into physical, morphological, thermal, and pasting properties of HHP-treated starches: effect of starch type and industry-scale concentration. Starch – Stärke 73(7–8):2000179. https://doi.org/10.1002/star.202000179

Hii SL, Tan JS, Ling TC, Ariff AB (2012) Pullulanase: role in starch hydrolysis and potential industrial applications. Enzyme Res 2012(1):921362. https://doi.org/10.1155/2012/921362

Hoover R, Ratnayake WS (2002) Starch characteristics of black bean, chick pea, lentil, navy bean and pinto bean cultivars grown in Canada. Food Chem 78(4):489–498. https://doi.org/10.1016/S0308-8146(02)00163-2

Hu A, Li Y, Zheng J (2019) Dual-frequency ultrasonic effect on the structure and properties of starch with different size. LWT 106:254–262. https://doi.org/10.1016/j.lwt.2019.02.040

Huang D, Men K, Li D, Wen T, Gong Z, Sunden B, Wu Z (2020) Application of ultrasound technology in the drying of food products. Ultrason Sonochem 63:104950. https://doi.org/10.1016/j.ultsonch.2019.104950

Ikram-ul-Haq, Ashraf H, Qadeer MA, Iqbal J (2005) Pearl millet, a source of alpha amylase production by bacillus licheniformis. Bioresour Technol 96(10):1201–1204. https://doi.org/10.1016/j.biortech.2004.09.012

Jambrak AR, Herceg Z, Šubarić D, Babić J, Brnčić M, Brnčić SR, Bosiljkov T, Čvek D, Tripalo B, Gelo J (2010) Ultrasound effect on physical properties of corn starch. Carbohydr Polym 79(1):91–100. https://doi.org/10.1016/j.carbpol.2009.07.051

Ju A, Song KB (2019) Development of teff starch films containing camu-camu (Myrciaria dubia mc. Vaugh) extract as an antioxidant packaging material. Ind Crop Prod 141:111737. https://doi.org/10.1016/j.indcrop.2019.111737

Kam J, Puranik S, Yadav R, Manwaring HR, Pierre S, Srivastava RK, Yadav RS (2016) Dietary interventions for type 2 diabetes: how millet comes to help. Front Plant Sci 7. https://doi.org/10.3389/fpls.2016.01454

Kang N, Zuo YJ, Hilliou L, Ashokkumar M, Hemar Y (2016) Viscosity and hydrodynamic radius relationship of high-power ultrasound depolymerised starch pastes with different amylose content. Food Hydrocoll 52:183–191. https://doi.org/10.1016/j.foodhyd.2015.06.017

Karwasra BL, Kaur M, Gill BS (2020) Impact of ultrasonication on functional and structural properties of Indian wheat (Triticum aestivum L.) cultivar starches. Int J Biol Macromol 164:1858–1866. https://doi.org/10.1016/j.ijbiomac.2020.08.013

Kaur P, Annapure US (2023) Effects of pin-to-plate atmospheric cold plasma for modification of pearl millet (Pennisetum glaucum) starch. Food Res Int 169:112930. https://doi.org/10.1016/j.foodres.2023.112930

Kim H-Y, Park DJ, Kim J-Y, Lim S-T (2013) Preparation of crystalline starch nanoparticles using cold acid hydrolysis and ultrasonication. Carbohydr Polym 98(1):295–301. https://doi.org/10.1016/j.carbpol.2013.05.085

Kringel DH, El Halal SLM, da Zavareze ER, Dias ARG (2020) Methods for the extraction of roots, tubers, pulses, Pseudocereals, and other unconventional starches sources: A review. Starch – Stärke 72(11–12):1900234. https://doi.org/10.1002/star.201900234

Kumari K, Kashyap P, Chakrabarti P (2024a) Germination and probiotic fermentation: A way to enhance nutritional and biochemical properties of cereals and millets. Food Sci Biotechnol 33(3):505–518. https://doi.org/10.1007/s10068-023-01401-2

Kumari S, Kaur BP, Thiruvalluvan M (2024b) Ultrasound modified millet starch: changes in functional, pasting, thermal, structural, in vitro digestibility properties, and potential food applications. Food Hydrocoll 153:110008. https://doi.org/10.1016/j.foodhyd.2024.110008

Lang, S., Sui, C., & Wang, L. (2025). Recent development in ozone-based starch modification: From generation methods to film applications. Int J Biol Macromol 309(Part 2): 142780 https://doi.org/10.1016/j.ijbiomac.2025.142780

Lawal O (2009) Starch hydroxyalkylation: physicochemical properties and enzymatic digestibility of native and hydroxypropylated finger millet (Eleusine coracana) starch. Food Hydrocoll 23:415–425. https://doi.org/10.1016/j.foodhyd.2008.02.013

Liu H, Fan H, Cao R, Blanchard C, Wang M (2016) Physicochemical properties and in vitro digestibility of sorghum starch altered by high hydrostatic pressure. Int J Biol Macromol 92:753–760. https://doi.org/10.1016/j.ijbiomac.2016.07.088

Magallanes-Cruz PA, Flores-Silva PC, Bello-Perez LA (2017) Starch structure influences its digestibility: A review. J Food Sci 82(9):2016–2023. https://doi.org/10.1111/1750-3841.13809

Mahajan P, Bera MB, Panesar PS, Chauhan A (2021) Millet starch: a review. Int J Biol Macromol 180:61–79. https://doi.org/10.1016/j.ijbiomac.2021.03.063

Mahlow S, Orzechowski S, Fettke J (2016) Starch phosphorylation: insights and perspectives. Cell Mol Life Sci 73(14):2753–2764. https://doi.org/10.1007/s00018-016-2248-4

Malik MK, Kumar V, Singh J, Kumar P (2023) Efficiency of phosphorylated mandua starch in matrix tablet for targeted release of mesalamine in colon. J Drug Del Sci Technol 81:104251. https://doi.org/10.1016/j.jddst.2023.104251

Meko OA, Mudiaga-Ojemu BO, Ubah CP, Eraga SO, Iwuagwu MA (2018a) Investigation into the use of acid modified millet (Pennisetum glaucum) starch mucilage as tablet binder. J Sci Pract Pharm 5:267. https://api.semanticscholar.org/CorpusID:267686873

Meko OA, Mudiaga B, Ubah C, Eraga S, Iwuagwu M (2018b) Investigation into the use of acid modified millet (Pennisetum glaucum) starch mucilage as tablet binder. J Sci Pract Pharm 5. https://doi.org/10.47227/jsppharm.v5i2.5

Mun S, Kim Y-L, Kang C-G, Park K-H, Shim J-Y, Kim Y-R (2009) Development of reduced-fat mayonnaise using 4αGTase-modified rice starch and xanthan gum. Int J Biol Macromol 44(5):400–407. https://doi.org/10.1016/j.ijbiomac.2009.02.008

Nagarajaiah SB, Ramakrishna MG, Prakash J (2019) Nutritional, antioxidant and sensory properties of value added muffins. Curr Nutr Food Sci 15(5):454–466. https://doi.org/10.2174/1874609811666180223155034

Nigar S, Subrahmanyam K, Choudhury DB, Gul K, Sehrawat R (2024) Alginate-millet starch composite matrices as novel carriers for controlled delivery of grape seed polyphenols: preparation, characterization, and in vitro release behaviour. Food Struct 42:100385. https://doi.org/10.1016/j.foostr.2024.100385

Odeku OA, Alabi CO (2007) Evaluation of native and modified forms of Pennisetum glaucum (millet) starch as disintegrant in chloroquine tablet formulations. J Drug Del Sci Technol 17(2):155–158. https://doi.org/10.1016/S1773-2247(07)50024-9

Osman Z, Farah Y, Ali Hassan H, Elsayed S (2021) Comparative physicochemical evaluation of starch extracted from pearl millet seeds grown in Sudan as a pharmaceutical excipient against maize and potato starch, using paracetamol as a model drug. Ann Pharm Fr 79(1):28–35. https://doi.org/10.1016/j.pharma.2020.08.004

Padhi S, Gupta KK, Saraugi SS, Sehrawat R, Routray W (2024) Comparative analysis of non-thermal technologies and solvent systems for the extraction and characterization of phytochemicals and antioxidants in Pandanus amaryllifolius. Chem Eng Res Des 210:212–229. https://doi.org/10.1016/j.cherd.2024.08.036

Pederson S (1995) A better enzyme for cyclodextrins. ChemTech 25:19–25

Pejin DJ, Mojović LV, Pejin JD, Grujić OS, Markov SL, Nikolić SB, Marković MN (2012) Increase in bioethanol production yield from triticale by simultaneous saccharification and fermentation with application of ultrasound. J Chem Technol Biotechnol 87(2):170–176. https://doi.org/10.1002/jctb.2675

Pozo C, Rodríguez-Llamazares S, Bouza R, Barral L, Castaño J, Müller N, Restrepo I (2018) Study of the structural order of native starch granules using combined FTIR and XRD analysis. J Polym Res 25(12):266. https://doi.org/10.1007/s10965-018-1651-y

Pratiwi M, Faridah DN, Lioe HN (2018) Structural changes to starch after acid hydrolysis, debranching, autoclaving-cooling cycles, and heat moisture treatment (HMT): A review. Starch – Stärke 70(1–2):1700028. https://doi.org/10.1002/star.201700028

Preetham Kumar KV, Dharmaraj U, Sakhare SD, Inamdar AA (2016) Flour functionality and nutritional characteristics of different roller milled streams of foxtail millet (Setaria italica). LWT 73:274–279. https://doi.org/10.1016/j.lwt.2016.06.028

Punia S, Kumar M, Siroha AK, Kennedy JF, Dhull SB, Whiteside WS (2021) Pearl millet grain as an emerging source of starch: A review on its structure, physicochemical properties, functionalization, and industrial applications. Carbohydr Polym 260:117776. https://doi.org/10.1016/j.carbpol.2021.117776

Raghunathan R, Pandiselvam R, Kothakota A, Mousavi Khaneghah A (2021) The application of emerging non-thermal technologies for the modification of cereal starches. LWT 138:110795. https://doi.org/10.1016/j.lwt.2020.110795

Sahoo S, Padhi S, Sehrawat R, Routray W (2023) Microwave and atmospheric cold plasma aided debittering of giloy (Tinospora cordifolia Miers.) juice: effect on bioactive compound content. J Appl Res Med Aromatic Plants 37:100520. https://doi.org/10.1016/j.jarmap.2023.100520

Saini S, Saxena S, Samtiya M, Puniya M, Dhewa T (2021) Potential of underutilized millets as Nutri-cereal: an overview. J Food Sci Technol 58(12):4465–4477. https://doi.org/10.1007/s13197-021-04985-x

Sandhu KS, Siroha AK (2017) Relationships between physicochemical, thermal, rheological and in vitro digestibility properties of starches from pearl millet cultivars. LWT Food Sci Technol 83:213–224. https://doi.org/10.1016/j.lwt.2017.05.015

Sandhu KS, Siroha AK, Punia S, Nehra M (2020) Effect of heat moisture treatment on rheological and in vitro digestibility properties of pearl millet starches. Carbohydr Polym Technol Appl 1:100002. https://doi.org/10.1016/j.carpta.2020.100002

Sandhu KS, Kaur M, Punia S, Ahmed J (2021a) Rheological, thermal, and structural properties of high-pressure treated litchi (Litchi chinensis) kernel starch. Int J Biol Macromol 175:229–234. https://doi.org/10.1016/j.ijbiomac.2021.02.030

Sandhu KS, Siroha AK, Punia S, Sangwan L, Nehra M, Purewal SS (2021b) Effect of degree of cross linking on physicochemical, rheological and morphological properties of sorghum starch. Carbohydr Polym Technol Appl 2:100073. https://doi.org/10.1016/j.carpta.2021.100073

Seo D-H, Yoo S-H, Choi S-J, Kim Y-R, Park C-S (2020) Versatile biotechnological applications of amylosucrase, a novel glucosyltransferase. Food Sci Biotechnol 29(1):1–16. https://doi.org/10.1007/s10068-019-00686-6

Shaikh M, Ali TM, Hasnain A (2017) Effects of different modification reagents on functional properties of pearl millet starches. Starch – Stärke 69(3–4):1600142. https://doi.org/10.1002/star.201600142

Sharma M, Singh AK, Yadav DN (2017) Rheological properties of reduced fat ice cream mix containing octenyl succinylated pearl millet starch. J Food Sci Technol 54(6):1638–1645. https://doi.org/10.1007/s13197-017-2595-7

Sharma V, Tomar MS, Sahoo S, Pradhan RC (2024) Development of barnyard millet flour rich in bioactive compounds with enhanced functional properties by application of multipin atmospheric cold plasma. J Food Process Eng 47(6):e14667. https://doi.org/10.1111/jfpe.14667

Singh DK, Kurichh R, Bist Y, Kheto A, Kumar Y, Sharma R, Shikha D, Saxena DC, Ganjloo A (2024) Effect of microwave roasting on the physicochemical, functional, rheological, and antioxidant properties of Kodo millet flour. Journal of Food Processing and Preservation 2024(1). https://doi.org/10.1155/2024/6861190

Siroha AK, Bangar SP (2024) Millet-based food products: an overview. Curr Food Sci Technol Rep 2:213. https://doi.org/10.1007/s43555-024-00031-x

Siwatch M, Yadav RB, Yadav BS (2022) Annealing and heat-moisture treatment of amaranth starch: effect on structural, pasting, and rheological properties. J Food Meas Charact 16(3):2323–2334. https://doi.org/10.1007/s11694-022-01325-1

Sonkar S, Jaddu S, Pradhan RC, Dwivedi M, Seth D, Goksen G, Sarangi PK, Lorenzo JM (2023) Effect of atmospheric cold plasma (pin type) on hydration and structure properties of kodo-millet starch. LWT 182:114889. https://doi.org/10.1016/j.lwt.2023.114889

Su Q, Cai S, Duan Q, Huang W, Huang Y, Chen P, Xie F (2024) Combined effect of heat moisture and ultrasound treatment on the physicochemical, thermal and structural properties of new variety of purple rice starch. Int J Biol Macromol 261:129748. https://doi.org/10.1016/j.ijbiomac.2024.129748

Subrahmanyam K, Gul K, Sehrawat R, Allai FM (2023) Impact of in-package cold plasma treatment on the physicochemical properties and shelf life of button mushrooms (Agaricus bisporus). Food Biosci 52:102425. https://doi.org/10.1016/j.fbio.2023.102425

Subrahmanyam K, Gul K, Sehrawat R, Tiwari BK, Sahoo S (2024) Cold plasma-mediated inactivation of microorganisms for the shelf-life extension of animal-based foods: efficiency, mechanism of inactivation, and impact on quality attributes. Food Control 162:110464. https://doi.org/10.1016/j.foodcont.2024.110464

Sun Q, Gong M, Li Y, Xiong L (2014) Effect of dry heat treatment on the physicochemical properties and structure of proso millet flour and starch. Carbohydr Polym 110:128–134. https://doi.org/10.1016/j.carbpol.2014.03.090

Sun X, Saleh ASM, Lu Y, Sun Z, Zhang X, Ge X, Shen H, Yu X, Li W (2022) Effects of ultra-high pressure combined with cold plasma on structural, physicochemical, and digestive properties of proso millet starch. Int J Biol Macromol 212:146–154. https://doi.org/10.1016/j.ijbiomac.2022.05.128

Tester RF, Morrison WR, Schulman AH (1993) Swelling and gelatinization of cereal starches. V. Risø mutants of Bomi and Carlsberg II barley cultivars. J Cereal Sci 17(1):1–9. https://doi.org/10.1006/jcrs.1993.1001

Thangaraju S, Shankar M, Buvaneshwaran M, Natarajan V (2023) Effect of processing on the functional potential of bioactive components. In: Thakur M, Belwal T (eds) Bioactive components: A sustainable system for good health and Well-being. Springer Nature, pp 183–207. https://doi.org/10.1007/978-981-19-2366-1_12

Van der Maarel MJEC, van der Veen B, Uitdehaag JCM, Leemhuis H, Dijkhuizen L (2002) Properties and applications of starch-converting enzymes of the α-amylase family. J Biotechnol 94(2):137–155. https://doi.org/10.1016/S0168-1656(01)00407-2

Wang S, Li C, Copeland L, Niu Q, Wang S (2015) Starch Retrogradation: A comprehensive review. Compr Rev Food Sci Food Saf 14(5):568–585. https://doi.org/10.1111/1541-4337.12143

Wang R, Chen C, Guo S (2017) Effects of drying methods on starch crystallinity of gelatinized foxtail millet (α-millet) and its eating quality. J Food Eng 207:81–89. https://doi.org/10.1016/j.jfoodeng.2017.03.018

Wang W, Hostettler CE, Damberger FF, Kossmann J, Lloyd JR, Zeeman SC (2018) Modification of cassava root starch phosphorylation enhances starch functional properties. Front Plant Sci 9. https://doi.org/10.3389/fpls.2018.01562

Wang C, McClements DJ, Jiao A, Wang J, Jin Z, Qiu C (2022) Resistant starch and its nanoparticles: recent advances in their green synthesis and application as functional food ingredients and bioactive delivery systems. Trends Food Sci Technol 119:90–100. https://doi.org/10.1016/j.tifs.2021.11.025

Wei Y, Li G, Zhu F (2023) Impact of long-term ultrasound treatment on structural and physicochemical properties of starches differing in granule size. Carbohydr Polym 320:121195. https://doi.org/10.1016/j.carbpol.2023.121195

Wu Y, Lin Q, Cui T, Xiao H (2014) Structural and physical properties of starches isolated from six varieties of millet grown in China. Int J Food Prop 17(10):2344–2360. https://doi.org/10.1080/10942912.2013.803119

Yadav S, Mishra S, Pradhan RC (2021) Ultrasound-assisted hydration of finger millet (Eleusine Coracana) and its effects on starch isolates and antinutrients. Ultrason Sonochem 73:105542. https://doi.org/10.1016/j.ultsonch.2021.105542

Yang S-Y, Cao L, Kim H, Beak S-E, Song KB (2018) Utilization of foxtail millet starch film incorporated with clove leaf oil for the packaging of Queso Blanco cheese as a model food. Starch - Stärke 70(3–4):1700171. https://doi.org/10.1002/star.201700171

Yashini M, Khushbu S, Madhurima N, Sunil CK, Mahendran R, Venkatachalapathy N (2024) Thermal properties of different types of starch: A review. Crit Rev Food Sci Nutr 64(13):4373–4396. https://doi.org/10.1080/10408398.2022.2141680

Zeng F, Li T, Gao Q, Liu B, Yu S (2018) Physicochemical properties and in vitro digestibility of high hydrostatic pressure treated waxy rice starch. Int J Biol Macromol 120:1030–1038. https://doi.org/10.1016/j.ijbiomac.2018.08.121

Zheng M, Xiao Y, Yang S, Liu H, Liu M, Yaqoob S, Xu X, Liu J (2020) Effects of heat–moisture, autoclaving, and microwave treatments on physicochemical properties of proso millet starch. Food Sci Nutr 8(2):735–743. https://doi.org/10.1002/fsn3.1295

Zhu F (2014) Structure, physicochemical properties, and uses of millet starch. Food Res Int 64:200–211. https://doi.org/10.1016/j.foodres.2014.06.026

13 Value Addition in Millets: Exploring Novel Technological Approaches

V. R. Bharathvaj, Praveena B, R. B. Ramyaa, and P. Nandavarman

13.1 Introduction

Millets are a diverse group of small-grained crops, considered to be the first crop cultivated by humans. It is known for their resilience and can thrive in some of the most challenging climates on Earth, earning them the reputation of "famine crops" in the semiarid tropics and dry regions of Africa and Asia. The key millet varieties cultivated in these regions include sorghum, pearl millet, finger millet, foxtail millet, barnyard millet, kodo millet, proso millet, and little millet.

Although millets have a rich nutritional profile, certain inherent characteristics such as a hard seed coat, the presence of antinutritional factors, poor digestibility, and low bioavailability of micronutrients pose significant challenges in their processing and cooking. The lack of advanced primary, secondary, and tertiary processing technologies has limited the availability of value-added, ready-to-eat millet products in the market, hindering dietary diversification. However, increasing awareness of the health benefits of millets has created opportunities for their marketing and branding as well-processed, value-added products (Deshpande and Nishad 2021).

Value addition is a key process to improve the acceptability of consumers with the high nutritional content in a less quantity. The value-added products in millets are in a large number, and the role of the novel technologies in value addition will be in the preprocessing step of the production. The novel technologies discussed in this chapter are microwave, 3D food printing, ohmic heating, infrared radiation, pulsed electric field (PEF), irradiation, hydrodynamic cavitation, and cold plasma.

V. R. Bharathvaj · B. Praveena · R. B. Ramyaa
National Institute of Food Technology, Entrepreneurship and Management, Thanjavur, Tamil Nadu, India

P. Nandavarman (✉)
Agricultural Engineering College and Research Institute, Tamil Nadu Agricultural University, Coimbatore, Tamil Nadu, India

S. Jaddu, R. Chandra Pradhan (eds.), *Innovative Millet Processing*,
https://doi.org/10.1007/978-3-031-99864-5_13

13.2 Microwave

Microwave technology is an electro-heat method that transforms electrical energy into thermal energy within a frequency range of 300 MHz to 300 GHz, with common food applications using frequencies of 915 MHz and 2.45 GHz. This electro-heat technique makes microwaves a more energy-efficient and quicker process compared to conventional methods like open pan boiling, which rely on indirect heat transfer. Microwave cooking is recognized as a fast household cooking method compared to traditional pan cooking. It enables the achievement of high temperatures in a short time through rapid vibration of polar molecules, providing uniform heating, high heating rates, and safe handling (Kumar et al. 2020). Microwaves themselves do not produce heat; instead, the heat arises primarily from intermolecular frictional forces. When food is exposed to microwave radiation, polar molecules, such as water, transition from a disordered state to an ordered structure aligned in a specific direction of applied electric filed. The rapid movement of these excited molecules generates friction, which in turn produces heat as shown in Fig. 13.1. (Hu et al. 2021).

Traditional methods were used to process millets to enhance its digestibility, nutritional value, and taste. Recently, microwave processing has emerged as a method to further improve the accessibility of micro- and macronutrients by increasing dehulling efficiency and reducing antinutritional factors. Due to its quicker drying rate, consistent and rapid heating, shorter processing time, precise control, and lower operational costs and time, microwave has gained attention among the researchers in the field of value addition. Physical properties like tapped density, bulk density, Carr's index, and Hausner ratio didn't show any difference and maintained the alterations during microwave treatment (Rao et al. 2021). Physiochemical properties such as color, water absorption index (WAI), and water solubility index (WSI) have higher values for microwave cooking than pan cooking method, and bulk density was lower for microwave cooking method (Kumar et al. 2020).

The millet flour pretreated with microwave had increased solubility and water absorption capacity while maintaining oil absorption capacity, swelling power, and color of flour within the acceptable limit. Microwave heating as a pretreatment effectively enhanced the functional properties of foxtail millet flours without

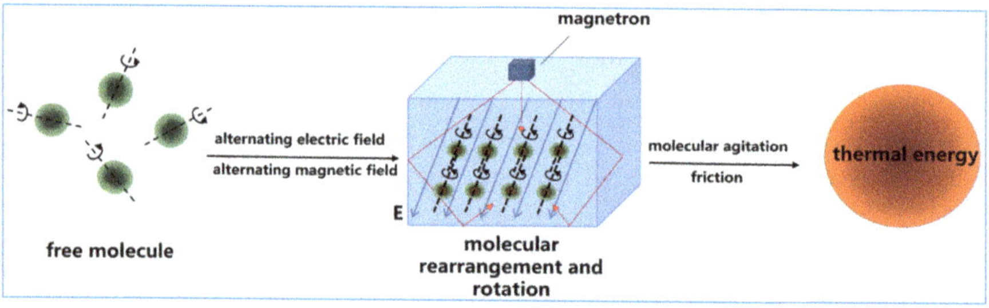

Fig. 13.1 Changes in the orientation of molecules during microwave treatment. (Hu et al. 2021)

affecting its physical parameters (Rao et al. 2021). The functional properties of proso and little millet such as water absorption capacity, oil absorption capacity, and water solubility index were higher in microwave-cooked millet compared to pan-cooked millet. The foaming capacity and foaming stability were lower for both cooking methods. Microwave cooking formed firm gel at higher concentration (Kumar et al. 2020). The microwave-treated foxtail millet flour incorporated with the flax seed improved the functional properties such as bulk density, water absorption index, water solubility index, water holding capacity, and oil absorption index and had good organoleptic score (Chowdhury and Maji 2024).

When proteins are denatured by heat, their molecules unfold, revealing polar side chains. This process causes the flour to absorb more water and reduces its ability to absorb oil. The reduced oil absorption capacity can be attributed to a lower concentration of hydrophobic amino acids in the sample (Rao et al. 2021). The stability and creation of foams rely on the interfacial film formed by proteins, which helps maintain air bubbles in suspension and slows down the rate of coalescence. Additionally, carbohydrates contribute by increasing the viscosity of the suspending medium (Kumar et al. 2020). The high fat content and lipase activity in pearl millet (*Pennisetum typhoides*) flour result in a limited shelf life when stored at room temperature (Yadav et al. 2012). Microwave treatment of pearl millet grains with 18% moisture for 80 s significantly reduced lipase activity in the flour. The pasting properties also notably declined with longer microwave exposure. Storage stability of flour treated with microwaves remained acceptable for up to 30 days when stored in LDPE pouches at ambient temperatures (15–35 °C) (Yadav et al. 2012). Microwave heating has been used to inactivate lipase in grain flours due to its thermal effect and, in a way, increase their shelf life (Xu et al. 2016).

The microwave treatment of proso millet starch increased the amylose and resistant starch content and reduced the structural recrystallization and bulk density. It resulted in a thinner stromal wall, with a tidier and more uniform network structure (Zheng et al. 2020). The microwave-modified pearl millet starch decreased amylose content, storage modulus, and swelling capacity. It increased the transition temperature and solubility of the film (Punia Bangar et al. 2021). Microwave treatment leads to a significant increase in the content of rapidly digestible starch and slowly digestible starch while decreasing the resistant starch content. Moreover, the treatment reduces the relative crystallinity and short-range order of the starch (Zhi et al. 2022). Microwave treatment (MT) can break down starch crystallinity before the particles expand, leading to more molecular material being leached at higher temperatures. Gelatinization under pressure can cause the amylopectin molecular structure to fracture or result in the hydrolysis of long-chain amylose molecules into shorter chains (Zheng et al. 2020).

Microwave treatment decreased the polyphenolic content of barnyard millet and increased in foxtail and proso millet, whereas the phenolic acid decreased in foxtail millet (Wang et al. 2022). Microwave treatment increased the germination rate, which subsequently boosted the total phenolic content, total flavonoid content, and antioxidant activity in the germinated millets (Sunil et al. 2024). A greater bulk density is influenced by the presence of lipids, which function as adhesives, promoting the agglomeration of carbohydrate and protein molecules (Kumar et al. 2020).

The ready-to-eat breakfast cereals is gaining popularity due to the change in lifestyle, urbanization, convenience, and health consciousness. The conventional process affects the digestibility and resistant starch formation. In the puffing process, starchy foods are expanded by either applying sudden heat at atmospheric pressure or by rapidly dropping the pressure in a high-pressure chamber at a high temperature (Pawar et al. 2015). Puffed ready-to-eat oil-free product was devolved from barnyard and finger millet using microwave treatment at a low processing time, low temperature, and increased the storage stability for 6 months (Dhumal et al. 2014). The microwave puffed product from steamed, cold-extruded composite millet flours involves first using convective heating at 196.0 °C for 132 s, followed by microwaving at 520 W for 173 s. The optimally microwave puffed product achieves a moisture content of 0.0304 kg/kg dry matter, hardness of 1315.4 g, crispness with 15.125 positive peaks, an expansion ratio of 1.96, and a color measurement of 43.9 (Pawar et al. 2015).

Microwave cooking gains popularity for its quickness, convenience, efficiency, and eco-friendliness; there is a growing need for research on its impact on the nutrient composition of indigenous staple foods, like spiced millet porridge. This research would provide valuable consumer education, as similar information is already available for many other cereal and seed-based meals. The spiced millet porridge treated with microwave increased the fat content and reduced the protein content due to higher temperature of microwave at 600 W. pH of the porridge was maintained at 4.5 that aids antimicrobial activity and helps in preserving the porridge. The relatively higher temperatures involved in microwave cooking may have caused more protein denaturation, resulting in less measurable protein as microwave power increased. While denaturation alters the protein's functional structure, the amino acids, which are more resistant to temperature due to their strong peptide bonds, remain unaffected and still provide nutritional benefits. The increased microwave power leads to higher temperatures, causing thermal decomposition and breaking the bonds holding fatty acids together (Mensah et al. 2022).

Milling is an important physical process of removing the outer layer of millets to remove antinutritional factors and to improve cooking. The microwave pretreatment for foxtail millet increases the dehulling efficiency, reduces the dehulling time, reduces cooking time, and increases percentage of brokens, and a reduction in head yield was observed with the increase in microwave power and exposure time (Sunil and Neeraja Chacko 2018). The greater reduction in moisture content during microwave treatment, which shortens de-husking time, leads to an increase in the percentage of broken grains and a decrease in head yield. The cooking time also decreases because the thermal effects of microwaves expand intermolecular spaces by breaking down bonds, allowing more water to penetrate and speeding up the cooking process (Sunil and Neeraja Chacko 2018).

13.3 3D Food Printing

3D food printing is the process which involves deposition of edible material (food ink) in layer by layer on a printing platform. 3D food printing is based on additive manufacturing technique; when food is involved, it is called as food layered manufacturing (FLM) (Nachal et al. 2019). Based on the method of printing, 3D food printing can be classified into four types:

1. Extrusion printing
2. Binder jetting
3. Selective laser sintering (SLS)
4. Inkjet printing

13.3.1 Extrusion Printing

The edible food material is held in a cartridge and forced through a nozzle, where it can be heated and deposited layer by layer on the printer's platform. Rheological and mechanical properties related to the food are to be considered for successful printing, as these affect the binding of layer, flowability, and stability of food structure after printing. Since some foods lack in maintaining the shape and structure post printing, food additives like guar gum, pectin, xanthan gum, starch, etc. are used to sustain the properties. In extrusion-based 3D printing, three types of mechanisms are present: screw-based extrusion, syringe-based extrusion, and air pressure-based extrusion as shown in Fig. 13.2. In screw-based extrusion, the food material is forced to extrude by the rotation of the screw. In syringe-based extrusion, a plunger inside the tube is used to force the food material through the nozzle, and in air pressure-based extrusion, the pressure of air is used for extrusion.

Based on the property of the material extruded, it is classified into soft material, melting, hydrogel-forming extrusion types. In soft material, extrusion involves mixing and deposition of self-supporting layers of materials such as dough, meat paste,

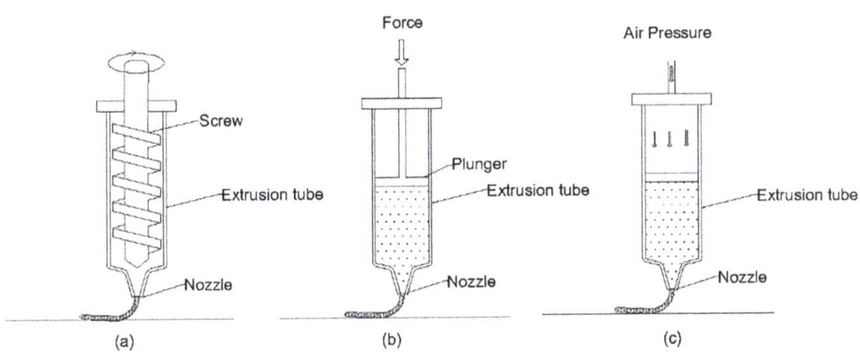

Fig. 13.2 (**a**) Screw-based extrusion; (**b**) syringe-based extrusion; (**c**) air pressure-based extrusion

and processed cheese. The food ink should have viscosity such that it makes it easier to flow through the fine nozzle (low viscosity) and to become solid once it got extruded (high viscosity). Melting extrusion is used to print food materials at a temperature ranging from 28 °C to 40 °C. It is being used in the printing of chocolate 3D objects. The formulations of chocolate self-supporting structures are challenging due to complex crystallization behavior exhibited by cocoa butter (Godoi et al. 2016). In hydrogel-forming type of extrusion, the hydrogel-forming materials are extruded in the printer which is highly dependent on the rheological properties and gel-forming mechanism. The gel should be formed post printing prior to the deposition of the next layer over it, and gelation of materials inside the printer is undesired and a temporal control over the gelation mechanism must be taken. Gel-forming mechanisms are classified into three types, namely:

1. Chemical cross-linking
2. Ionotropic cross-linking
3. Complex coacervate cross-linking

Chemical cross-linking is not used in food industries due to its harmful reagents used. Ionotropic cross-linking is used in food industries, especially in microencapsulation (Godoi et al. 2016). The use of hydrocolloids as an alternative with other food ingredients is suggested to create printable foods composed of starch, protein, etc. (Vesco and Lipson 2009).

For the printing of materials, various factors are important, one of the important factors that influences the quality of 3D printed food is the pressure in the nozzle while printing. For getting precise results, the pressure must be kept constant and stable while printing. For printing soft foods like fish, beet puree, egg white foam, etc., the pressure range must be between 20 and 50 kPa (Siacor et al. 2021) and stiffer pastes (e.g., vegemite, marmite, etc.) in the range of 100 to 170 kPa (Kouzani et al. 2017). The composition of food ink and temperature of printing affect the rheological properties of food ink and which in turn affects the texture of food (García-Segovia et al. 2020). High-resolution 3D products are developed with small nozzle diameter which also provides good precision, although it increases the printing time, decreasing the productivity.

13.3.2 Powder Binding Deposition

Powder binding deposition is divided into three subtypes:

1. Selective laser sintering (SLS)
2. Selective hot air sintering and melting (SHASAM)
3. Liquid binding (LB)

13.3.2.1 Selective Laser Sintering (SLS)

In selective laser sintering, the powder particles are fused together by the laser. The laser is focused at specific areas according to the predesigned 3D model. After fusion of powder particles in one layer, the powder bed is lowered by one layer thickness, and a new layer is spread over the fused layer, and laser is subjected; this process is repeated until the final product is made. In this method, each layer can be made of different food material components (Dias et al. 2016).

13.3.2.2 Selective Hot Air Sintering and Melting (SHASAM)

In selective hot air sintering and melting, a narrow, directed, low velocity beam of hot air is used to sinter the powder food particles by melting the powder.

13.3.2.3 Liquid Binding (LB)

In liquid binding (LB), a liquid binder is applied on the powder bed by a drop-on-demand print head. The liquid binder joins the nearby particles forming a layer. The liquid binder is ejected at specific places according to the 3D model. After a layer is formed, the powder is spread over it, and again, the liquid binder is ejected; this process is continued until the 3D model is achieved (Godoi et al. 2016).

13.3.3 Inkjet Printing

This method is mostly used for food decoration purposes and surface filling. The small food ink droplets are deposited onto the surface of the foodstuff, such as a cake, cookie, or candy forming an image (Varvara et al. 2021). Inkjet printers use thermal or piezoelectric print heads. In thermal inkjet printer, print head is electrically heated to produce pulses of pressure that pushes the droplets from the nozzle. Piezoelectric inkjet printers create pressure waves to push the droplets (Murphy and Atala 2014). This process can be done in two ways either continuous or drop on demand (DoD). This DoD uses pneumatic nozzle jets which forms tiny drops onto a moving object forming an image, surface fill, or cavity deposition (Food Jet 2021). Inkjet printers use low viscosity materials; the typical materials used are chocolate, liquid dough, sugar icing, meat paste, cheese, jams, gels, etc.,

Millets along with other ingredients can be mixed and 3D printed; the millets with other ingredients must be comminuted to small size as larger size materials clog nozzle during the printing process. The addition of water to the comminuted flour matters a lot in deciding flowability and extrusion. A fiber-rich snack has been developed by a mixture of barnyard millet 42%, green gram 42%, fried gram 14%, and ajwain seeds 2%. All the components are initially milled to a size of 0.2 mm, and then, water is added to form batter; this batter exhibited shear-thinning behavior, i.e., the apparent viscosity decreases with the increase in shear rate which is a very important factor for extrusion-based 3D printing (Krishnaraj et al. 2019). Pearl millet flour fortified idli batter has shown good structural stability and creep recovery (Raja et al. 2023). Post-processing is crucial for stabilizing 3D printed foods made with pearl millet. A comparison of baking and steaming, considering sensory attributes, structural integrity and physicochemical features, it has been reported that baking to be the most effective post-processing method for the pearl millet–banana pulp blend (Santhoshkumar et al. 2024). A millet-based food enriched with algal biomass *Azolla filiculoides* microalgae has been printed to create a gluten-free fortified millet product. The incorporation of algal biomass at 15% gave optimal results, and the firmness of product increased with increase in algal biomass incorporation, exhibiting high protein and carbohydrate content (Hassan et al. 2025). With the above studies and its results, it has been known that millets have a good potential to get printed along with other ingredients in a blend.

13.4 Ohmic Heating

Ohmic heating is also called as Joule heating or direct resistance heating. In this process, the electric current is passed through the food matrix, and volumetric heating occurs by the resistance offered by each component present in the food. Electrical conductivity plays a major role in this process. The passing of electric current also produces a nonthermal effect that kills microorganisms present in food. Ohmic heating can be done only for pumpable foods containing liquid medium in it. Ohmic heating has been reported as one of the powerful and economically effective techniques to inactivate lipase oxidation in peal millets. Dias-Martins et al. (2019) investigated the ohmic heating process at 98 °C with 0.1 M salt solution of whole pearl millet and decorticated pearl millet. The study found that due to small size of pearl millet, the electrical conductivity was low. As compared to the whole millet, the decorticated millet had slightly high electrical conductivity, which may be due to removal of pericarp which exposed the nutrients and leaching of intracellular materials. The cooking yield obtained by ohmic heating for whole millets was less compared to conventional cooking and which was because of the presence of pericarp, acting as a physical barrier for electrical conductivity, whereas in decorticated millets, the difference was very less between ohmic heating and conventional cooking method. The OH-decorticated millets showed higher lightness L* value than conventional cooking.

13.5 IR Radiation

Infrared radiation transfers the thermal energy to the food components in the form of electromagnetic waves. IR can be classified into three classes, namely, near-infrared (750 nm to 1400 nm), mid-infrared (1400 to 3000 nm), and far-infrared (3000 to 10^6nm) (Sakai and Hanzawa 1994). Whenever the electromagnetic radiation impinges on a surface, changes in rotational, vibrational, and electronic states of atoms and molecules occur. Changes at the atomic scale occur depending on the energy absorbed by the material. Radiation in the wavelength range of 200 nm to 700 nm (UV-visible range) induces electronic transitions, while vibrational and rotational transitions typically correspond to wavelengths of 2500 nm to 1 mm and greater than 1 mm, respectively. The millet flour has good nutrition value despite being underutilized which is due to the rancidity that occurs during storage; different types of IR treatments decreased the rancidity. Pearl millet flour subjected to hydrothermal and near-infrared (NIR) processing exhibits reduced activities of lipase, lipoxygenase, peroxidase, and polyphenol oxidase, without affecting its starch and protein digestibility properties (Vinutha et al. 2022).

It was reported that the IR processing of sorghum millet at 120 °C for 8.5 min was found best to obtain low free fatty acid and lipase activity, increasing the storage stability of the millet (Swaminathan et al. 2015). The IR treatment does not affect the color of the flour (Kumari and Jayadeep 2024).

13.6 Pulsed Electric Field (PEF)

PEF is the application of short duration pulses of electric field in the food between the two electrodes. PEF can be performed only for liquid and semisolid food. The electric field intensity used ranges from 100–300 V/cm to 20–80 kV/cm. This application inactivates the microorganisms in the food by rupturing the cell membrane without damaging the color, flavor, taste, and nutrients of food. PEF is also used in the extraction of active constituents, modification of biomacromolecules, enhancement of chemical reactions, and even in accelerating the aging process in fermented foods (Niu et al. 2020). Electroporation happens during PEF treatment, which states that when the membrane of a cell is exposed to high-intensity electric field pulses, destabilization at specific regions of the cell occurs; during this process, the cell membrane is highly permeable to exogenous molecules present in the surrounding media (Chang et al. 2012). The independent process parameters are voltage applied, inter-electrode gap, number and frequency of pulses, food resistivity, and PEF application time in food. Pulsed electric field set at optimized parameter values aids in releasing the bound phenolic and antioxidants in millets especially sorghum (Lohani and Muthukumarappan 2016). The hydrophobicity and solubility of foxtail millet prolamin, which is a dominant class of seed storage protein in many cereals, which contain proline and amide nitrogen, increased and also promoted an increase in soluble aggregates, resulting in an increase in protein particle size by the application of pulsed electric field (Zhang et al. 2023).

13.7 Irradiation

Millets are often processed at home, leading to limited availability of vital nutrients for the general population. A variety of millet is consumed in India. Products include porridge, chapati, laddu, machula, dubkee, churkani, and regional specialties. One of the most prevalent reasons for millet grain's limited use is its antinutritional content. However, insufficient processing knowledge limits crop consumption, especially in unconventional areas. Novel technological approaches like irradiation have been explored for value addition in millets, aiming to enhance their nutritional profile and functional characteristics. Food irradiation refers to the process of exposing food to ionizing energy, typically by the use of gamma photons generated by 60Co (or by 137Cs) radioisotopes, machine-generated X-rays ("Bremsstrahlung") of max. 5 MeV, or accelerated electrons with max. 10 MeV kinetic energy. Electromagnetic radiation from the first two sources has high penetration, while accelerated electrons have minimal penetration (Farkas and Mohácsi-Farkas 2011). Irradiation in food preservation offers numerous benefits, including enhanced food safety by limiting the growth of microorganisms and pathogens (BN 2023). Irradiation principles involve exposing food or tissues to ionizing radiation for various purposes. In the context of food preservation, irradiation is a nonthermal process that creates ions and free radicals, leading to chemical events that enhance food safety (Sruthi et al. 2023). This method helps disinfect food from insects, inhibiting sprouting and extending the shelf life of various food products (Sarkar et al. 2023).

Sources of irradiation include a wide range of radiation-generating systems as encompassing natural and artificial sources such as radioisotopes, solar radiation, accelerators, X-ray generators, lasers, and nuclear reactors (Tanabe 2022). Advanced radiation source systems can generate multiple types of radiation using electron guns, accelerators, and target systems that selectively produce various radiation types based on parameters like position and rotation angle. Gamma irradiation, a common form of irradiation, is a nonthermal technology that effectively reduces microbial growth without compromising the nutritional quality of the food. This process is economically feasible and safe and has a strong antimicrobial effect, making it a reliable method for treating large volumes of food products. Additionally, irradiation can enhance the phenolic content and antioxidant activity of food, contributing to its overall quality and safety (Shams et al. 2023). Overall, irradiation technology plays a crucial role in ensuring food security and safety and extending the shelf life of various food commodities, benefiting both producers and consumers. Studies have demonstrated that gamma irradiation of millet grains at doses above 0.5 kGy significantly reduces fungal incidence, free fatty acids, tannin, and phytic acid content while increasing protein solubility and in vitro protein digestibility (Wani et al. 2022). Numerous research have shown that irradiation can enhance the functionality and quality of millet flour. The study (Mahmoud et al. 2016) found that a dose-dependent irradiated pearl millet loses antinutrients like tannin and phytic acid due to free radical destruction. Irradiation alters protein structure, leading to denaturation and improved digestibility in vitro (Sruthi and Rao 2021). Lipoxygenase activity is a serious concern for the food sector. Hydroperoxides produced by the LOX process serve as a substrate for downstream enzymes, leading to the creation of volatile chemicals such as hexanal, which are associated with unpleasant flavors in millets. The study by Mahmoud et al. (2016) found that doses above 0.5 kGy significantly reduced both fungal incidence and FFA. The drop in FFA can be attributed to reduced lipase in the grain, resulting in improved shelf life.

Gamma irradiation breaks down starch molecules into smaller dextrin fragments, which can be electrically charged. Alternatively, free radicals are uncharged. Macronutrients (carbohydrates, proteins, and lipids) are relatively stable to

Table 13.1 Irradiation process in millets and its properties changes

Types of millet	Improved properties	Reference
Finger millet	Physicochemical, functional, proximate, and microstructural characteristics, structural modifications	Gowthamraj et al. (2021)
White finger millet	Quality improvement during storage	Shobha et al. (2023)
Proso millet starch content	Structure and physicochemical properties	Liu et al. (2023)
Millet flour	Enhance the functionality	Mahmoud et al. (2016)

irradiation doses up to 10 kGy. A decrease in tannin content might be the result of chemical degradation by the action of free radicals formed by the radiation. However, gamma irradiation affects proteins, causing conformational changes, oxidation of amino acids, rupturing of covalent bonds, and formation of protein-free radicals (Issa et al. 2011). Microbes are effectively killed by irradiation through various mechanisms. The radiation resistance of microbes, such as *Bacillus pumilus* spores, is overcome by gamma rays, X-rays, and electron beams, leading to microbial inactivation (McEvoy et al. 2023). Irradiation reduces the total microbial load on food products, with a linear decrease observed as the absorbed dose increases, ultimately causing the "death" of a significant portion of microbes (Klemo et al. 2015). The sterilization of tissues using ionizing radiation ensures a high sterility assurance level by directly and indirectly killing all microbes present, influenced by factors like radiation dose rate and the type of microbe (Yusof and Hilmy 2014). Ionizing radiation in the food industry is particularly effective in controlling microbiological contaminants, with different types of pathogens showing varying sensitivity to irradiation, ultimately preserving food safety (Belbe and Tofana 2010). The effect of irradiation on the physicochemical and functional properties of various millet varieties is shown in Table 13.1.

13.8 Hydrodynamic Cavitation

In current times, food industries demand the development of new and emerging technologies in food processing according to the consumers' needs in terms of safer, healthier, and minimally processed foods (Gavahian et al. 2021; Ramírez-Jiménez and Castro-Muñoz 2021). Cavitation is a phenomenon caused by variable velocity and a sudden change in pressure inside a liquid flow system. It involves the formation and rapid collapse of vapor bubbles due to sudden pressure changes within milliseconds, releasing intense localized energy (1000–10,000 K and 100–5000 bar) that facilitates chemical reactions capable of breaking down environmentally and biologically hazardous substances. Several chemical reactions can make use of cavitation to break down materials that are hazardous to the environment in addition to people. The huge release of energy is due to the cavitation effect in the form of temperature and high pressure (between 100 and 5000 bars) 1000–10,000 K (Suslick

1990). Hydrodynamic cavitation is created when a liquid is forced through a narrow opening or constriction, increasing its kinetic energy. Nucleation, bubble expansion, and bubble collapse are all involved in the phenomenon. The vena contracta is where the liquid's flow area is at its lowest. There should be enough throttling to cause the pressure decrease to dip below the pressure threshold at the vena contracta causing millions of cavities to be released. Afterward, as the pressure decreases, the cavities disintegrate as the liquid swells. A powerful shock wave of energy is discharged, and it is released into the environment. This shock wave produces scale-free heating, a potent microscopic mixing effect, and controllable rotor/liquid friction (Gogate and Pandit 2001, 2005; Gogate 2011). The process is similar to that of acoustic cavitation. There is just a variation in terms of pressure and temperature. The collapse of individual cavities is considerably less intense than compared with the ultrasound-induced cavitation; however, it can be enhanced by adjusting operating parameters, equipment configuration and energy levels. Therefore, it is possible to regulate the quantity of cavitation events successfully. Moreover, HC can be produced mechanically spinning a liquid at a particular speed. One way to do that is to use a rotor that has holes in it referred to as a hydrodynamic cavitator. The rotating element of the rotor action generates HC inside the holes away from the metal surface. Hydrodynamic cavitation (HC) is a unique nonthermal technology that has recently come under investigation. It can be used to extract, homogenize, emulsify, and sterilize liquid foods, including milk, juices, and water while minimizing the loss of bioactive ingredients. This innovative method may mitigate the detrimental effects of traditional heat treatments on the sensory, physio-chemical, and nutritional qualities of liquid meals. This is because the HC's cavities form, swell, and collapse in milliseconds, releasing a lot of energy that physically stresses the microbial cell and sterilizes it without raising the temperature significantly and destroying the nutrients in liquid foods. It is important to evaluate new food processing methods to see if they can reduce energy usage, achieve microbiological lethality at lower temperatures, and preserve the freshness of the final product (Arya et al. 2020). Millets, being rich in starch, can benefit from hydrodynamic cavitation for starch modification, improving their binding, thickening, and stabilizing capabilities. This process involves the use of nonthermal physical methods like ultrasonication and high-pressure processing to extract and modify starch from millets, ultimately enhancing their techno-functional properties for various applications (Heena et al. 2024). The application of hydrodynamic cavitation can play a crucial role in developing innovative millet-based functional foods with improved nutritional, medicinal, and therapeutic properties, catering to the expanding consumer interest in plant-based products (Heena et al. 2024). By leveraging hydrodynamic cavitation, millet processors can unlock the full potential of these nutritious grains and meet the growing demand for value-added millet products in the market.

Hydrodynamic cavitation plays a crucial role in enhancing millet nutritional value by reducing the antinutritional factors in it. Studies have shown that high-intensity ultrasound treatment, a form of hydrodynamic cavitation, can significantly improve the nutritional quality of millet by reducing antinutrient levels such as phytates and tannins, thus increasing the bioavailability of nutrients (Dubey and

Tripathy 2024). Additionally, this process can increase the amount of freely available bioactive compounds from millet bran, leading to higher antioxidant activity and total phenolic content, which are essential for health benefits (Voučko et al. 2019). Furthermore, the combination of sonication with germination has been found to enhance the functional properties of millet milk, increasing phenolic activity and antioxidant activity, and reducing antinutrient concentrations, ultimately improving the overall nutritional profile of millet-based products (Saxena et al. 2023). Therefore, hydrodynamic cavitation techniques like high-intensity ultrasound can be a valuable tool in maximizing the nutritional value of millet and its by-products for the food industry. Research demonstrated that high-intensity ultrasound treatment on millet bran increased the antioxidant activity by 15% and the total phenolic content by 16%, ultimately enhancing the nutritive value of millet. Additionally, the treatment led to a 38% increase in water-soluble and ethanol-insoluble dietary fiber, crucial for functional cereal product development (Voučko et al. 2019). The study also highlighted improvements in water retention and limited browning of the sample, showcasing the potential of hydrodynamic cavitation to optimize millet processing for bakery and pasta industries. By utilizing hydrodynamic cavitation, millet's bioactive compounds can be more readily available, enhancing its overall value and applicability in the food industry.

13.9 Cold Plasma

Cold plasma is a novel nonthermal technology that finds its use in foods in increasing the shelf life by reducing the microbial growth in the foods. Plasma is an ionized gas containing reactive oxygen species, reactive nitrogen species, ultraviolet radiation, charged particles, and free radiation (Laroque et al. 2022). Cold plasma treatment reduced the antinutritional factors tannin and phytic acid in pearl and barnyard millets; also, a decrease in the phenolic content in the millets was noticed which was due to the direct interaction of the energetic ions with phenolic compounds leading to their breakdown. The iron content in the millets increased with cold plasma treatment, which can be taken as the ability of old plasma to increase the bioavailability of iron from millets (Charu 2024). Treatment of kodo millet starch with atmospheric cold plasma improved the hydration capacity of the starch, which improved the application of starch as a better thickening and bulking agent than the native millet starch. The modification and the effect of cold plasma are over the surface; the changes that occur in the surface of the foods alter the microstructure; pore formations are mostly responsible for the major improvement in functional properties (Sonkar et al. 2023). Little millet flour treated with atmospheric multipin cold plasma improved the functional properties of the flour by the breakdown of starch granules, and relative crystallinity decreased with the depolymerization of starch caused by reactive oxygen species and reactive nitrogen species. The treated flour due to its improved functional properties could become an important ingredient in bakery, porridges, and frozen foods (Jaddu et al. 2022). The cold plasma treatment improved the formation of protein gels, which is due to the formation of disulfide bonds. Water holding capacity of the cold plasma-treated millet flour increased, which is due to the depolymerization of starch in the millets; gel formation is an important functional property related to the development of value-added products (Jaddu et al. 2024).

13.10 Synergistic Approach of Different Treatments

Dey et al. (2024) reported that the synergistic effect of germination and ultrasonication treatment on kodo millets resulted in a number of positive benefits as it reduced the antinutritional factors—phytates and tannins—which is due to the activation of phytase enzyme during the germination treatment, and ultrasonication ruptured the cells and exposed these phytate to phytase enzyme. The antioxidant content increased after the treatment which is due to the generation of enzymes after the germination, which were responsible for release of phenol content. According to Raghav et al. (2025), the combined use of ultrasound and high-pressure techniques significantly enhances protein extraction efficiency from pearl millet, along with improving its functional properties. This synergistic effect results in a lower least gelation concentration and better gelation behaviour of the extracted protein. These effects are associated with the disruption of the globular structure, increased exposure of hydrophobic and sulfhydryl groups, and conformational changes indicated by a decrease in α-helix content and an increase in β-sheet structures. Different treatments in combined form are needed to be performed and analyzed and optimization be done in the future to get a clear detail about the changes in functional and other properties.

13.11 Conclusion

Malnutrition is still being one of the major problems in some parts of the world, where the availability of grains and food is low, causing an imbalance in the growth and other functions of the body. Value addition of products and extension of shelf life using novel technologies play a major role in mitigating the unavailability of foods and give nutritional security to the people. The improvement in the functional properties like water holding capacity, gel strength, solubility, and swelling power by the action of different treatments like cold plasma, hydrodynamic cavitations, and microwave increases the bioavailability of nutrients and decreases the antinutritional factors present in the raw form of the millets. Therefore, there is a dual benefit of improving the sensorial and textural properties as well as increasing the bioavailability of the nutrients. Different novel technologies and their effect on millets and the interaction of proteins, starch, and lipids when subjected to different treatments and how the functional properties are changed are some of the important areas to be focused. The optimization of novel treatments to the millet in making different value-added products is crucial for the development of nutrient-rich and highly accepted products. The demand for millet is steadily increasing, and the commercial production of various value-added millet products is being bolstered by the availability of suitable processing machinery and equipment. Advances in millet processing technologies are creating new opportunities, helping millets to compete more effectively as a staple food.

References

Arya SS, Sawant O, Sonawane SK et al (2020) Novel, nonthermal, energy efficient, industrially scalable hydrodynamic cavitation—applications in food processing. Food Rev Intl 36:668–691. https://doi.org/10.1080/87559129.2019.1669163

Belbe GM, Tofana M (2010) Effects of ionizing radiation on microbiological contaminants of foods. Bulletin of University of Agricultural Sciences and Veterinary Medicine Cluj-Napoca Agriculture 67:. https://doi.org/10.15835/buasvmcn-agr:5052

BN E (2023) Food irradiation—an effective technology for food safety and security. Food Sci Nutr Technol 8:1–6. https://doi.org/10.23880/fsnt-16000312

Chang DC, Saunders JA, Chassy BM, Sowers AE (2012) Overview of electroporation and electrofusion. Guide to Electroporation and Electrofusion 1–6. https://doi.org/10.1016/B978-0-08-091727-6.50004-6

Chowdhury AR, Maji M (2024) Study on functional properties of protein-rich composite flour utilizing foxtail millet and flaxseed. MOJ Food Process Technol 12:27–32. https://doi.org/10.15406/mojfpt.2024.12.00295

Deshpande SD, Nishad PK (2021) Technology for millet value-added products. Millets and Millet Technology 293–303. https://doi.org/10.1007/978-981-16-0676-2_14

Dey S, Saxena A, Kumar Y, et al (2024) Synergistic effects of germination and ultrasonication on nutritional and structural characteristics of Kodo (Paspalum scrobiculatum) and little (Panicum sumatrense) millet. 2024. https://doi.org/10.1155/2024/4951196

Dhumal CV, Pardeshi IL, Sutar PP et al (2014) Optimization of process parameters for development of microwave puffed product. J Ready to Eat Food 1:111–119

Dias JV, van Bommel KJC, Noort MW-J et al (2016) Method for the production of edible objects using sls and food products. Google Patents

Dias-Martins AM, Cappato LP, da Costa Mattos M et al (2019) Impacts of ohmic heating on decorticated and whole pearl millet grains compared to open-pan cooking. J Cereal Sci 85:120–129. https://doi.org/10.1016/j.jcs.2018.11.007

Dubey A, Tripathy PP (2024) Ultrasound-mediated hydration of finger millet: effects on antinutrients, techno-functional and bioactive properties, with evaluation of ANN-PSO and RSM optimization methods. Food Chem 435:137516. https://doi.org/10.1016/j.foodchem.2023.137516

Farkas J, Mohácsi-Farkas C (2011) History and future of food irradiation. Trends Food Sci Technol 22:121–126. https://doi.org/10.1016/j.tifs.2010.04.002

FoodJet (2021) FoodJet printing systems—web page. https://www.foodjet.com/

García-Segovia P, García-Alcaraz V, Balasch-Parisi S, Martínez-Monzó J (2020) 3D printing of gels based on xanthan/konjac gums. Innov Food Sci Emerg Technol 64:102343. https://doi.org/10.1016/j.ifset.2020.102343

Gavahian M, Mathad GN, Pandiselvam R et al (2021) Emerging technologies to obtain pectin from food processing by-products: a strategy for enhancing resource efficiency. Trends Food Sci Technol 115:42–54. https://doi.org/10.1016/j.tifs.2021.06.018

Godoi FC, Prakash S, Bhandari BR (2016) 3d printing technologies applied for food design: status and prospects. J Food Eng 179:44–54

Gogate PR (2011) Application of hydrodynamic cavitation for food and bioprocessing. Pp 141–173

Gogate PR, Pandit AB (2001) Hydrodynamic cavitation reactors: a state of the art review. Rev Chem Eng 17:1–85. https://doi.org/10.1515/REVCE.2001.17.1.1

Gogate PR, Pandit AB (2005) A review and assessment of hydrodynamic cavitation as a technology for the future. Ultrason Sonochem 12:21–27. https://doi.org/10.1016/j.ultsonch.2004.03.007

Gowthamraj G, Jubeena C, Sangeetha N (2021) The effect of γ-irradiation on the physicochemical, functional, proximate, and anti-nutrient characteristics of finger millet (CO14 & CO15) flours. Radiat Phys Chem 183:109403. https://doi.org/10.1016/j.radphyschem.2021.109403

Hassan I, Rasheed N, Gani A, Gani A (2025) Rice starch, millet flour supplemented with algal biomass for 3D food printing. Int J Biol Macromol 303:140604. https://doi.org/10.1016/j.ijbiomac.2025.140604

Heena, Kumar N, Singh R et al (2024) Application and functional properties of millet starch: wet milling extraction process and different modification approaches. Heliyon 10:e25330. https://doi.org/10.1016/j.heliyon.2024.e25330

Hu Q, He Y, Wang F et al (2021) Microwave technology: a novel approach to the transformation of natural metabolites. Chinese Medicine (United Kingdom) 16:87. https://doi.org/10.1186/s13020-021-00500-8

Issa F, Latip NAA, Bradley DA, Nisbet A (2011) Ge-doped optical fibres as thermoluminescence dosimeters for kilovoltage X-ray therapy irradiations. Nucl Instrum Methods Phys Res A 652:834–837. https://doi.org/10.1016/j.nima.2010.09.080

Jaddu S, Pradhan RC, Dwivedi M (2022) Effect of multipin atmospheric cold plasma discharge on functional properties of little millet (Panicum miliare) flour. Innovative Food Sci Emerg Technol 77:102957. https://doi.org/10.1016/j.ifset.2022.102957

Jaddu S, Sonkar S, Seth D et al (2024) Cold plasma: unveiling its impact on hydration, rheology, nutritional, and anti-nutritional properties in food materials—an overview. Food Chem X 22:101266. https://doi.org/10.1016/j.fochx.2024.101266

Klemo M, Hysa A, Sefa S (2015) Irradiation treatment of animal foods. AJIS. https://doi.org/10.5901/ajis.2015.v4n3s1p256

Kouzani AZ, Adams SJ, Whyte D et al (2017) 3D printing of food for people with swallowing difficulties. KnE Engineering 2:23. https://doi.org/10.18502/keg.v2i2.591

Krishnaraj P, Anukiruthika T, Choudhary P et al (2019) 3D extrusion printing and post-processing of fibre-rich snack from indigenous composite flour. Food Bioprocess Technol 12:1776–1786. https://doi.org/10.1007/s11947-019-02336-5

Kumar SR, Sadiq MB, Anal AK (2020) Comparative study of physicochemical and functional properties of pan and microwave cooked underutilized millets (proso and little). Lwt 128:109465. https://doi.org/10.1016/j.lwt.2020.109465

Kumari V, Jayadeep A (2024) Effect of infrared heatingon finger millet in unhydrated and hydrated condition on physical, functional, pasting and phenolic profile. J Cereal Sci 116:103834. https://doi.org/10.1016/j.jcs.2023.103834

Laroque DA, Seó ST, Valencia GA et al (2022) Cold plasma in food processing: design, mechanisms, and application. J Food Eng 312:110748. https://doi.org/10.1016/j.jfoodeng.2021.110748

Liu X, Liang W, Zheng J et al (2023) The role and mechanism of electron beam irradiation in glutaric anhydride esterified proso millet starch: multi-scale structure and physicochemical properties. Int J Biol Macromol 243:125246. https://doi.org/10.1016/j.ijbiomac.2023.125246

Lohani UC, Muthukumarappan K (2016) SC. Innov Food Sci Emerg Technol 35:29. https://doi.org/10.1016/j.ifset.2016.03.012

Mahmoud BSM, Nannapaneni R, Chang S et al (2016) Improving the safety and quality of raw tuna fillets by X-ray irradiation. Food Control 60:569–574. https://doi.org/10.1016/j.foodcont.2015.08.039

McEvoy B, Maksimovic A, Howell D et al (2023) Studies on the comparative effectiveness of X-rays, gamma rays and electron beams to inactivate microorganisms at different dose rates in industrial sterilization of medical devices. Radiat Phys Chem 208:110915. https://doi.org/10.1016/j.radphyschem.2023.110915

Mensah EE, Asamoah YD, Yunus S et al (2022) A comparative study of the effect of microwave and conventional heating methods on proximate composition of spiced millet porridge. Int J Food Sci Nutr 7:56–61

Murphy SV, Atala A (2014) 3D bioprinting of tissues and organs. Nat Biotechnol 32:773–785. https://doi.org/10.1038/nbt.2958

Nachal N, Moses JA, Karthik P, Anandharamakrishnan C (2019) Applications of 3D printing in food processing. Food Eng Rev 11:123–141

Niu D, Zeng XA, Ren EF et al (2020) Review of the application of pulsed electric fields (PEF) technology for food processing in China. Food Res Int 137:109715. https://doi.org/10.1016/j.foodres.2020.109715

Pawar SG, Pardeshi IL, Mohril MP et al (2015) Optimization of microwave puffing process for preparing ready-to-eat snack using composite millet. Flours 2:101–111

Punia Bangar S, Nehra M, Siroha AK et al (2021) Development and characterization of physical modified pearl millet starch-based films. Foods 10. https://doi.org/10.3390/foods10071609

Raghav K, Buvaneswaran M, Bhavana MR et al (2025) Effects of ultrasound and high-pressure assisted extraction of pearl millet protein isolate: functional, digestibility, and structural properties. Int J Biol Macromol 289:138877. https://doi.org/10.1016/j.ijbiomac.2024.138877

Raja V, Moses JA, Anandharamakrishnan C (2023) Effect of 3D printing conditions and post-printing fermentation on pearl millet fortified idli. J Sci Food Agric 103:2401–2412. https://doi.org/10.1002/jsfa.12410

Ramírez-Jiménez AK, Castro-Muñoz R (2021) Emerging techniques assisting nixtamalization products and by-products processing: an overview. Crit Rev Food Sci Nutr 61:3407–3420. https://doi.org/10.1080/10408398.2020.1798352

Rao MV, KG A, CK S et al (2021) Effect of microwave treatment on physical and functional properties of foxtail millet flour. Int J Chem Stud 9:2762–2767. https://doi.org/10.22271/chemi.2021.v9.i1am.11641

Sakai N, Hanzawa T (1994) Applications and advances in far-infrared heating in Japan. Trends Food Sci Technol 5:357–362. https://doi.org/10.1016/0924-2244(94)90213-5

Santhoshkumar P, Raja V, Priyadarshini SR, Moses JA (2024) Impact of baking and steaming of 3D printed pearl millet—Banana pulp blends on physicochemical and sensory attributes. Int J Gastron Food Sci 37:100962. https://doi.org/10.1016/j.ijgfs.2024.100962

Sarkar P, Tamili D, Bhattacharya S, Chakraborti PB (2023) Preservation of food by gamma-irradiation technology. In: Emerging Technologies in Food Preservation. CRC Press, Boca Raton, pp 199–260

Saxena S, Vasudevan H, Saini S, Sasmal S (2023) Comparative nutritional assessment of millet-based Milk produced by ultrasound, germination, and a combined approach. ACS Food Sci Technol 3:600–607. https://doi.org/10.1021/acsfoodscitech.2c00342

Shobha D, Shubhashree K, Kavali S, Badigannavar A (2023) Influence of Irradiation and Packaging on the Shelf Life of White Finger Millet (Ragi) [*Eleusine coracana* (L.)] Flour. Mysore Journal of Agricultural Sciences 57(2)

Shams R, Rizvi QUEH, Jan B, Manzoor A (2023) Role of gamma irradiation in the preservation of fruits and vegetables. In: Quality control in fruit and vegetable processing. Apple Academic Press, New York, pp 57–75

Siacor FDC, Chen Q, Zhao JY et al (2021) On the additive manufacturing (3D printing) of viscoelastic materials and flow behavior: from composites to food manufacturing. Addit Manuf 45:102043. https://doi.org/10.1016/j.addma.2021.102043

Sonkar S, Jaddu S, Pradhan RC et al (2023) Effect of atmospheric cold plasma (pin type) on hydration and structure properties of kodo-millet starch. Lwt 182:114889. https://doi.org/10.1016/j.lwt.2023.114889

Sruthi NU, Rao PS (2021) Effect of processing on storage stability of millet flour: a review. Trends Food Sci Technol 112:58–74. https://doi.org/10.1016/j.tifs.2021.03.043

Sruthi P, Srinivasa UM, Rastogi NK, Naidu MM (2023) Principles of irradiation and its equipment. In: Non-thermal Food Processing Operations. Elsevier, pp 205–220

Sunil CK, Neeraja Chacko (2018) Microwave assisted De-husking of foxtail millet. J Agric Eng (India) 55:21–33. https://doi.org/10.52151/jae2018554.1666

Sunil CK, Gowda NAN, Nayak N, Rawson A (2024) Unveiling the effect of processing on bioactive compounds in millets: implications for health benefits and risks. Process Biochem 138:79–96. https://doi.org/10.1016/j.procbio.2024.01.010

Suslick KS (1990) Sonochemistry. Science (1979) 247:1439–1445. https://doi.org/10.1126/science.247.4949.1439

Swaminathan I, Guha M, Hunglur UH, Rao DB (2015) Optimization of infrared heating conditions of sorghum flour using central composite design. Food Sci Biotechnol 24:1667–1671. https://doi.org/10.1007/s10068-015-0216-7

Tanabe T (2022) Sources of energetic quanta (EQ) (radiation sources). In: Radiation: an energy carrier. Springer Nature Singapore, Singapore, pp 57–76

Varvara RA, Szabo K, Vodnar DC (2021) 3D food printing: principles of obtaining digitally-designed nourishment. Nutrients 13

Vesco A, Lipson H (2009) Hydrocolloid printing. 807–817

Vinutha T, Kumar D, Bansal N et al (2022) Thermal treatments reduce rancidity and modulate structural and digestive properties of starch in pearl millet flour. Int J Biol Macromol 195:207–216. https://doi.org/10.1016/j.ijbiomac.2021.12.011

Voučko B, Novotni D, Drakula S et al (2019) Optimization of high intensity ultrasound treatment of Proso millet bran to improve physical and nutritional quality. Food Technol Biotechnol 57:183–190. https://doi.org/10.17113/ftb.57.02.19.6100

Wang H, Fu Y, Zhao Q et al (2022) Effect of different processing methods on the millet polyphenols and their anti-diabetic potential. Front Nutr 9. https://doi.org/10.3389/fnut.2022.780499

Wani HM, Sharma P, Gul K et al (2022) Influence of Y-irradiation and genotype on the structural and techno-functional properties and microbial quality of millet flour. Applied Food Research 2:100181. https://doi.org/10.1016/j.afres.2022.100181

Xu B, Wang LK, Miao WJ et al (2016) Thermal versus microwave inactivation kinetics of lipase and lipoxygenase from wheat germ. J Food Process Eng 39:247–255. https://doi.org/10.1111/jfpe.12216

Yadav DN, Anand T, Kaur J, Singh AK (2012) Improved storage stability of pearl millet flour through microwave treatment. Agric Res 1:399–404. https://doi.org/10.1007/s40003-012-0040-8

Yusof N, Hilmy N (2014) Radiation biology of tissue Radiosterilization. Comprehensive Biomedical Physics. Elsevier, pp 263–287

Zhang Y, Jing X, Chen Z, Wang X (2023) Effects of moderate-intensity pulsed electric field on the structure and physicochemical properties of foxtail millet (Setaria italica) prolamin. Cereal Chem 100:360–370. https://doi.org/10.1002/cche.10614

Zheng MZ, Xiao Y, Yang S et al (2020) Effects of heat–moisture, autoclaving, and microwave treatments on physicochemical properties of proso millet starch. Food Sci Nutr 8:735–743. https://doi.org/10.1002/fsn3.1295

Zhi W, Zhou Y, Wang R et al (2022) Effect of microwave treatment on the properties of starch in millet kernels. Starch/Staerke 74. https://doi.org/10.1002/star.202200063

C C, S V, DV C et al (2024) Impact of cold plasma on pearl and barnyard millets' microbial quality, antioxidant status, and nutritional composition. Food Human 2:100238. https://doi.org/10.1016/j.foohum.2024.100238

Nutritional Quality: A Comprehensive Analysis of the Impact of Conventional and Modern Millet Processing Methods on Nutritional Quality

14

Sneha Subramanian and Emi Grace Mary Gowshika Rajendran

14.1 Introduction

Millets, the nutri-cereals, are commonly grown in the arid and semiarid regions of the world. These tiny-seeded grasses are part of the *Poaceae* family. They contribute significantly to India's economic security. They are packed with macro- and micronutrients making them a game changer in combating food insecurity and malnutrition. "The major millets are foxtail millet, finger millet, and proso millet or white millet, and the minor millets include barnyard, kodo, little millet, guinea millet, and browntop millet" (Rana et al. 2023). Millets are a rich source of vitamins and minerals and a variety of phytochemicals that play numerous roles in health and disease prevention. Thus, millets can act as both supplementary nutrition and add therapeutic value owing to the presence of different phytochemicals such as flavonoids, phenolics, anthocyanidins, and other bioactive compounds that exert antioxidant effect (Jacob et al. 2024).

According to the Global Hunger Index (GHI) report 2023, India has a serious hunger problem, with a score of 28.3% and ranking 111 out of 125 countries. With growing interest in nutrition-sensitive agriculture, further enhancing the production of millet can aid in improving health status and nutrition security. The world's largest producer of millet is India. From 2012 to 2022, a significant rise in the productivity of millets was observed. Additionally, India's share of global millet production has remained constant, averaging 17% from 2012 to 2022 and increasing to 19% in that year (APEDA report 2023). Today's world, however, faces a plethora of global problems, including environmental contamination, water scarcity, rising food costs, changing climate patterns, and other socioeconomic effects. These unfavorable

S. Subramanian (✉)
SixD Healthcare Private Limited, Chennai, Tamil Nadu, India

E. G. M. G. Rajendran
Food Science and Research, Nerdy Grad e-Organization for Education, Chennai, Tamil Nadu, India

S. Jaddu, R. Chandra Pradhan (eds.), *Innovative Millet Processing*,
https://doi.org/10.1007/978-3-031-99864-5_14

elements may have an impact on local agricultural developments and decrease cereal production, which would raise food prices and cause serious global food security problems (Yousaf et al. 2021). The government's shift in emphasis from food security to nutritional security is anticipated to have a significant impact on millets moving forward. The changes in the trends have now created a pressing need for researchers to further explore the effective methods of millet processing to increase production without compromising on their nutritional benefits.

Both at the household level and industrial level, millets are subjected to multiple processing methods which could affect their physical and nutrient composition. However, weighing the potential benefits of these processing methods can help in ensuring better nutritional quality. Studies have found that fermentation alters the ratio of nutritional to antinutritive components in millets and increases the amount of biologically active ingredients (Wang et al. 2022). Germination, another traditional processing method, has been found to increase the total polyphenolic content of millets while reducing the antinutritional content. The absence of suitable technologies is a significant barrier to enhancing millet production and processing. Different millets have different-sized outer pericarps or hulls. The hull becomes challenging to remove because of its close attachment to the endosperm. Several modern processing techniques were developed to counteract this and guarantee higher nutritional value.

Millets offer food security benefits due to their high nutritive quality, drought resistance, and adaptability. However, poor shelf life, difficulty in processing, and inherently present antinutritional properties hinder their use. Various food processing techniques, including microwave, radiofrequency, infrared, and nonthermal methods, have been explored for their ease, time and energy saving, and reduced nutrient losses.

14.2 Nutritional Significance of Millets

Millets are a nutritious crop, loaded with protein, vitamins, and minerals. Millets do not contain gluten making it a perfect and sustainable choice for individuals with gluten sensitivity. The low glycemic index of foods ensures a safer diet for blood sugar control. With abundant phytochemicals, millet acts as a functional food to prevent lifestyle disorders and mitigate malnutrition. Millets such as finger millet are rich in iron and calcium (Kalsi et al. 2023) and add supplementary benefits in treating anemia among pregnant women, young children, and adolescents.

Increasing awareness about the nutritional benefits of millets among the masses has led to a considerable increase in demand for millets as food products. The largest fraction of the total grain weight of the millets are non-starchy polysaccharides and dietary fiber. The protein content of millet varies among the varieties. It commonly ranges between 6% and 13%. Among all types, proso and foxtail millet are the richest sources of protein (Jacob et al. 2024). Like cereal grains, millet contains many essential amino acids, such as methionine, lysine, and cysteine. The fat content of millet grains is usually moderate at 2–8% of grain weight, with pearl millet having the highest fat content. Because they contain unsaturated fatty acids like oleic and linoleic acid, millet fats have been shown to have beneficial effects on health and nutrition (Annor et al. 2015). The amount of dietary fiber in millet is similar to that of whole wheat and nearly twice that of rice. Finger millet and pearl millet contain good amounts of dietary fiber. Both soluble and insoluble fractions make up millet's dietary fiber. Soluble fibers include β-glucans, arabinoxylans, and pectins, while insoluble fibers include cellulose and hemicellulose (Gupta et al. 2015).

Apart from macronutrients, millets are packed with micronutrients. They are a good source of iron, zinc, calcium, phosphorous, thiamine, niacin, and riboflavin (Anitha et al. 2020). Millets contain abundant phenolics (phenolic acids, flavonoids, and tannins), phytosterols, and policosanols compound which makes them a potent antioxidant. The millet seed coat contains polyphenols such as tannins, flavonoids, and phenolic acids, which function as reducing agents by quenching singlet oxygen, metal chelators, and free radicals (Kumar et al. 2021).

Millets are claimed to be an ideal grain alternative in the management of diabetes mellitus. Diabetes is a global health problem characterized by elevated blood sugar levels brought on by either a complete lack of insulin or a reduction in its action. The presence of low-digestible starch and polyphenols in millets is attributed to the hypoglycemic effect (Krishnan et al. 2021).The potent antioxidants in millets aid in preventing oxidative damage thus reducing the risk of complications. Millets containing a good source of dietary fiber, proteins, vitamins, minerals, and antioxidants aid in the prevention and management of hypertension. The nutritional quality of millets falls under the criteria of DASH (Dietary Approach to Stop Hypertension) making them an excellent choice. With adequate dietary fiber, millets are a good source of prebiotics thus helping in ensuring better gut health and in turn prevention of several chronic diseases. *The several nutritional benefits of millet are given in* Table 14.1.

14.3 Importance of Millet Processing

Some of the phytochemical substances known as antinutrients are produced by plants as a defense mechanism. There is a decrease in nutrient bioavailability and utilization as a result of these antinutritional factors that impede nutrient absorption (Bora et al. 2019). Common antinutrients present in millets include tannins, phytic acid, oxalates, and inhibitors such as trypsin and chymotrypsin. Raw finger millet contains 301 mg/100 g of tannin and 425 mg/100 g of phytic acid. Raw pearl millet contains 450 mg/100 g of tannin and 550 mg/100 g of phytic acid (Mane et al. 2023). Proteins and minerals have the lowest bioavailability of nutrients due to antinutritional factors such as enzyme inhibition capacity and metal chelation.

However, due to their role in antioxidant properties, some of the antinutritional elements like polyphenolic compounds have been identified to have health benefits in recent years (N. Gowda et al. 2022). Several studies on conventional and modern processing of millets have demonstrated enhanced nutrient bioavailability and

Table 14.1 Therapeutic benefits of millets

Millet variety	Nutritional significance	References
Pearl millet	Hinders enzymes involved in postprandial hyperglycemia Manages hepatic glucose production	Krishnan et al. (2022)
Foxtail millet	Studies on diabetic rats showed improvement in lipid contents and the effective metabolism of glucose In diabetic mice, it alleviated insulin resistance, enhanced liver function, and modulated serum metabolic profiles It was found to lower the risk of colonic inflammation and colorectal cancer in rats	Wang et al. (2023); Zhang et al. (2020)
Barnyard millet	In diabetic rats, it reduced diabetes-related complications and HbA1C levels	Ito et al. (2022)
Finger millet	Polyphenols shown to increase the growth of beneficial bacteria in the gut and suppress pathogenic bacteria such as *salmonella* sp., and *Escherichia coli* Antioxidant activity reduces postprandial hyperglycemia and prevents oxidative stress	Devi et al. (2014); Antony and Chandra (1997)
Sorghum	Animal studies on sorghum-based diets suggested proper thyroid gland regulation and promotion of weight loss through fat metabolism	Kangama (2017); Al-Mamary et al. (2001)
Kodo millet	Phenolic compounds in kodo millet demonstrated effectiveness in reducing the proliferation of human adenocarcinoma cells	Chandrasekara and Shahidi (2011)
Proso and barnyard	Vanillin extracted from these millets has shown inhibition of the proliferation of colon cancer cell lines	Ramadoss and Sivalingam (2020)

better protein and starch digestibility. A study on finger millets showed that germination for 48 hours at 30 °C reduced the antinutrient significantly and also increased its protein digestibility (Hejazi and Orsat 2016). This emphasizes the importance of antinutrient elimination through various processing techniques.

14.4 Conventional Millet Processing: Impact on Nutritional Quality

Like any other cereals, millets are processed to make it fit for consumption. This processing also improves their shelf life and sensory and nutritional attributes. The processing of millet is done through two stages. Processing techniques in the primary phase include dehulling, germination, soaking, roasting, drying, polishing, and milling. Secondary processing techniques include fermenting, parboiling, cooking, puffing, popping, malting, baking, extrusion, etc. These methods are commonly used for the formulation and development of millet-based value-added food products (Singh et al. 2024).

Processing of millets into different value-added products can increase its consumption and utilization and also lower the antinutritional content such as phytate and tannins. It also increases the bioavailability of minerals and amino acids and enhances macronutrient (starch and protein) digestibility (Singh et al. 2024). Overall processing of millets can exert both positive and negative effects on their nutritional properties. Thus, considering the effect of different processing techniques on millet is highly necessary to maximize nutrient availability and increase consumption.

14.4.1 Dehulling by Traditional Methods

Dehulling or decortication is the process of removal of the outer layer or hull that covers the endosperm. The hull is indigestible and contains comparatively less nutritional content. Owing to the smaller size of the hull and its variability among different millets, developing appropriate technologies for hull removal without causing damage to the endosperm has gained significant attention among researchers.

Removal of the hull is an important processing phase; however, removal is suitable only up to 12–30%. Transcending this limit can result in the loss of bioactive compounds, dietary fiber, and other macro- and micronutrients (Goudar and Sathisha 2016).

Dehulling and polishing of grains improve the bioavailability of nutrients and enhance their digestibility. The outer pericarp is the concentrated source of antinutrients which can be decreased only through this process (Babiker et al. 2018).

At the domestic level, millets were dehulled by pounding using a mortar and pestle. This is a common processing technique in India owing to the variability of the hull of different millets. Traditionally equipment that utilizes mechanical forces for dehulling and polishing was utilized. Most of these equipment were based on paddy processing technologies. The efficiency of this equipment was demonstrated for both dehulling effectiveness and nutrient bioavailability. The impact of traditional dehullers based on abrasion, compression, and mechanical force on nutritional quality is given in Table 14.2.

Upon decortication, several studies have shown an increase in their in vitro protein digestibility attributed to the elimination of polyphenols and antinutrients (Rani et al. 2018; Urooj 2017). Another invest igation found that along with an increase in protein digestibility, improvement in the estimated glycemic index of the millet. This reduction might be due to decreased levels of insoluble dietary fiber (Annor et al. 2017).

The current equipment's drawbacks include their high maintenance needs, need for skilled labor, noise, high investment costs, high power consumption, etc. To combat these limitations, suitable processing equipment that is easy to use, affordable for farmers, and environmentally friendly must be created for all varieties of millets. In addition to meeting consumer demand, the advancement and promotion of processing machinery will contribute to a rise in millet consumption.

Table 14.2 Impact of traditional dehullers on nutritional quality

Dehuller	Type of millet	Nutritional quality	References
Rubber roll dehuller	Foxtail millet	Decreased antinutrient contents Increased digestibility of nutrients	Mohamad Gesudaraj (2012)
Abrasive polisher	Pearl millet	Decreased activity of phytates	Lestienne et al. (2007)
Traditional Stone dehuller (6NS-33, India)	Pearl millet	Decreased protein, crude fiber, and fat contents Reduced tannin and phytic acid levels Among minerals, decreased iron, calcium, and phosphorous	Elfadil Babiker et al. (2018)
Pedal-operated sorghum dehuller	Sorghum	Improved dehulling efficiency of jumbo and Dionje varieties of sorghum of up to 90.01% and 83.40%, respectively	El. Lazaro et al. (2013)
Centrifugal impeller	Minor millets	Improved dehulling efficiency of foxtail, little, proso, kodo, and barnyard up to 92.7%, 91.00%, 93.5%, 80.90%, and 74.40, respectively	R. B. Gowda et al. (2020)

14.5 Germination

Germination or malting is an age-old processing technique generally used at the household level. Several studies have concluded that the germination of millets significantly reduces the antinutritional factors while increasing protein and starch digestibility.

A study comparing the antinutritional factors in finger millet demonstrated that values of tannin and phytic acid reduced significantly after germination. The raw finger millet contained 301 mg/100 g of tannin and 425 mg/100 g of phytic acid; after germination, the values were decreased to 168 mg/100 g and 270 mg/100 g, respectively. The decrease in tannins upon germination was suggested to be caused by increased enzymatic activity during germination. The hydrolysis of phytate due to the increased activity of phytase was concluded to be the reason for the decrease in phytic acid. Leaching of the polyphenols and phytic acid was also attributed as a common reason. The proximate analysis of the germinated millets revealed improved quantities of crude fiber, iron, and zinc (Mane et al. 2023).

Mineral analysis of finger millet after germination confirmed increased levels of calcium due to the reduction of oxalic acid. A significant increase in the iron content of germinated finger millet flour was demonstrated. Other minerals such as phosphorous were found to decrease significantly post-germination (Singh Chauhan and Sarita 2018). Concerning macronutrients a significant increase and decrease in protein and carbohydrates were found, respectively. This can be a result of the production of good-quality protein by mobilizing the stored nitrogen for plant development. Sprouting of the millet flour also enhanced the vitamin C levels. Other than phytates and tannins, germination also reduced the trypsin inhibitor activity from 4188 ± 8.00 U/g to 2001 ± 5.10 U/g (Singh Chauhan and Sarita 2018).

The ability of millet flour to scavenge free radicals has been greatly enhanced by germination. This may be because biofunctional substances are produced during germination as a result of the breakdown of high molecular weight polymers. Of the raw and germinated millet flours, little millet had the highest antioxidant properties, followed by barnyard and foxtail millet flour (Kulla et al. 2021). Germination also had a significant effect on protein and starch digestibility. Sprouted millet showed increased in vitro protein digestibility (IVPD) after 48-hour germination, attributed to enhanced intrinsic proteases, removal of protease inhibitors, and enhanced protein solubility, as demonstrated by their IVPD values. Also, an increase in starch digestibility was also seen, attributed to the increased availability of starch granules to the enzymes and increased alpha-amylase activity (Sharma and Gujral 2020).

Utilizing techniques like germination in food processing can enhance nutrient availability, reduce antinutrient levels, and can be used to produce a range of nutrient-dense value-added food products that are sustainable and would also meet dietary needs.

14.5.1 Soaking

Soaking of millets is a traditional method commonly performed before decortication or germination. For decortication, the temperature at which the millets have been soaked could highly affect the dehulling efficiency. Higher temperature leads to better water absorptivity by the hull, followed by a drying process, and then, the millets are dehulled. This preprocessing technique generates multiple changes in the physiochemical and functional properties.

A study by Abioye VF found that the phenolic content of finger millet reduced significantly from 40.29 mg/100 g to 16.00 mg/100 g as the soaking hour increased (Abioye et al. 2022). Several studies have demonstrated similar results with decreased antinutritional properties on soaking owing to the leaching of phytates and tannins.

The iron and zinc contents of millet grains were found to decrease when they were soaked in water; this could be attributed to the leaching of the minerals (Yousaf et al. 2021).

A comparative study on five millets (little millet, foxtail millet, pearl millet, finger millet, and kodo millet) by 24 h soaking, followed by germination, suggested enhanced antioxidant properties and reduced tannin content (Bhuvaneshwari et al. 2020). Another study on high-pressure soaking showed that soaking foxtail millet grains under high pressure dramatically increased their water uptake. This, in turn, increased the flour's degree of starch gelatinization, which reached a maximum value of 64.93%. Moreover, the antinutrient levels (phytic acid and tannin) dropped with high-pressure soaking, proving that this technique can enhance the quality of foxtail millet grains and flour (N. Sharma et al. 2018). Overall, soaking can be attributed as an effective pretreatment method which could potentially reduce the antinutritional properties and enhance bioavailability.

14.5.2 Fermentation

One of the oldest techniques for preserving food is fermentation, which is usually used to process grains like millet and cereal. Several studies on fermented food products have shown that they enhance the nutritional value of foods while also helping to improve their sensory qualities, such as taste, flavor, and texture. Owing to these special advantages, fermentation has emerged as a major processing technique among millets and cereals especially in terms of production of value-added food products. During fermentation, the millet undergoes several physiochemical changes such as the breakdown of sugar, softening, and changes in starch leading to improved nutritional quality and reduced antinutrients (Gupta et al. 2015).

A study reported that fermentation of pearl millet flour significantly improved the phytochemical content. When the parameters for fermentation such as yeast concentration and temperature were optimized, an increase in total phenolic content was noted. The treated millet samples provided the maximum yield of different phytochemicals such as ferulic acid (33.66 μg/mg), catechin (20.10 μg/mg), and vanillic acid (5.67 μg/mg) (R. Kumari et al. 2022). It was attributed to the increase in acidity during fermentation altering the structure of the grain. Moreover, the release of various enzymes during fermentation could also aid in the release of phytochemicals (Adebo and Gabriela Medina-Meza 2020). It was also found that a further increase in the fermentation interval and temperature resulted in a decline in phenolic contents (R. Kumari et al. 2022).

The presence of microorganisms using the available compounds for growth might have caused the reduction (Adetuyi and Ibrahim 2014). A study by Purewal et al. (2019) has shown that using *Rhizopus azygosporus* as a starting culture for 10 days of fermentation greatly increased the bioactive constituents present in pearl millet from 6.6 to 21.8 mg GAE/g for the total phenolic content. Another research on pearl millet reported similar changes with increased polyphenols and further reduced phytic acid content upon fermentation for 72 hours at varying temperatures such as 20, 25, and 30 °C (Urooj 2017).

With respect to the proximate composition, a study on pearl millet has found an increase in the protein content from 15.25 ± 0.21 g/100 g to 15.35 ± 0.35 g/100 g on 24 hours of fermentation. Moreover, only negligible changes were found with fat and ash content. At 24 hours of fermentation, the carbohydrate content decreased from 72.63% to 70.97% with a corresponding increase in soluble sugars. A significant decrease among amino acids such as glycine, lysine, and arginine was noted (Osman 2011). Another study on the same millet has found an increase in the level of free amino acids such as serine, histidine, alanine, and tryptophan and a higher crude protein concentration for the same duration of 24 hours (R. Kumari et al. 2022). A recent study by Akinola et al. (2017) reported an increase in protein and crude fat content of pearl millet from 10.99% to 13.65% and 1.83% to 3.71%, respectively. However, a similar decrease in carbohydrate content from 75.75% to 73.76% was shown.

The fiber content of finger millet on fermentation with different microorganisms resulted in a similar reduction. The value of fiber content decreased from 5.67% to

4.40% when *Saccharomyces cerevisiae* L. was employed as the fermentation activator, and it decreased to 4.13% when yeast and ammonium sulfate were combined. This decrease was attributed to the metabolizing action of the fermenting microorganisms. Among the antinutrients, treatment with yeast reduced the phytate content from 629.00 mg/100 g to 521.33 mg/100 g, whereas when combined with ammonium sulfate, the phytate was further reduced to 416.62 mg/100 g. Upon fermentation of finger millet using *Saccharomyces cerevisiae* L., a significant increase in minerals such as copper by 22.3%, iron by 22.14%, zinc by 18%, and manganese content by 28.37% was reported (Jan et al. 2022).

Upon natural fermentation for 72 hours at different temperatures, an increase in protein and starch digestibility in pearl millet was reported by Urooj (2017). At the highest temperature of 30 °C, the digestibility increased from 17.85% to 110.1%. This improved starch digestibility was a result of the conversion of starch to oligosaccharides and decreased phytate content. Several studies on protein digestibility found similar improvement upon fermentation, which could be potentially due to the decrease in phytate content, proteolytic enzyme secretion by microbes, and the breakdown of complex proteins into simple proteins (Mohammed Nour et al. 2018). Thus, fermentation could be viewed as an effective traditional processing technique resulting in improved nutritional properties of millets.

14.5.3 Milling and Sieving

One of the crucial initial stages in processing millets is milling or grinding, which includes removing the coarse bran or seed coat and reducing the grain's size. Generally, milling is done either manually or with the help of machinery. This could potentially alter the bioactive compounds and other chemical compositions due to the removal of the bran (Rathore 2016).

Usually, the process of milling aids in the elimination of antinutrients. Studies have shown a significant reduction of polyphenols and phytate phosphorus of up to 74.7% and 39.8%, respectively, in finger millet (Ramashia et al. 2019). However, this benefit comes with a compromise of other important minerals and vitamins (Yousaf et al. 2021). Milling of browntop millet, an underutilized millet known for several health benefits, has shown considerable reduction in phytic acid, saponins, and other indigestible oligosaccharides (Sunagar and Sreerama 2024).

A study on finger millet has shown a significant decrease in protein from 7.15 to 6.33 g, fat from 1.78 to 1.29 g, and insoluble dietary fiber from 20.23 to 12.15 g upon milling. Along with this, the iron, zinc, and calcium content was reduced from 6.52 to 3.29 mg, 2.50g to 1.98 mg, and 404.3 to 294.8 mg, respectively. Other nutrients such as thiamine and riboflavin were also reduced significantly. On the other hand, the phytate and tannin contents were also remarkably reduced from 628.2 to 432.0 mg and 851.2 to 563.9 mg, respectively (Oghbaei and Prakash 2016).

Improved protein and starch digestibility can be achieved through milling due to the elimination of antinutrients (Yousaf et al. 2021). Considering the potential loss of important nutrients along with antinutrients, milling is a less convenient technology for the processing of millets. Further optimization with modern processing for the removal of bran and production of flour could improve the nutritional quality.

14.5.4 Roasting

Roasting is a traditional food processing method that browns the food's surface and intensifies its flavor. It's also possible for roasting to expand the range of aroma from the millets. This is a common household technique that improves the digestibility of the millet without any loss of nutrients. This method could potentially remove most antinutritional factors from the millet. A study on roasting of foxtail millets showed a decrease in moisture content from 8.9% to 3.7%. This was attributed to the dry heat processing technique. Compared to carbohydrates and fat, the protein content of the millet reduced significantly from 12.1% to 10.8%. The antinutritional factors including polyphenols, tannins, and phytic acid were reduced from 14.5 to 7.8 mg GAE/100 g, 221.1 to 92.4 mg CAE/100 g, and 306 to 180.5 mg/100 g, respectively (A. P. Khapre et al. 2016).

Analysis of roasting on different temperatures and durations (120 °C, 140 °C for 5, 10, and 15 min and 160 °C, 180 °C for 3, 5, and 10 min) of pearl millet flour has shown improved functional properties such as an increase in water solubility index and oil absorption capacity. With increasing temperature, a decrease in phenolic content was noted (Obadina et al. 2016). With barnyard millet, the carbohydrate content increased by 9.18%, protein by 0.3%, and dietary fiber by 0.53%. The increase in protein content was attributed to the roasting temperature of the millet which was set at a lower range of 110 °C for 10 minutes which increased the protein containing secondary metabolites. Thus, the ideal range of roasting for improved availability of protein in the millet was at a lower temperature. The total phenolic content of the barnyard millet increased from 0.031 mg GAE/g to 0.042 mgGAE/g.

The heat transfer during roasting might have favored the hydrolysis of C-glycosyl flavones leading to the release of phenolic compounds from barnyard millet (Mohd et al. 2023). A similar study on proso millet has also found an increase in total phenols (Kalam Azad et al. 2019), whereas contrasting results were reported about phenols in the case of pearl and finger millet. Roasting of pearl millet has resulted in a major reduction in phenolic content from 169.85 to 90.60 mg/100 g (Obadina et al. 2016) and finger millet from 314.24 to 223.31 mg/100 g. This decrease was attributed to the breakdown or leaching of the phenolics. In finger millet, this reduction was coupled with a decrease in antioxidant activity from 89.9% to 86.78% (N.Singh et al. 2018).

Another study has found an increase in protein digestibility from 22.3% to 60.1% on roasting at 140 degrees Celsius (Tumwine et al. 2019). Upon heating, the nitrogen levels might have improved leading to increased protein hydrolysis (N. Singh et al. 2018). This might be helpful in the development of millet-based composite flour for value-added products. Thus, roasting could aid in improving the nutritional qualities of the millets; however, optimization of roasting temperature is an important factor.

14.5.5 Puffing/Popping

Popping is the simplest, least expensive, and fastest traditional method of cooking millets. This process involves applying dry heat to seed kernels until the internal moisture expands, which is then used to make weaning foods and nutritious, ready-to-eat snack products. Home-processed (popping) ready-to-eat snacks have a lot of potential as value-added health products. This is because consumers are shifting to more convenient, nutrient-dense, and less refined and polished grains (Singh Chauhan and Sarita 2018). Moreover, it was noted that popped millet grains had a high mineral and phenolic compound content, which made them advantageous to use as a food product (Yousaf et al. 2021).

A study on the popping of finger millet flour found an increase in carbohydrate and protein content from 6.3 ± 0.20 g/100 g and 71.9 ± 1.9 g/100 g to 7.1 ± 0.30 g/100 g and 75.73 ± 0.98 g/100 g, respectively. Also, a significant reduction in the crude fiber, fat, and moisture content was reported. The thermal processing might have resulted in reduced fat and increased lipolytic action causing fat reduction. Among minerals, a considerable decrease and increase in calcium and iron were found, respectively. The calcium and iron content in raw finger millet changed from 342.4 ± 1.36 mg/100 g and 3.7 ± 0.06 mg/100 g to 338.3 ± 1.00 mg/100 g and 5.1 ± 0.10 mg/100 g, respectively. Popping has significantly reduced the tannin, phytic acid, and oxalate contents and trypsin inhibitor activity from 870.8 ± 1.05 mg/100 g, 851.4 ± 1.6 mg/100 g, 45.8 ± 3.5 mg/100 g, and 4188 ± 8.00 U/g to 610.2 ± 2.10 mg/100 g, 333.1 ± 1.07 mg/100 g, 32.2 ± 1.7 mg/100 g, and 3090 ± 4.50 U/g, respectively (Singh Chauhan and Sarita 2018). In a study on proso millet by different presoaking methods and popping, it has been shown to enhance the nutritional profile and exhibit better health benefits. Following popping, it was discovered that the amounts of tannin and phytic acid, which ranged from 115.78 to 74.00 mg/100 g and 1307.1 to 656.78 mg/100 g, respectively, were lower. This reduction was attributed to the impact of popping on the outer layer (bran) that contains these antinutritional elements (Nithyashree and Vijayalaxmi 2022). Puffing/popping can also significantly affect the digestive properties of millet. Studies have shown improvement in the digestibility of starch and protein in millets after popping. This increase was attributed to the removal of tannin and inhibition of trypsin inhibitor which otherwise would hinder the proteolytic enzyme activity (Huang et al. 2018, Kirti Jalgaonkar et al. 2015). These studies indicate that the process of puffing can significantly impact the millet's nutritional quality and improve its adaptation in the production of nutritious millet-based food products. The impact of conventional millet processing on nutritional quality is summarized in Table 14.3.

Table 14.3 Impact of conventional millet processing on the nutritional quality

Conventional processing method	Millet variety	Nutritional impact	References
Germination	Finger millet	Decreased phytic acid and tannins Improved quantities of crude fiber, iron, and zinc Increased levels of calcium and iron Increased protein content and decreased carbohydrate content	Mane et al. (2023); Singh Chauhan and Sarita (2018)
Soaking	Finger millet	Reduced phenolic content	Abioye et al. (2022)
Soaking	Little millet, foxtail millet, pearl millet, finger millet, and kodo millet	Enhanced antioxidant properties	Bhuvaneshwari et al. (2020)
Fermentation	Pearl millet	Increase in phytochemicals such as ferulic acid, catechin, and vanillic acid	R. Kumari et al. (2022); Purewal et al. (2019); Urooj (2017)
	Pearl millet	Increase in protein and decrease in carbohydrates Increase in the level of free amino acids	Osman (2011); R. Kumari et al. (2022); Akinola et al. (2017)
	Finger millet	Reduction in antinutrients and fiber content Increase in minerals such as copper, iron, zinc, and manganese	Jan et al. (2022)
	Pearl millet	Increase in starch and protein digestibility	Mohammed Nour et al. (2018); Urooj (2017)
Milling	Finger millet	Reduction in polyphenols and phytates	Ramashia et al. (2019)
	Browntop millet	Reduction in phytic acid, saponins, and other indigestible oligosaccharides	Sunagar and Sreerama (2024)
	Finger millet	Reduction in protein, fat, crude fiber, iron, zinc, calcium, thiamine, and riboflavin Decrease in phytate and tannin content	Oghbaei and Prakash (2016)
Roasting	Foxtail millet	Reduced moisture and protein content Reduction in antinutrients	A. P. Khapre et al. (2016)
	Pearl millet flour	Reduction in phenolic content	Obadina et al. (2016)
	Barnyard millet	Increase in carbohydrates, protein, and crude fiber Increase in total phenolic contents	Mohd et al. (2023)
	Proso millet	Increase in total phenols	Kalam Azad et al. (2019)
	Pearl and finger millet	Decrease in total phenolic compounds	Obadina et al. (2016)

14.6 Modern Millet Processing: Impact on Nutritional Quality

With increasing demands, millet is a better alternative for major cereals both economically and nutritionally. However, their utilization is restricted due to their poor shelf life, inherent antinutrients, and difficulty in processing. Promising millet processing techniques are now being explored to combat these issues and ensure better physical and functional properties. The use of innovative thermal and nonthermal technologies to preserve and enhance nutritional quality is crucial these days due to the drawbacks of conventional processing techniques. Some of the innovative millet processing technologies and their impact on nutritional quality are discussed below. The effect of modern millet processing on different nutritional parameters is listed in Table 14.4.

14.6.1 Thermal Methods

14.6.1.1 Microwave Treatment

Over the years, conventional processing methods were predominantly used for dehulling millets. In recent times, improved dehulling efficiency through microwave processing has led to an improvement in the physicochemical availability of both macro- and micronutrients and also in the reduction of antinutrients. This promising technique could potentially aid in proper heating, low operational cost, better storage stability, and lesser processing time of the millet (Ananthu et al. 2023).

Several studies using microwave treatment have demonstrated better physical and functional properties of millets. Microwave pretreatment of foxtail millet flour using a domestic microwave model with a maximum input power of 900 W has shown reduced oil absorption capacity and increased water absorption capacity. This was explained as a result of the heat-induced denaturation of the protein. This change in physical property would potentially affect the overall texture of cooked products by playing a major role in starch gelatinization. The flour obtained from microwave-treated foxtail millet can be utilized for various food processing applications, thereby contributing to the value addition of the millet (Rao et al. 2021). The treatment also lessens the antinutrients like trypsin, tannins, and others that are present in millet. It also increases the amount of water available inside the grain, which facilitates the process of gelatinization during steaming (S. M. Singh et al. 2024). Another study evaluating the effect of kodo millet has found an increase in the total phenolic content at maximum power of 720 W. The total phenolic content surged from 3.26 ± 0.09 mg GAE/100 g to 4.23 ± 0.71 mg GAE/100 g at 720 W for 270 seconds. It was concluded that the protein-phenolic component could have resulted in the alteration and degradation of tannin due to thermal effects. On the other hand, a decrease in the antioxidant activity of the microwave-treated millets was noted (Gopal and Bhuvana 2021).

Millets have poor storage stability related to their fat hydrolytic rancidity. Multiple studies evaluated the effect of microwave treatment on improving the

Table 14.4 Impact of modern millet processing on the nutritional quality

Modern processing method	Millet variety	Nutritional impact	References
Microwave treatment	Foxtail millet	Decreased antinutrient	Singh et al. (2024)
	Kodo millet	Increase in total phenolic content	Gopal and Bhuvana (2021)
	Pearl millet and foxtail millet	Improved storage stability and enhanced shelf life	Yadav et al. (2021)
	Millet kernels	Increase in rapidly and slowly digestible starch	Zhi et al. (2022)
Infrared treatment	Finger millet	Increased losses of phenolic content	Kumari and Jayadeep (2024)
	Pearl millet	Reduction in enzymes causing rancidity	Vinutha et al. (2022)
Ultrasound treatment	Finger millet	Decrease in total phenolic content Decreased tannin content	S. Yadav et al. (2021)
	Millet-based probiotic beverage	Increased total flavonoid content Increase in antioxidant activity Increase in total phenolic content	Meena et al. (2023)
Cold plasma treatment	Pearl millet flour	With higher voltage and duration, a decrease in phenols was reported, whereas with low intensity, the total phenolic contents were increased	Kheto et al. (2023)
	Pearl and barnyard millet	Decreased antioxidant activity Decreased phenolic contents Decreased tannins and phytic acid Increased iron content Decreased calcium and phosphorous content	Charu et al. (2024)
Ozonation	Foxtail millet	Decrease in tannin and total flavonoid content Increase in total phenols and phytic acid content	Sharma et al. (2022)
	Pearl and sorghum starches	Decrease in starch digestibility	Thakur et al. (2023)

storage stability of millets. A study on pearl millet showed decreased lipase activity and a simultaneous increase in moisture level. The subjective evaluation of the flour showed an enhanced shelf life of up to 30 days without unpleasant odor and bitter taste (D. N. Yadav et al. 2012). A similar study on foxtail millet flour reduced lipase activity and better storage stability (Rao et al. 2023).

Enhancing the millet's starch properties may have an impact on how it's used in food preparation. Microwave treatment has been shown to affect the physical and chemical properties of starch in the millet kernels. Microwave treatment for

150 seconds has been shown to significantly increase the rapidly and slowly digestible starch which leads to the reduction in resistant starch. This effect can be a useful tool for the alteration of starch in millet for desired food products (Zhi et al. 2022).

14.6.1.2 Infrared Treatment

Infrared treatment causes heating which could be utilized in the dehulling of the grains. These radiations are electromagnetic waves, which range from 0.78 to 1000.00 μm. The grain experiences a sudden internal heating as a result of its wavelike transmission. When radiation strikes the grain, it causes vibrations in the grain's molecules, which in turn causes intermolecular friction and heats the grain. Compared to conventional methods of heating, infrared treatment saves time, increases energy efficiency, and results in the production of higher-quality products. This processing technique as a pretreatment for removing the hull can improve the dehulling efficiency (S. M. Singh et al. 2024). Infrared heating might also potentially alter the physical and functional properties of millets. A study on hydrated finger millet found higher losses of soluble and bound phenolic contents on infrared heating of the millet (V. Kumari and Jayadeep 2024). A study documented the combined hydrothermal infrared treatment on pearl millet flour and found a significant decrease in several enzymes such as lipase, polyphenol oxidase, and lipoxygenase that are responsible for rancidity. It was found that the flour's functional qualities were enhanced during storage by this combined infrared treatment (Vinutha et al. 2022). While infrared treatment holds potential in the areas of improving the physical and functional properties of millet, more investigations on the impact of nutritional quality have to be conducted.

14.6.2 Nonthermal Methods

14.6.2.1 Ultrasound-Mediated Hydration

During processing, hydrating the crop aids in lowering antinutrient levels and improving nutrient availability. Additionally, it softens the grain texture of the cereal to facilitate easy extraction and grinding (Park and Han 2016). Traditional methods of hydration for a longer time could pose more risks than benefits such as off-odor or microbial contamination. While thermal methods of hydration can reduce the processing time, still it could affect the nutritional properties of the grains. Thus, ultrasound hydration was developed as a much-equipped nonthermal method of hydration (S. Yadav et al. 2021). Studies have revealed that ultrasound hydration is an effective processing technique with lesser time and energy consumption and better nutrient qualities.

An investigation assessing the impact of ultrasound hydration on finger millet has found improved hydration related to a significant increase in moisture content. The increase in moisture content was directly proportional to the increase in amplitude and time. Also, with higher amplitude and time, the total phenolic content of the millet grain was significantly decreased. The tannin content of the millet was affected by the level of moisture. The total tannin content between the range of

35.83 and 187.83 mg TAE/100 g was reduced with higher water content. This was explained by the ultrasound's capacity to change tannic acid that is hydrolyzable into gallic acid. Additionally, it causes the sample's condensed tannin to leach out, lowering the tannin content overall. This processing technique was also optimized to the level of "66% of ultrasound amplitude, 26 min of treatment time, and 1:3 of grain to water ratio" for better quality and hydration in the study by Yadav et al. (2021).

Further, ultrasound treatment has also shown a significant impact on value-added products such as millet-based probiotic beverages. A study in this context by Meena et al. (2023) has found reduced fermentation time due to better nutrient uptake by bacterial cells and lactic acid productivity. The ultrasound treatment after the inoculation of bacteria showed an increase in total flavonoid content from 3.64 ± 0.32 (mg QE/100 g) to 9.74 ± 1.58 (mg QE/100 g) and an increase in antioxidant activity. Flavonoids are known for their health-promoting effects such as anti-inflammatory, anticancer, and antioxidant. Thus, maximizing the advantages of these bioactive compounds can be facilitated by high-intensity ultrasound treatment.

Another investigation on similar high-intensity ultrasound pretreatment on proso millet bran has found a 15% increase in antioxidants and a 16% increase in total phenolic content. With increasing time and amplitude of the treatment, the highest increase of 38% in dietary fiber was recorded. Thus, optimized treatment using ultrasound can improve the availability of bioactive compounds present in millets and also reduce the antinutrient level significantly.

14.6.2.2 Cold Plasma Treatment

The cold plasma treatment is a nonthermal technique that is relatively newer and offers multiple benefits to the food industry in the aspects of processing. A partially or completely ionized state of matter existing as a gas is referred to as plasma. In this state, free electrons and ions are present because some or all of the gas particles have gained or lost electrons.

The process of ionization entails the electrons being detached from atoms or molecules creating a plasma state (Jaddu et al. 2024).

Generally, swelling capacity, solubility, and water absorption capacity are important parameters that are altered during plasma treatment resulting in better outcomes for food products. Cold plasma treatment results in increased water holding capacity of cereal grains due to the starch hydrolytic depolymerization. Several studies have demonstrated improved hydration properties of millets using the cold plasma technique. This treatment has increased parameters such as swelling capacity, solubility, water absorption capacity, and oil absorption capacity that were noted in pearl millet flour (Lokeswari et al. 2021; Sarkar et al. 2023).

Due to the nonthermal behavior of cold plasma, it could induce only a negligible effect on the nutritional properties of the millet. However, plasma treatment might result in the decomposition of water leading to the formation of free oxygen radicals causing reduced moisture content of the food item (Jaddu et al. 2024). A similar study on barnyard and pearl millet has also reported decreased moisture content in cold plasma treatment (Charu et al. 2024).

A study on pearl millet flour using atmospheric pressure for plasma discharge found a decrease in moisture level and fat which later increased with an increase in time. Also, a reduction in fiber content and no change in protein content were noted (Lokeswari et al. 2021). On the other hand, the discharge of plasma with a multipin electric discharge by Sarkar et al. (2023) on the same millet flour has shown a reduction in moisture and CHO with an increase in fat and protein. The increase in fat content was attributed to the formation of secondary products derived from reactive oxygen species that resulted in lipid oxidation. The majority of results showed that the protein and carbohydrate (CHO) contents of foods treated with plasma had no or very little effects (Jaddu et al. 2024).

A prominent impact of cold plasma treatment on biologically active compounds was demonstrated in several studies. The total phenolic content of pearl millet flour was significantly affected by the application of plasma treatment. This change was based on the voltage of the treatment. With higher voltage and duration, a decrease in phenols was reported, whereas with low intensity, the total phenolic contents were increased (Kheto et al. 2023). A similar report by Charu et al. (2024) shows that the antioxidant activity of pearl and barnyard millet reduced after cold plasma application. The treatment has also been found to influence the total phenolic contents of the respective millets. The unprocessed pearl and barnyard millets have 84.03 ± 0.075 mg/100 g and 82.28 ± 0.03 mg/100 g of phenolic content, respectively. On treatment, these values were reduced to 2% in pearl and 19% in barnyard millet. The generation of highly energetic electrons leading to the breakdown of phenolic compounds and conversion of aromatic phenolic structures to aliphatic molecules disrupting phenols were attributed as the common reason for this reduction.

Concerning antinutrients, such as tannin and phytic acid, a reduction in their content was reported which was affected by the treatment duration and power (Sarkar et al. 2023). A study by Charu et al. (2024) also showed a significant reduction in total tannin and phytic acid content after treatment. The initial concentrations of tannin of pearl and barnyard millet were 236 ± 12.53 mg/100 g and 232.33 ± 9.29 mg/100 g, respectively. On treatment, a decrease of 18% in pearl millet and 24% in barnyard millet were reported.

According to reports, cold plasma treatment can be potentially utilized in the process of decontamination of microorganisms. Barnyard and pearl millet were enumerated with microbial populations and later treated with cold plasma which showed a significant reduction of all three populations: *Enterococcus* spp., bacterial, *E. coli*, and yeast populations. On the other hand, cold plasma treatment also had a significant impact on the mineral content. In pearl and barnyard millet, iron content increased, whereas phosphorous and calcium levels decreased. The increase in bioavailability was attributed to the increase in the breakdown of iron-binding compounds, thus increasing the concentration.

These results imply that cold plasma treatment is a useful method for improving millet's nutritional value. To fully realize its potential, more research and treatment condition optimization are required. Nonetheless, it is imperative to recognize specific constraints and offer direction for subsequent investigations. Further research should concentrate on the cold plasma technology's wider effects, scalability, and economic viability in the millet sector.

14.6.2.3 Ozonation

Ozone is a potent oxidant that can be used in various food applications to guarantee food safety. Ozone treatment is regarded as an economical and environmentally beneficial food processing method. Ozone affects seed germination and is frequently used to control pests in grain. There are still many unanswered questions regarding the application of this treatment for millet processing, and the available studies are scant.

A study on the ozonation of foxtail millet flour has found a significant decrease in the tannin content from 64.83 mg/100 g to 112.43 mg/100 g, and also, a reduction in total flavonoid content was reported. However, a significant rise in total phenolic content and phytic acid was noted. This treatment has shown significant changes in the functionality of the flour improving the water and oil absorption, solubility, and swelling power. These outputs made this treatment an efficient processing method for improved functionality; however, it slightly compromises the total flavonoid and phenol content (R. Sharma et al. 2022).

Several studies analyzing the impact of ozone treatment in terms of shelf life and microbial contamination were done on cereal grains. According to a study on wheat bran, the optimized ozone treatment reduced the microbial contents by up to 90%. And noodles prepared from the treated wheat bran had improved storage capacity (Liu et al. 2020). Another study on the characterization of ozonated pearl millet and sorghum starches demonstrated a significant reduction in in vitro starch digestibility (Thakur et al. 2023).

Ozone treatment has been shown in numerous studies to be an effective fumigant and to inactivate microorganisms for the processing and storage of grains. This unique benefit of ozone treatment as an alternative to harmful pesticides to improve the quality of cereal grains can be potentially leveraged for millets. This opens a new avenue for further exploration (Dubey et al. 2022).

14.7 Integrated Approaches in Millet Processing Technologies

Millet processing techniques have evolved from conventional methods to modern approaches, aiming to maximize nutritional benefits and extend shelf life. Traditional processes like mechanical processing, fermentation, germination, and soaking have been widely used but often lead to alterations in nutritional and functional properties. On the other hand, modern processing techniques, both thermal and nonthermal, have shown promising results in enhancing millet quality. Interestingly, a synergistic approach combining different processing methods has shown potential for achieving enhanced results.

Recent studies have demonstrated the effectiveness of combining ultrasonication and germination as integrated processing techniques to enhance the nutritional and structural properties of minor millets. One such investigation on kodo (*Paspalum scrobiculatum*) and little millet (*Panicum sumatrense*) reported significant reductions in antinutritional factors like tannins and phytates, alongside increased antioxidant activity and improved nutritional bio-accessibility. Notably, in vitro protein digestibility exceeded 82% in treated samples (Dey et al. 2024). These findings highlight the synergistic potential of ultrasonication and germination in improving the functional and nutritional value of millets. A synergistic approach, combining multiple processing methods, may provide the best results in terms of enhancing nutritional quality and addressing the challenges associated with millet processing, such as rapid rancidity and nutrient loss. Further research into combined processing methods for millets could lead to more effective and sustainable solutions for the food industry.

14.8 Future Scope

Future research on millet processing should focus on optimizing the integration of traditional and modern techniques to enhance both nutritional and sensory qualities. Although emerging technologies offer improved nutritional outcomes, they may adversely affect sensory attributes such as taste, aroma, and texture, necessitating a balanced approach. The integration of multiple processing methods may help overcome these challenges, providing synergistic benefits that maximize nutrient retention while maintaining consumer acceptability.

Further investigation is required to address persistent issues such as rancidity and off-flavors, which can impact the shelf life and palatability of millet-based products. The development of nutrient-rich millet-based composite flours incorporating other health-promoting ingredients is also recommended. With the rising demand for minimally processed, functional foods, continued innovation and interdisciplinary research in millet processing technologies will be critical to unlocking the full potential of millets in addressing global nutritional challenges.

14.9 Conclusion

Millets are an important source of nutrition loaded with antioxidant properties. This helps in combating chronic diseases and aids in the prevention of malnutrition. With a growing population, millets are our best bet in terms of nutritional and agricultural-sensitive crops. However, the ambiguity in millet processing and the presence of antinutrients are the major drawbacks in increasing their production and consumption. Conventional methods such as milling, malting, fermentation, and heat treatments have shown effectiveness in reducing antinutritional factors. For instance, germination for 48 hours significantly increased protein content, protein digestibility, and total sugars while decreasing phytate content in millet grains. Modern processing techniques, including cold plasma technology, high-pressure processing, and novel thermal treatments, have demonstrated potential in improving the functional attributes and nutritional quality of millet products. These methods offer advantages in terms of microbial decontamination and enzyme inactivation while minimizing nutrient loss and quality changes. However, it is important to note that processing methods can have varying effects on different nutritional components. For example, roasting at high temperatures may increase protein digestibility but can also reduce protein and total sugar content. The wide range of millet grain varieties and their untapped nutritional potential make it imperative to concentrate on their production, processing, and increasing the efficiency of consumption to enhance food and nutritional security.

References

Abioye V, Olatunde S, Ogunlakin G, Abioye O (2022) Effect of soaking conditions on chemical composition, antioxidant activity, total phenols, flavonoids and anti-nutritional contents of finger millet. Afr J Food Agric Nutr Dev 22(112):20942–20956. https://doi.org/10.18697/ajfand.112.20480

Adebo OA, Gabriela Medina-Meza I (2020) Impact of fermentation on the phenolic compounds and antioxidant activity of whole cereal grains: a mini review. Molecules 25(4):927. https://doi.org/10.3390/molecules25040927

Adetuyi FO, Ibrahim TA (2014) Effect of fermentation time on the phenolic, flavonoid and vitamin C contents and antioxidant activities of okra (Abelmoschus esculentus) seeds. Niger Food J 32(2):128–137. https://doi.org/10.1016/S0189-7241(15)30128-4

Agricultural and Processed Food Products Export Development Authority (APEDA). (2023, March 19). Annual Account Report 2022–23. New Delhi: APEDA.

Al-Mamary M, Molham A-H, Abdulwali A-A, Al-Obeidi A (2001) In vivo effects of dietary sorghum tannins on rabbit digestive enzymes and mineral absorption. Nutr Res 21(10):1393–1401. https://doi.org/10.1016/S0271-5317(01)00334-7

Ananthu R, Buvaneswaran M, Meena L, Sunil CK (2023) Microwave treatment effect on functional and pasting properties, and storage stability of white finger millet. J Food Process Eng 46(7). https://doi.org/10.1111/jfpe.14341

Anitha S, Govindaraj M, Kane-Potaka J (2020) Balanced amino acid and higher micronutrients in millets complements legumes for improved human dietary nutrition. Cereal Chem 97(1):74–84. https://doi.org/10.1002/cche.10227

Annor GA, Marcone M, Corredig M, Bertoft E, Seetharaman K (2015) Effects of the amount and type of fatty acids present in millets on their in vitro starch digestibility and expected glycemic index (eGI). J Cereal Sci 64:76–81. https://doi.org/10.1016/j.jcs.2015.05.004

Annor GA, Tyl C, Marcone M, Ragaee S, Marti A (2017) Why do millets have slower starch and protein digestibility than other cereals? Trends Food Sci Technol 66:73–83. https://doi.org/10.1016/j.tifs.2017.05.012

Antony U, Chandra TS (1997) Microbial population and biochemical changes in fermenting finger millet (Eleusine coracana). World J Microbiol Biotechnol 13(5):533–537. https://doi.org/10.1023/A:1018561224777

Akinola SA, Badejo AA, Osundahunsi OF, Edema MO (2017) Effect of preprocessing techniques on pearl millet flour and changes in technological properties. Int J Food Sci Technol 52(4):992–999

Babiker E, Abdelseed B, Hassan H, Adiamo O (2018) Effect of decortication methods on the chemical composition, antinutrients, Ca, P and Fe contents of two pearl millet cultivars during storage. World J Sci Technol Sustain Dev 15:278

Bhuvaneshwari G, Nirmalakumari A, Kalaiselvi S (2020) Impact of soaking, sprouting on antioxidant and anti-nutritional factors in millet grains. J Phytol 62:–66. https://doi.org/10.25081/jp.2020.v12.6384

Bora P, Ragaee S, Marcone M (2019) Characterisation of several types of millets as functional food ingredients. Int J Food Sci Nutr 70(6):714–724. https://doi.org/10.1080/09637486.2019.1570086

Chandrasekara A, Shahidi F (2011) Inhibitory activities of soluble and bound millet seed phenolics on free radicals and reactive oxygen species. J Agric Food Chem 59(1):428–436. https://doi.org/10.1021/jf103896z

Charu C, Vignesh S, Chidanand DV, Mahendran R, Baskaran N (2024) Impact of cold plasma on pearl and barnyard millets' microbial quality, antioxidant status, and nutritional composition. Food Human 2:100238. https://doi.org/10.1016/j.foohum.2024.100238

Devi PB, Vijayabharathi R, Sathyabama S, Malleshi NG, Priyadarisini VB (2014) Health benefits of finger millet (Eleusine coracana L.) polyphenols and dietary fiber: a review. J Food Sci Technol 51(6):1021–1040. https://doi.org/10.1007/s13197-011-0584-9

Dey S, Saxena A, Kumar Y, Maity T, Tarafdar A (2024) Synergistic effects of germination and ultrasonication on nutritional and structural characteristics of Kodo (*Paspalum scrobiculatum*) and little (*Panicum sumatrense*) millet. J Food Qual 2024(1). https://doi.org/10.1155/2024/4951196

Dubey P, Singh A, Yousuf O (2022) Ozonation: an evolving disinfectant technology for the food industry. Food Bioprocess Technol 15(9):2102–2113. https://doi.org/10.1007/s11947-022-02876-3

Gesudaraj M (2012) Development and performance evaluation of dehusker for foxtail millet (Setaria italica L.)

Gopal DS, Bhuvana S (2021). Effect of microwave treatment on phytochemical, functional and rheological property of Kodo millet (*Paspalum scrobiculatum*). *Pharma Innovation 10(11):397–401*

Goudar G, Sathisha GJ (2016) Effect of processing on ferulic acid content in foxtail millet (Setaria Italica) grain cultivars evaluated by HPTLC. Orient J Chem 32(4):2251–2258. https://doi.org/10.13005/ojc/320458

Gowda NAN, Satishkumar, Taj F, Subramanya S, Ranganna B (2020) Development a table top centrifugal dehuller for small millets. Agric Mech Asia Afr Lat Am 51(1)

Gowda NAN, Siliveru K, Prasad PVV, Bhatt Y, Netravati BP, Gurikar C (2022) Modern processing of Indian millets: a perspective on changes in nutritional properties. Foods 11(4):499. https://doi.org/10.3390/foods11040499

Gupta RK, Gangoliya SS, Singh NK (2015) Reduction of phytic acid and enhancement of bioavailable micronutrients in food grains. J Food Sci Technol 52(2):676–684. https://doi.org/10.1007/s13197-013-0978-y

Hejazi SN, Orsat V (2016) Malting process optimization for protein digestibility enhancement in finger millet grain. J Food Sci Technol 53(4):1929–1938. https://doi.org/10.1007/s13197-016-2188-x

Huang R, Pan X, Lv J, Zhong W, Yan F, Duan F, Jia L (2018) Effects of explosion puffing on the nutritional composition and digestibility of grains. Int J Food Prop 21(1):2193–2204. https://doi.org/10.1080/10942912.2018.1514634

Ito Y, Suzuki A, Nasukawa H, Miyaki K, Yano A, Nagasawa T (2022) Ameliorative effects of Japanese barnyard millet (Echinochloa esculenta H.Scholz) bran supplementation in streptozotocin-induced diabetic rats. Food Sci Technol Res 28(5):FSTR-D-22-00079. https://doi.org/10.3136/fstr.FSTR-D-22-00079

Jacob J, Krishnan V, Antony C, Bhavyasri M, Aruna C, Mishra K, Nepolean T, Satyavathi CT, Visarada KBRS (2024) The nutrition and therapeutic potential of millets: an updated narrative review. Front Nutr 11:1346869. https://doi.org/10.3389/fnut.2024.1346869

Jaddu S, Sonkar S, Seth D, Dwivedi M, Pradhan RC, Goksen G, Kumar Sarangi P, Režek Jambrak A (2024) Cold plasma: unveiling its impact on hydration, rheology, nutritional, and anti-nutritional properties in food materials—an overview. Food Chem X 22:101266. https://doi.org/10.1016/j.fochx.2024.101266

Jalgaonkar K, Jha SK, Sharma DK (2015) Effect of thermal treatments on the storage life of pearl millet (Pennisetum glaucum) flour. Indian J Agric Sci 86

Jan S, Krishan Kumar KK, Yadav AN, Ahmed N, Thakur P, Chauhan D, Hyder Rizvi Q-U-E, Dhaliwal HS (2022) Effect of diverse fermentation treatments on nutritional composition, bioactive components, and anti-nutritional factors of finger millet (Eleusine coracana L.). J Appl Biol Biotechnol:46–52. https://doi.org/10.7324/JABB.2022.10s107

Kalam Azad MO, Jeong DI, Adnan M, Salitxay T, Heo JW, Naznin MT, Lim JD, Cho DH, Park BJ, Park CH (2019) Effect of different processing methods on the accumulation of the phenolic compounds and antioxidant profile of broomcorn millet (Panicum miliaceum L.) flour. Foods 8(7):230. https://doi.org/10.3390/foods8070230

Kalsi R, Bhasin J, Goksen G, Kashyap P (2023) Exploration of nutritional, pharmacological, and the processing trends for valorization of finger millet (*Eleusine coracana*): a review. Food Sci Nutr 11(11):6802–6819. https://doi.org/10.1002/fsn3.3659

Kangama CO (2017) Importance of sorghum bicolor in African's cultures. J Agric Environ Sci 6(2):10.15640/jaes.v6n2a16

Khapre AP, Shere DM, Deshpande HW (2016) Studies on effect of roasting on nutritional and anti-nutritional components of foxtail millet (Setaria Italica). The Bioscan

Kheto A, Mallik A, Sehrawat R, Gul K, Routray W (2023) Atmospheric cold plasma induced nutritional & anti-nutritional, molecular modifications and in-vitro protein digestibility of guar seed (Cyamopsis tetragonoloba L.) flour. Food Res Int 168:112790. https://doi.org/10.1016/j.foodres.2023.112790

Krishnan V, Awana M, Singh A, Goswami S, Vinutha T, Kumar RR, Singh SP, Sathyavathi T, Sachdev A, Praveen S (2021) Starch molecular configuration and starch-sugar homeostasis: key determinants of sweet sensory perception and starch hydrolysis in pearl millet (Pennisetum glaucum). Int J Biol Macromol 183:1087–1095. https://doi.org/10.1016/j.ijbiomac.2021.05.004

Krishnan V, Verma P, Saha S, Singh B, Vinutha T, Kumar RR, Kulshreshta A, Singh SP, Sathyavathi T, Sachdev A, Praveen S (2022) Polyphenol-enriched extract from pearl millet (Pennisetum glaucum) inhibits key enzymes involved in post prandial hyper glycemia (α-amylase, α-glucosidase) and regulates hepatic glucose uptake. Biocatal Agric Biotechnol 43:102411. https://doi.org/10.1016/j.bcab.2022.102411

Kulla S, Hymavathi TV, Kumari BA, Reddy RG, Rani CVD (2021) Impact of germination on the nutritional, antioxidant and antinutrient characteristics of selected minor millet flours. Ann Phytomed 10(1):10.21276/ap.2021.10.1.18

Kumar A, Rani M, Mani S, Shah P, Singh DB, Kudapa H, Varshney RK (2021) Nutritional significance and antioxidant-mediated antiaging effects of finger millet: molecular insights and prospects. Front Sustain Food Syst 5. https://doi.org/10.3389/fsufs.2021.684318

Kumari V, Jayadeep A (2024) Effect of infrared heatingon finger millet in unhydrated and hydrated condition on physical, functional, pasting and phenolic profile. J Cereal Sci 116:103834. https://doi.org/10.1016/j.jcs.2023.103834

Kumari R, Bhatt S, Agrawal H, Dadwal V, Gupta M (2022) Effect of fermentation conditions on nutritional and phytochemical constituents of pearl millet flour (Pennisetum glaucum) using response surface methodology. Appl Food Res 2(1):100055. https://doi.org/10.1016/j.afres.2022.100055

Lazaro EL, Benjamin Y, Mpanduji SM (2013) Development of a low cost machine for improved sorghum Dehulling efficiency, vol 12(2). Faculty of Agriculture Sokoine University of Agriculture, Morogoro, Tanzania

Lestienne I, Buisson M, Lullien-Pellerin V, Picq C, Trèche S (2007) Losses of nutrients and anti-nutritional factors during abrasive decortication of two pearl millet cultivars (Pennisetum glaucum). Food Chem 100(4):1316–1323. https://doi.org/10.1016/j.foodchem.2005.11.027

Liu C, Zhang Y, Li H, Li L, Zheng X (2020) Effect of ozone treatment on processing properties of wheat bran and shelf life characteristics of noodles fortified with wheat bran. J Food Sci Technol 57(10):3893–3902. https://doi.org/10.1007/s13197-020-04421-6

Lokeswari R, Sharanyakanth PS, Jaspin S, Mahendran R (2021) Cold plasma effects on changes in physical, nutritional, hydration, and pasting properties of pearl millet (*Pennisetum Glaucum*). IEEE Trans Plasma Sci 49(5):1745–1751. https://doi.org/10.1109/TPS.2021.3074441

Mane RP, Kshirsagar RB, Agarkar BS, Patil BM, Katke SD (2023) Effects of germination on antinutritional, colour, functional and physicochemical properties of millets. Pharma Innov 12(1):350–355. https://doi.org/10.22271/tpi.2023.v12.i1d.18008

Meena L, Malini B, Byresh TS, Sunil CK, Rawson A, Venkatachalapathy N (2024) Ultrasound as a pre-treatment in millet-based probiotic beverage: Its effect on fermentation kinetics and beverage quality. Food Chemistry Advances 4:100631 https://doi.org/10.1016/j.focha.2024.100631

Mohammed Nour AA, Mohamed AR, Adiamo OQ, Babiker EE (2018) Changes in protein nutritional quality as affected by processing of millet supplemented with Moringa seed flour. J Saudi Soc Agric Sci 17(3):275–281. https://doi.org/10.1016/j.jssas.2016.05.006

Mohd MS, Shafie NH, Ahmad F, Shafie SR (2023) Effect of roasting on whole grain barnyard millet to the proximate composition, amino acid profile, total phenolic content and antioxidant activity. Malays J Med Health Sci 19:69

Nithyashree K, Vijayalaxmi KG (2022) Effect of popping on nutrients and anti-nutrients in Proso millet. Mysore J Agric Sci

Obadina A, Ishola IO, Adekoya IO, Soares AG, de Carvalho CWP, Barboza HT (2016) Nutritional and physico-chemical properties of flour from native and roasted whole grain pearl millet (Pennisetum glaucum [L.]R. Br.). J Cereal Sci 70:247–252. https://doi.org/10.1016/j.jcs.2016.06.005

Oghbaei M, Prakash J (2016) Effect of primary processing of cereals and legumes on its nutritional quality: a comprehensive review. Cogent Food Agric 2(1). https://doi.org/10.1080/23311932.2015.1136015

Osman MA (2011) Effect of traditional fermentation process on the nutrient and antinutrient contents of pearl millet during preparation of Lohoh. J Saudi Soc Agric Sci 10(1):1–6. https://doi.org/10.1016/j.jssas.2010.06.001

Park D-J, Han J-A (2016) Quality controlling of brown rice by ultrasound treatment and its effect on isolated starch. Carbohydr Polym 137:30–38. https://doi.org/10.1016/j.carbpol.2015.10.045

Purewal SS, Sandhu KS, Salar RK, Kaur P (2019) Fermented pearl millet: a product with enhanced bioactive compounds and DNA damage protection activity. Journal of Food Measurement and Characterization 13(2):1479–1488. https://doi.org/10.1007/s11694-019-00063-1

Ramadoss DP, Sivalingam N (2020) Vanillin extracted from Proso and Barnyard millets induce apoptotic cell death in HT-29 human colon cancer cell line. Nutr Cancer 72(8):1422–1437. https://doi.org/10.1080/01635581.2019.1672763

Ramashia SE, Anyasi TA, Gwata ET, Meddows-Taylor S, Jideani AIO (2019) Processing, nutritional composition and health benefits of finger millet in sub-saharan Africa. Food Sci Technol 39(2):253–266. https://doi.org/10.1590/fst.25017

Rana SS, Tiwari S, Gupta N, Tripathi MK, Tripathi N, Singh S, Bhagyawant SS (2023) Validating the nutraceutical significance of minor millets by employing nutritional–antinutritional profiling. Life 13(9):1918. https://doi.org/10.3390/life13091918

Rani S, Singh R, Sehrawat R, Kaur BP, Upadhyay A (2018) Pearl millet processing: a review. Nutr Food Sci 48(1):30–44. https://doi.org/10.1108/NFS-04-2017-0070

Rao MV, Akhil KG, Sunil CK, Venkatachalapathy N, Jaganmohan R (2021) Effect of microwave treatment on physical and functional properties of foxtail millet flour. Int J Chem Stud 9(1):2762–2767. https://doi.org/10.22271/chemi.2021.v9.i1am.11641

Rao MV, Sunil CK, Venkatachalapathy N (2023) Effect of microwave and hot air radiofrequency treatments on physicochemical and functional properties of foxtail millet flour and its protein isolate. J Cereal Sci 114:103774. https://doi.org/10.1016/j.jcs.2023.103774

Rathore S (2016) Millet grain processing, utilization and its role in health promotion: a review. Int J Nutr Food Sci 5(5):318. https://doi.org/10.11648/j.ijnfs.20160505.12

Sarkar A, Niranjan T, Patel G, Kheto A, Tiwari BK, Dwivedi M (2023) Impact of cold plasma treatment on nutritional, antinutritional, functional, thermal, rheological, and structural properties of pearl millet flour. J Food Process Eng 46(5). https://doi.org/10.1111/jfpe.14317

Sharma N, Goyal SK, Alam T, Fatma S, Chaoruangrit A, Niranjan K (2018) Effect of high pressure soaking on water absorption, gelatinization, and biochemical properties of germinated

and non-germinated foxtail millet grains. J Cereal Sci 83:162–170. https://doi.org/10.1016/j.jcs.2018.08.013
Sharma R, Bhandari M, Kaur K, Singh A, Sharma S, Kaur P (2022) Molecular interactome and starch–protein matrix, functional properties, phytochemical constituents, and antioxidant activity of foxtail millet (*Setaria italica*) flour as influenced during gaseous ozonation. Cereal Chem 99(5):1101–1111. https://doi.org/10.1002/cche.10559
Sharma B, Gujral HS (2020) Modifying the dough mixing behavior, protein & starch digestibility and antinutritional profile of minor millets by sprouting. International Journal of Biological Macromolecules 153:962–970. https://doi.org/10.1016/j.ijbiomac.2019.10.225
Singh Chauhan E, Sarita S (2018) Effects of processing (germination and popping) on the nutritional and anti-nutritional properties of finger millet (Eleusine Coracana). Curr Res Nutr Food Sci J 6(2):566–572. https://doi.org/10.12944/CRNFSJ.6.2.30
Singh N, David J, Thompkinson DK, Seelam BS, Rajput H, Morya S (2018) Effect of roasting on functional and phytochemical constituents of finger millet (Eleusine coracana L.). Pharma Innov J
Singh SM, Jayasree Joshi T, Rao PS (2024) Technological advancements in millet dehulling and polishing process: an insight into pretreatment methods, machineries and impact on nutritional quality. Grain Oil Sci Technol 7:186. KeAi Communications Co. https://doi.org/10.1016/j.gaost.2024.05.007
Sunagar RR, Sreerama YN (2024) Impact of milling on the nutrients and anti-nutrients in browntop millet (*Urochloa ramosa*) and its milled fractions: evaluation of their flour functionality. J Sci Food Agric 104(9):5504–5512. https://doi.org/10.1002/jsfa.13382
Thakur K, Sharma S, Sharma R, Bhardwaj R, Sohu RS, Singh B (2023) Characterization of ozonated pearl millet and sorghum starches: pasting profile, functional properties, in vitro-digestibility, molecular interactions, and morphology. Starch-Stärke 75(11–12). https://doi.org/10.1002/star.202300020
Tumwine G, Atukwase A, Tumuhimbise GA, Tucungwirwe F, Linnemann A (2019) Production of nutrient-enhanced millet-based composite flour using skimmed milk powder and vegetables. Food Sci Nutr 7(1):22–34. https://doi.org/10.1002/fsn3.777
Urooj A (2017) Impact of household processing methods on the nutritional characteristics of pearl millet (Pennisetumtyphoideum): a review. MOJ Food Process Technol 4(1):10.15406/mojfpt.2017.04.00082
Vinutha T, Kumar D, Bansal N, Krishnan V, Goswami S, Kumar RR, Kundu A, Poondia V, Rudra SG, Muthusamy V, Rama Prashat G, Venkatesh P, Kumari S, Jaiswal P, Singh A, Sachdev A, Singh SP, Satyavathi T, Ramesh SV, Praveen S (2022) Thermal treatments reduce rancidity and modulate structural and digestive properties of starch in pearl millet flour. Int J Biol Macromol 195:207–216. https://doi.org/10.1016/j.ijbiomac.2021.12.011
Wang H, Fu Y, Zhao Q, Hou D, Yang X, Bai S, Diao X, Xue Y, Shen Q (2022) Effect of different processing methods on the millet polyphenols and their anti-diabetic potential. Front Nutr 9. https://doi.org/10.3389/fnut.2022.780499
Wang H, Zhao Q, Fu Y, Liu Z, Zhu Y, Wang C, Shen Q (2023) Foxtail millet (Setaria italica) alleviates non-alcoholic fatty liver disease in high-fat diet/streptozotocin-induced diabetic mice through gut microbiota modulation. Food Biosci 53:102797. https://doi.org/10.1016/j.fbio.2023.102797
Yadav DN, Anand T, Kaur J, Singh AK (2012) Improved storage stability of pearl millet flour through microwave treatment. Agric Res 1(4):399–404. https://doi.org/10.1007/s40003-012-0040-8
Yadav S, Mishra S, Pradhan RC (2021) Ultrasound-assisted hydration of finger millet (Eleusine Coracana) and its effects on starch isolates and antinutrients. Ultrason Sonochem 73:105542. https://doi.org/10.1016/j.ultsonch.2021.105542
Yousaf L, Hou D, Liaqat H, Shen Q (2021) Millet: a review of its nutritional and functional changes during processing. Food Res Int (Ottawa, Ont) 142:110197. https://doi.org/10.1016/j.foodres.2021.110197

Zhang B, Xu Y, Liu S, Lv H, Hu Y, Wang Y, Li Z, Wang J, Ji X, Ma H, Wang X, Wang S (2020) Dietary supplementation of foxtail millet ameliorates colitis-associated colorectal cancer in mice via activation of gut receptors and suppression of the STAT3 pathway. Nutrients 12(8):2367. https://doi.org/10.3390/nu12082367

Zhi W, Zhou Y, Wang R, Wang M, Wang W, Hu A, Zheng J (2022) Effect of microwave treatment on the properties of starch in millet kernels. Starch - Stärke 74(9–10). https://doi.org/10.1002/star.202200063

Index

S. Jaddu, R. Chandra Pradhan (eds.), *Innovative Millet Processing*,
https://doi.org/10.1007/978-3-031-99864-5

www.ingramcontent.com/pod-product-compliance
Lightning Source LLC
Chambersburg PA
CBHW071953170426
43348CB00003B/24
* 9 7 8 3 0 3 1 9 9 8 6 3 8 *